AF893893

Enforcing fire safety in the catering industry

Leiden University Press

Lay-out: AlphaZet prepress, Waddinxveen

Leiden University Press is een imprint van Amsterdam University Press

ISBN 978 90 8728 061 1
e-ISBN 978 90 4850 604 0

Enforcing fire safety in the catering industry

An economic analysis

PROEFSCHRIFT

ter verkrijging van de graad van Doctor aan de Universiteit Leiden,
op gezag van de Rector Magnificus prof. mr. P.F. van der Heijden,
volgens besluit van het College voor Promoties
te verdedigen op woensdag 10 december 2008
klokke 13.45 uur

door

Guido Suurmond

geboren te Gouda in 1979

Promotiecommissie:

Promotor:	prof. dr. K.P. Goudswaard
Copromotor:	dr. B.C.J. van Velthoven
Referent:	prof. dr. M.G. Faure LLM (Universiteit Maastricht en Erasmus Universiteit Rotterdam)
Overige leden:	Prof. dr. mr. W. Huisman (Vrije Universiteit Amsterdam)
	Prof. dr. mr. E.R. Muller
	Prof. dr. J.J.M. Theeuwes (Universiteit van Amsterdam)
	Prof. dr. W.J.M. Voermans

Content

1 Introduction

1.1 Enforcement deficits

On New Year's Eve 2001 a fire occurred in café "'t Hemeltje" in the municipality of Volendam, the Netherlands. As a result of the Christmas decorations catching fire, fourteen youngsters died and approximately 300 people were wounded. It appeared that the fire safety in the café was far below standard. The proprietor (Mr. Veerman) was responsible for inadequate emergency exits, decorations that cause a fire hazard, too many people present and not being in possession of the prescribed license for the use of the establishment. Although the primary responsibility rested with the proprietor, the government was also to blame for failing to grant and/or inspect the necessary licenses. The municipality of Edam-Volendam was consciously not granting the use licenses. They did not employ personnel for inspecting bars or restaurants on compliance with fire safety regulations. The national and provincial government failed to supervise the municipality. Put differently, since several governmental agencies had failed to supervise the licenses, the proprietor did not feel the responsibility to take the regulations seriously and an accident could therefore end in a disaster. This weak enforcement practice of municipalities was common at the time in most municipalities. The use licenses had been granted for less than 30% of all buildings (IOOV, 2002*). Moreover, inspection of the licenses was exceptional. Where violations were found, enforcement hardly ever took place. These observations caused a major increase in municipal effort concerning fire safety enforcement. Municipalities began to eliminate the backlogs in use licenses and to carry out regular inspections in the licensed buildings. Due to the Volendam disaster, bars and restaurants were labeled as having the highest priority.

Other countries have similar experiences. For example, on 30 October 1998, 63 youngsters died in a great fire in a discothèque in Goteborg (Sweden). In December 1994, 15 people died and 164 were wounded at a New Year's Eve party in the Switel-hotel in Antwerp (Belgium), because a candle set fire to a Christmas tree. In Buenos Aires (Argentina), a fire in a discothèque caused the death of 193 persons in December 2004. In February 2003, a fire in a nightclub in Warwick, Rhode Island in the United States killed 100 people. In these accidents, similar problems to those identified above arose with respect to the fire safety of the building and the enforcement of the fire safety by the municipality. A report by the U.S. Fire Administration (USFA, 2000, p.1) mentions that, in response to the Rhode Island nightclub fire, "many local jurisdictions across the country reviewed their fire safety codes or increased inspections of local nightclubs to enforce existing codes. These typically included sufficient egress, illuminated exits signs, occupancy limits, and requirements for sprinkler systems for clubs based on size and occupancy. (...) Several local news stories reported cases of clear disregard for safety on the part of some nightclub owners. Examples often include exit doors that were blocked by equipment or storage, locked exit doors (often to keep patrons from sneaking in to avoid cover charges or entrance lines), exit hallways used as storage, missing or unlit exit signs, or owners who routinely allowed the club to exceed occupancy limits." These observations are much in line with those for the Netherlands following the Volendam disaster.

Before the Volendam disaster, there was already academic and governmental concern about inappropriate enforcement – accumulated in the report of the Michiels Committee (1998*). This committee pointed out that there are serious enforcement deficits in many domains, including fire safety. With the help and pressure of the national government, municipalities began to professionalize administrative enforcement. A program such as "High Quality Enforcement" was launched, expert centers were formed, policy instruments developed and national Inspectorates professionalized.

The need for a professionalization of enforcement also arose because of the large increase in regulation. New problems such as the environment, occupational health and customer safety, have led to much new regulation for firms and require corresponding enforcement efforts. As a result, there has been a major increase in enforcement efforts. It is estimated that the costs of administrative enforcement by national regulatory agencies in the Netherlands has risen from €280 million in 1989 to €800 million in 2005.[1] At the same time, there is a constant pressure to reduce the burden of regulation for firms by withdrawing contradictory or unnecessarily burdening regulations.

The question is which ideas and theories underlie these initiatives. Is there any attention for properly balancing the costs and benefits of enforcement? The double trend to both intensify and reduce regulatory enforcement requires an economic perspective that analyzes the effectiveness and desirability of enforcement. Are the recent enforcement efforts able to improve compliance with the regulations, and thereby to reduce harm to citizens? Does the magnitude of this effect justify its costs? Is it possible to improve enforcement policies and instruments by reducing their costs or increasing their effects? For instance, should we emphasize inspections or sanctions? Which sanctions or which magnitude of sanctions should we impose? Should we rely on enforcement after harm has been done or on enforcement of activities prior to harm? Should we apply a strategy of strict punishment or a strategy of cooperation and negotiations?

I will consider these questions by analyzing the case of the enforcement of fire safety in bars and restaurants in the Netherlands.

---

1 Leeuw and Willemsen (2006*).

1.2 MAIN QUESTION

The main question of this thesis is: What is an effective and efficient enforcement policy for fire safety in the catering industry?

I generally take the law and the regulation as given, and concern myself with their enforcement. The enforcement of fire safety can basically be organized in three ways (or a combination of these). With private enforcement, enforcement is left to the victims of the harm. Liability rules enable customers to require compensation after damage has occurred. Under administrative enforcement, regulation is enforced by municipalities through the granting of licenses, inspections, the imposition of administrative sanctions, etc.[2] These administrative actions are based on the expected harm. Municipalities check whether proprietors satisfy the requirements before actual harm takes place. Finally, public prosecutors can prosecute an offense, so that a criminal law sanction can be imposed on proprietors who fail to meet the requirements.

The question for each of the enforcement methods is whether and when it is effective in inducing compliance with the fire safety standards. Furthermore, I investigate whether enforcement results in a higher level of social welfare: what is an efficient enforcement policy? Most attention will be given to administrative enforcement. What are the benefits of a policy of advice, persuasion and warnings? How should the recent attention for administrative enforcement be evaluated? However, it is important to study not only administrative or public law enforcement, but also private enforcement. First, for enforcing fire safety, all three enforcement methods are observed in practice. All three are a potentially interesting alternative. Whether an enforcement policy is efficient, can only be decided by considering the relevant alternatives. Secondly, the enforcement of safety regulation is often, as in case of fire safety, combined with private enforcement. Therefore, their joint use has to be examined. This requires that the effects of private enforcement are understood.

In answering the main question I will apply the theoretical strengths of the law and economics literature to a real enforcement problem. This enables me to show that the law and economics perspective is useful in analyzing actual enforcement policies. But it also enables me to contribute to the development of law and economics by confronting it with seemingly paradoxical enforcement policies and with criticism expressed in other social sciences.

---

2 I define administrative enforcement broadly as every action by a governmental agency that is directed towards compliance of the rules it has laid down. As such the enforcement process covers granting permits, public campaigns, inspections, agreements, sanctions, etc. This definition corresponds to definitions by Hoitink and Michiels (1993*, p.3) and Blomberg and Michiels (1997*, p.29). It contrasts with definitions of enforcement in which enforcement is understood to only cover actions following a detected violation. Here, the detection of violations is also seen as part of the enforcement process.

1.2.1 *The state of the art in the literature*

The economics of law has outgrown its childhood. There is now a large branch of literature on the economics of contracts, torts, property rights, crime etc.[3] This thesis is concerned with the economics of law enforcement, both private and public. The enforcement of crime began with the analysis of Becker (1968). Calabresi (1970) initiated the analysis of liability rules. Both inspired many scholars to further extend the analysis, so that by now many aspects of these legal problems have been analyzed.

In short, the analysis inspired by Becker (1968) goes as follows. Because obedience of law is not taken for granted, public and/or private resources are required to prevent offenses and apprehend offenders. Moreover, punishment, sometimes severe punishment of convicted offenders is needed because conviction itself is generally considered an insufficient solution. This raises the important (normative) question of how many resources are optimal to spend on enforcing legislation. An activity is inefficient if it causes more harm to the victims and society than benefits to the criminal. If enforcement is costless and harm is always larger than the gain, eliminating crime is socially desirable. Generally, enforcement is costly so that the costs of enforcement should be balanced with the benefits of deterrence.

If the individual is risk-neutral, an expected fine equal to the personal gain will deter him from committing the crime. If apprehension is costly and imposing a sanction is not, enforcement costs are minimized by a probability of apprehension arbitrarily close to zero and a maximal fine sufficient to deter. Fines are far less costly than imprisonment because they merely involve a wealth transfer (income redistribution from the fined individual to the tax payers or other beneficiaries), while under imprisonment, socially important time is lost and resources are spent on prisons and guards. Therefore, imposing fines is preferred above imprisonment. Under a system of fines, it is optimal to impose the highest possible fine (constrained by for example the wealth of an individual) and realize a probability of apprehension that suffices to induce compliance. If the costs of imposing a sanction are non-negligible, it should be considered whether these costs, if realized, justify the deterrence that is gained by imposing this sanction.

This analysis has also been applied to the enforcement of administrative regulation, such as regulation of markets, the environment, occupational health, customer safety, etc. There exists a huge body of theoretical research into these areas as well as a lot of empirical studies. However, the analysis of

3 For an extensive summary, see Shavell (2004). See also the New Palgrave Dictionary of Economics and the Law (Newman, 1998) and the Encyclopedia of Law and Economics (Bouckaert and De Geest, 2000). The economic literature on law enforcement is summarized in Suurmond and Van Velthoven (2004*).

administrative law enforcement[4] as a distinct method of enforcement has so far been rather limited.[5] The existing literature usually studies criminal law enforcement of the regulation, or at least the imposition of fines. It mainly considers cases of the United States or other Anglo-Saxon countries. And it concerns national regulatory agencies. But many regulations are enforced by lower level governments like municipalities who are not empowered to impose fines. The (empirical) literature usually concerns the most serious violations that firms commit. Of course, this is of interest, but it neglects important elements of the enforcement problem. Less serious offenses are the bulk of violations and may, in fact, impose non-negligible threats to citizens. In addition, before regulatory agencies impose sanctions, (especially in the field of social regulation) there is often a time consuming process in which the regulatory agency tries to make violations undone. Only a small and non-representative part of the violations ends up in sanctions.

The neglect of administrative enforcement in the economic literature is unfortunate in view of the fact that the practice of administrative law enforcement remains quite puzzling. The standard economic model of law enforcement stresses the importance of punishing individuals and firms in order to deter them from committing a violation. However, in many (European) countries the sanctions that administrative bodies like municipalities are empowered to impose, do not include fines or imprisonment, but penal sums, administrative coercion and withdrawal of licenses, which grant firms time to restore compliance. Usually sanctions are preceded by (informal) warnings. From an economic perspective these practices seem to harm the deterrent effect of enforcement. However, the non-economic literature in particular argues that such an enforcement style, also called a *compliance strategy*, is much more effective than the *deterrence strategy* promoted by economists. Under a compliance strategy the enforcement official does not act as a 'policeman' but as an 'advisor'. Enforcement is carried out not by sanctioning, but by persuading the individual or firm to comply and by bargaining about the level of precautions. Compliance is the result of cooperative behavior on both sides. Formal sanctions are reserved for persistent violators. A compliance strategy does not assume that firms or individuals make conscious decisions, but instead recognizes that non-compliance is often the result of mistake or misinformation. People are willing to comply, but possibly incompetent to do so or of the opinion that compliance is unreasonable.

---

4 By administrative enforcement I mean enforcement by governmental (administrative) bodies like municipalities, Ministers, regulatory agencies, etc. In the Netherlands, administrative enforcement is a separate method of law enforcement, empowered in a special law (the *Algemene Bestuurswet*). The term should not be confused with administrative law and economics, which deals with the position of the government (the administration) relative to that of courts, with the optimal level of discretion etc.

5 Michiels (2006*, p.50) points to the fact that there are only 8 pages of bibliographical information on administrative law in the 4196 pages of the Encyclopedia of Law and Economics (Bouckaert and De Geest, 2000). Similar conclusions hold for other encyclopedic works.

The challenge is to examine whether these policies are really puzzling. Can the economic model explain the observed enforcement policies in the field of regulatory compliance of firms? Why do enforcement officials choose a cooperative enforcement style, and are they right in doing so? What are the effects of administrative law enforcement on compliance?

1.2.2 *The contribution of the theoretical part*

In part I of this thesis I analyze the economic literature on enforcement of safety standards, especially public enforcement of public regulation. It describes the conditions from the literature under which private, administrative and criminal law enforcement are effective and efficient for fire safety in the catering industry. This provides a framework for the empirical analysis in part II.

In particular, I examine the desirability of the compliance strategy described above. There are few systematic economic analyses of the debate on enforcement strategies. However, various articles and contributions contain arguments that can be related to this discussion. Therefore, in the first place, this literature review integrates the literature on public enforcement by considering the question whether and when a compliance strategy can be favorable. What are the benefits of an enforcement policy of advice, persuasion and warnings? By doing so, I provide an extensive review concerning the enforcement of regulatory crime from an economic perspective. To my knowledge, such a review is not available.[6] As said, the literature is rather fragmented in articles in different journals. Most theories concern common crime by individuals or crimes such as fraud, tax evasion and antitrust. Instead, I focus on non-compliance with safety standards by firms. The theoretical part results in a general description of the optimal enforcement method that can be used for analyzing any specific sector in this area.

It is important to analyze the enforcement strategy from an economic perspective for two reasons. First of all, in practice, many enforcement agencies seem to apply a compliance strategy. Therefore, it is important to know whether and when this can be beneficial. It is important to understand what explains the discrepancy between the economic model and the actual enforcement policies: the unrealistic nature of the economic model or the inefficiency of enforcement authorities? Secondly, studying the enforcement strategies may narrow the gap between economics and other social sciences on the recommended enforcement policies. Adding the economic analysis may improve our understanding of the enforcement strategies. I do not claim to provide a definitive and full answer on the optimal enforcement strategy. However, I want to address some important elements of the discussion from an economic perspective.

---

6 Heyes (2000) offers the most important, yet limited, overview.

1.2.3 The contribution of the empirical part

In part II (empirical analysis) I address the question of this thesis by considering the actual data on the enforcement of fire safety in the catering industry in the Netherlands. This is necessary because the optimal enforcement policy depends on the relevant cost functions. What is the optimal enforcement budget in the concrete? Part I shows that there are arguments in favor of both a deterrence and a compliance strategy. Which strategy, or: which mix of strategies, should be applied can only be answered by considering an actual case. Similarly, the literature presents several conditions for an effective use of private enforcement by liability rules or of criminal enforcement by fines and imprisonment. Whether these conditions are satisfied can only be determined by considering the facts and the actual costs and consequences. In part II I analyze whether the optimal enforcement policy derived in part I is properly applied in the enforcement policy of fire safety in the catering industry.

Such an empirical analysis is relevant because it improves our understanding of actual enforcement policies and its effects.[7] Theories are only useful if they enable us to explain or investigate the real world. Examining the enforcement of fire safety in the catering industry opens up part of this real world. This is the more important since by now the theoretical economic analysis of law is rather well developed. Today, the challenge is to demonstrate that this theory is useful in practice, in studying actual social problems.

The field of fire safety in the catering industry is a useful case to study. First, it is an important social problem that received much attention of both citizens and the government due to the Volendam disaster. Secondly, the catering industry is a properly defined group of firms, which is large enough to be a relevant study group and to allow for data collection. Thirdly, fire safety is a problem that can be mitigated by different enforcement methods and thus allows comparing these methods. It is also a field in which, on first view, compliance strategies seem to be applied and justified. Moreover, fire safety is a field that is not often studied in the literature.

1.2.4 The economic perspective

In my analysis I adopt the perspective of an economist. The literature on enforcement and compliance is much broader. Regulatory compliance by firms has also been examined in several other disciplines, such as criminology, (organizational) psychology and sociology, and the science of law and

7 The knowledge of actual effects is limited as Leeuw and Willemsen (2006*) and the National Court of Audit (2005*) conclude for the Netherlands, and Macrory (2006b) for the United Kingdom.

public administration (also known as socio-legal studies).[8] I do not discuss these branches of literature.

At this point in time, it is in my view more important to demonstrate the relevance of the economic perspective on compliance and enforcement than to try to integrate the different disciplines. Often, critics have made few efforts to offer a clear picture of the economic model of enforcement, but instead immediately begin to criticize, for instance because of the rational choice approach. A usual argument is that a deterrence strategy applies to rational offenders and a compliance strategy to non-rational offenders. I regard it as my task to show that the economic theories involved are much broader and more nuanced than the standard models usually referred to. Moreover, they are able to explain the use of both administrative law enforcement and of compliance strategies.

In addition, I hope to demonstrate that adding the economic perspective to both the academic and the policy debate on enforcement makes for an interesting analysis. It requires attention for the constant balancing of the benefits and costs of enforcement. In addition – as opposed to most other disciplines – it systematically focuses on the enforcement policy. It does not stop at examining which factors play a role in deciding to comply, but questions how these affect the (optimal) enforcement policy.

That this thesis is written for non-economists too, implies that the analysis assumes that the reader is not an expert in the economic analysis of law enforcement. If necessary I summarize arguments that may be well-known for this expert, but may help the uninformed reader to catch on. To be sure, as discussed above, the analysis of administrative law enforcement and of an enforcement style that is different from systematically punishing offenders, is also of interest for economists themselves. Moreover, this thesis examines the practical relevance of the economic models. What are the actual effects of enforcement actions? Studying a concrete, actual enforcement problem enables me to show that economic analysis is not only of academic interest, but also an important guide for dealing with real enforcement problems. Ultimately, the objective is not to provide theories, but to explain and solve problems in the real world. This is, or should be, the common interest of both economists and non-economists.

---

8 See Huisman and Beukelman (2007*) for an overview. An example of an interdisciplinary approach to compliance and enforcement is Elffers et al. (2006).

1.3 METHODOLOGY

The main question of this thesis is: what is an effective and efficient method of enforcing fire safety in the catering industry?

1.3.1 *The analysis of the economic literature*

Part I of this thesis provides an overview of the economic literature on law enforcement as it applies to the enforcement of social regulation of firms. The leading question in this part of the thesis is whether the economic theory provides an optimal economic policy for the enforcement of fire safety regulation in the catering industry. I discuss the conditions for efficient enforcement through private, administrative or criminal law enforcement. I only outline the analysis of safety standards through private law enforcement, because this analysis is well-known, at least for law and economics scholars.[9] In detail the optimal public enforcement policy is examined, especially in relationship to the optimal enforcement strategy. The literature discussed is selected by considering whether it provides an argument for the imposition of small sanctions, or even for the use of advice, persuasion and warnings or other enforcement tools. This selection covers many different themes. I do not, and do not intend to, analyze all the literature on themes such as intrinsic motivation, legal procedures or discretion. The literature is selected and discussed in relationship to the question how it affects the optimal enforcement policy, in particular the enforcement strategy.

This literature review only concerns the theoretical literature. To limit the scope of this thesis, I pay only limited attention to the empirical literature concerned with testing the theories. Furthermore, I focus on the literature on the optimal enforcement policy and do not broadly examine the social and economic factors that explain non-compliant behavior by firms.[10] Moreover, as mentioned above, I only review the economic literature, although I do attempt to relate it to the general debate on deterrence and compliance strategies.

Since the focus of this thesis lies on the analysis of optimal enforcement of fire precautions in the catering industry, I will focus on the enforcement of safety standards for firms, i.e. so-called regulatory crime. Regulatory or business crime differs from organized crime.[11] Organized crime is crime committed by illegal organizations with a criminal motive, and aimed at supplying illegal goods or services (such as drugs, prostitution, gambling, or human

---

9 Shavell (1987) is the most well-known textbook. See also Shavell (2004).

10 For the explanation of criminal behavior, interested readers are referred to the reviews in Eide (2000) or Ehrlich (1996). See further Heineke (1978), Eide (1994) and Van Velthoven (1994*). For firms overviews can be found in Lott (2000) and Huisman and Beukelman (2007*).

11 A readable introduction to economic crime is found in Sjögren and Skogh (2004).

trafficking). Business crime is crime by corporations in order to gain extra profit within an otherwise legal business (such as environmental or occupational violations). Society does not want to restrain the activity or the level of activity. Contributions on business crime tend to focus on (tax) fraud, antitrust and environmental non-compliance. In common crimes by individuals, as well as in fraud, tax evasion or antitrust violations, the standard often concerns a forbidden activity. In fire safety regulation the standards concern a level of precautions that must be satisfied. If the firm chooses a level of precautions lower than the standard, it can be punished. There are costs of compliance instead of benefits of violation. The literature I discuss mainly originates from the analysis of the enforcement of environmental, safety and health regulation for firms.[12]

1.3.2 *Methodology of the empirical part*

Part II of this thesis focuses on analyzing the enforcement policy for fire safety regulation in the catering industry in the Netherlands with the actual data. I evaluate the currently observed enforcement policies, in particular the administrative enforcement efforts following the Volendam disaster, and try to determine which enforcement policy is efficient. Several methods are used to carry out this evaluation. The problem with analyzing actual enforcement policies is to find or collect the right data on compliance and compliance benefits. I try to do so in a number of ways.

First, I conduct interviews with enforcement officials in thirteen municipalities to acquire insight into actual enforcement policies. These interviews provide mainly qualitative data. These data, together with the available information in policy documents, in (management) reports, on websites and the like, are used to examine whether the theoretical predictions on the optimal enforcement strategy are applied. This analysis yields insight into the question whether the administrative enforcement policy that is currently applied in almost all municipalities, is the most effective enforcement policy.

Secondly, I combine bits and pieces of available evidence into a cost-benefit analysis of the enforcement policies following the Volendam disaster. This information is obtained from national government agencies and the branch organization of the catering industry. A cost-benefit analysis tries to list all effects of a policy intervention. Monetizing these effects allows them to be compared and to evaluate whether the intervention leads to a positive or negative contribution to social welfare. I use a cost-benefit analysis to answer the question whether the increase in administrative enforcement efforts due to the Volendam disaster has been a socially desirable investment. This yields

12 So called social regulation. Different types of regulation are categorized in Faure (2000), den Hertog (2000) and Ogus (2001).

insight into the question whether administrative law enforcement is an efficient enforcement method for fire safety in the catering industry.

Thirdly, I analyze the available information about the settlement of claims for compensation of victims following the Volendam disaster. This case study is of particular interest since it received relatively much attention and is therefore likely to have a relatively large influence on proprietors' future expectations. Moreover, there are simply not enough compensation claims and not enough knowledge concerning compensation claims to carry out a more general analysis. Information about the case of the Volendam disaster can be found in reports and messages of victim interest groups, the municipality and other relevant organizations, in parliamentary documents and in newspapers. With these sources it is possible to reconstruct the payments that the different parties have made to the victims. The main question in this context is: who is actually paying for the harm caused? I analyze these payments by considering whether they provide a sufficient incentive to take precautions. This yields insight into the question whether private enforcement through liability rules is effective.

Finally, I carry out a simulation of different enforcement policies for a representative, hypothetical municipality. The construction of this municipality is based on all available data. Assuming that the proprietor of a bar or restaurant is a risk-neutral wealth maximizer I analyze the effects of different enforcement policies. Not only are the currently observed policies examined. This simulation allows me to analyze alternative enforcement policies and to examine which policy might lead to the highest level of social welfare. Of course, the disadvantage of such an analysis is that it abstracts from reality. However, without it, it becomes impossible to further analyze the effects and the social welfare consequences of enforcement on compliance.

The available data and the methods applied with these data do not provide unambiguous answers to the question of the efficient enforcement policy. There are no 'hard', quantitative data on the effects of enforcement. This is not a good reason to refrain from an empirical analysis. Precisely because these data are absent, the methods used are necessary to improve our understanding of actual enforcement policies.

1.4 Definitions and concepts

It may be helpful for non-economic readers to describe the meaning of several concepts that are used. I also introduce the outlines of the case study analyzed.

1.4.1 *Enforcement as social welfare maximization*

The main question of this thesis refers to the (economic) concepts of effectiveness and efficiency. Enforcement is effective if it leads to a higher level of compliance. Efficiency refers to the desirability of enforcement. From an economic point of view, the social objective is to maximize social welfare, which is an aggregate of the welfare of all individuals involved. I assume that all individuals (and firms) are equally important, and thus ignore the income distribution. Income distribution is not the objective of the regulation of fire accidents. Society may, in fact, be interested in the (income) position of actual victims. However, there are better measures to help these victims than regulation and enforcement, such as the social security system, public health care, and/or a collective damage fund (Kaplow and Shavell, 2001; Shavell, 2004).[13] Moreover, people can insure themselves against potential losses.

So, the social objective is the maximization of total well-being of all individuals in society. Because in regulation and enforcement the interests of the injurers and victims are generally opposed, the different effects of enforcement have to be compared. Put differently, because we are unable to find actual Pareto improvements, we have to look for potential Pareto improvements (Hicks-Kaldor criterion). According to the Hicks-Kaldor criterion, welfare is improved if the gain in utility of the winning party exceeds the decline in utility of the losing party. Therefore we must be able to compare the utility of the different parties, for example by estimating utility in income or wealth. Utility can be measured in monetary expressions by determining the willingness to pay for and/or to accept changes in starting positions.[14]

Enforcement is called efficient if social welfare is maximized. Spending an extra euro on enforcement is efficient, if it leads to an at least equal reduction in the monetary value of expected damage and compliance costs. Judging the efficiency of enforcement implies that the effectiveness of enforcement is known. After determining the extent to which enforcement increases compliance, it is possible to determine the benefits of this increase in compliance and to compare it with its costs. In this respect, analyzing efficient enforcement includes analyzing effective enforcement. For clarity both are explicitly included in the main question.

1.4.2 *Rationality*

In economics, it is assumed that individuals and firms behave rationally, i.e. they make choices in ways that maximize their expected utility. They evaluate the expected outcomes of the relevant alternatives and choose the most promising one, given the available information and the constraints of wealth, time and other relevant factors.

This notion of rational behavior can easily be misunderstood. Therefore, it seems useful to shortly describe what the meaning of the rational choice

---

13 See also Van Velthoven and Suurmond (2003*) on a discussion of the social welfare (efficiency) criterion and the possibilities to include the income distribution.

14 E.g. Shavell (1987).

approach is.[15] It means that behavior is purposeful, intended to fulfill the ends with the available means. An individual reaches a decision by systematically balancing the costs and benefits. Rationality only deals with the 'procedure' of making a decision (choosing the alternative that seems to fit the personal goals best), not with the 'rationality' of preferences itself. According to economic theory individuals pursue their own interest, but the nature of an individual's interest is not restricted in any way. Genetic endowment may influence both preferences and constraints. Someone who pities hungry people in Africa behaves rationally by becoming a relief worker. Someone who is jealous of his neighbor and cannot stand the differences in income, may behave rationally by rejecting an offer in which he is granted €100 and his neighbor €200.

In the same way, since Becker (1968), criminal behavior is studied in terms of purposeful behavior, without imposing restrictions on the purposes. Since rationality means nothing more or less than choosing the alternative which serves the preferences of an individual best, complying with the rules on social or moral concerns (voluntary compliance) can be viewed as a rational choice. Hence, voluntary, spontaneous compliance is also possible. Similarly, non-compliance due to a preference for unlawful behavior or the influence of group behavior does not necessarily imply that an individual is irrational. It is only when an individual is not influenced by any risk of sanctions at all that we can say that this individual is irrational, since he does not take all costs and benefits into account.[16]

A better founded criticism on rational choice theory is that individuals do not (always) make conscious, well-considered, well informed decisions, and do not always choose what is in their own best interest. Information is not necessarily perfect, and hence 'incorrect' decisions may occur, because information and deliberation costs are too high. Individuals may have important cognitive problems to update and process all available information. This is incorporated in the research founded by Nobel Laureate Herbert Simon on *bounded rationality*. Because maximization is sometimes too difficult, individuals choose an alternative that is 'good enough'.[17] They are sensitive to the way alternatives and information is presented (*framing*) and the process

---

15 A clear exposition of rational choice is given in Schäfer and Ott (2004, chapter 4). Eide (1994) in particular provides a handsome discussion of the use of rational choice theory for crime. "The norm-guided rational choice model, comprising personal norms and wants, feasible sets of activities, environmental characteristics, and individual outcomes, provides a suitable framework for discussing various theories of crime, including characteristics of individuals and circumstances. The framework allows for a simultaneous consideration of many possible determinants of crime. The abstract model is a means of gaining insight into the elements of rational behavior, and it permits filling bits of information into a broader context" (p.45). Elffers (2005*) also discusses how rational choice theory can be used to analyze crime.

16 Although even this might be a rational choice if the gains of non-compliance are regarded as infinitely large for the individual concerned.

17 See Schäfer and Ott (2004, p.59 and further).

of decision-making is based on simple rules (*anchoring*). Individuals have a *status quo bias* (reference effect). They have trouble estimating the correct outcome associated with an alternative. For instance, individuals tend to overestimate their own abilities and have difficulties assessing the correct opportunity costs. They also have difficulties estimating the correct probability of events. Individuals have difficulties assessing the ordered sequence of events, so that they underestimate unlikely scenarios. More generally, they have problems estimating small probabilities; they tend to set them to zero. But if large losses are involved, the impact of the event is over-estimated. People also make use of hindsight (i.e. overestimate a probability if an event has (recently) happened), because it influences the availability heuristic. The consequences of bounded rationality for public enforcement will be shortly discussed in section 3.1.4.

1.4.3 *The case of fire safety in the catering industry*

The enforcement of fire safety regulation in the catering industry in the Netherlands serves as my case study. It is therefore useful at this point to describe this case study's most important characteristics.

Horeca establishments

The catering industry will be labeled 'horeca'. *Horeca* usually stands for *ho*tels, *re*staurants and *ca*fés, as well as all similar establishments where food and drinks are supplied. The term is sometimes used in a wider sense, including all tourist and/or recreational businesses. In this study, I use the term horeca to indicate a limited group of these businesses, namely those where food and/or drinks are served that are consumed on the premises: restaurants, cafés, bars, pubs, discothèques and party centers. Similar prescriptions apply to all these establishments. Not included are hotels, which have to fulfill stronger requirements (like sprinklers) because they offer the opportunity to spend the night, a time when people are more vulnerable to fire. Moreover, in contrast to cafés and restaurants, hotel guests are spread over many rooms. Snack bars and other small take-away restaurants are also excluded because they do not have to take precautions to help people escape in case of fire since there are only few people present.

All these horeca firms are characterized by a rather simple organization with few hierarchy levels. The manager is usually the person who receives the profits of exploiting the horeca establishment and who can directly monitor the behavior of his (limited number of) personnel.[18] As a consequence, principal agency problems are ignored.

---

18 There are a few larger chains, notably fast-food restaurants, hotel-restaurants or roadhouses (e.g. McDonald's), which are characterized by a professional and more complex organization.

The use requirements

Fire safety prescriptions are divided in requirements with respect to construction, installations, and use. I focus on the requirements concerning the use of an establishment. These requirements include that (1) escape-routes and emergency exits should be kept free; (2) furnishings and decorations should be correctly hung and impregnated; (3) escape-route signs should be clearly visible and burning; (4) fire extinguishers should be clearly indicated, ready to hand and annually certified; (5) candle lights, trash, ash-trays etc. should be used safely; (6) the maximum number of persons allowed should not be exceeded; (7) remaining use requirements, for example those concerning technical devices or evacuation plans, should be obeyed.

With respect to fire safety, two important developments were recently initiated by the Ministry of Housing, Spatial Planning and Environment. First, a process of simplifying regulations was initiated. In this context, unnecessary regulations have been abolished and other regulations harmonized and nationalized. This includes the harmonization of all building and fire safety regulations into one national Use Decree (now planned for mid 2008). As a consequence, exploiters of a building no longer have to possess a use license. However, they still have to satisfy the requirements described above. Under the Use Decree, these requirements can be enforced directly. Moreover, a general duty of care will be introduced, so that the exploiter of a building is responsible for its sound use. Secondly, a single spatial permit will be introduced in 2008, which will replace all licenses in the field of physical safety, such as the environmental license, building license, use license etc. Firms have to apply for a single license at a single governmental agency. The Fire Departments are expected to play an important role in granting the license and inspecting fire safety.

International comparison

The regulation and enforcement practice of the Netherlands is similar to that in most other (Northern) European countries.[19] Most European countries have established national fire safety standards in a Building Code or specific Fire Code. Local government officials are generally responsible for enforcement. Local fire brigades or building officials have the opportunity to inspect buildings and are entitled (if necessary) to carry out enforcement measures, such as a notice to comply, closure of the business, and – if violations are severe – fines or the start of criminal proceedings.[20] The major differ-

19 See Meijer and Visscher (2004*) and the reports on which their article is based. The authors have compared the system of building control in eight European countries, to name: Belgium, Denmark, Germany, England (and Wales), France, Netherlands, Norway and Sweden. The comparison is based on the formal description of the systems, not on what the systems bring about in practice.

20 This information can be found on www.feucare.org (25-07-2006), a website that provides information on legislation of fire safety, including prevention, for most European countries.

ence is that in the Netherlands exploiters of a property require a use license, whereas most other countries enforce this type of regulations directly. In the near future, this will also be the case in the Netherlands. As mentioned above, other countries have experienced disasters similar to the one in Volendam and have observed similar enforcement failures.

1.5 Plan of this thesis

This thesis consists of two parts. In part I, I will provide an overview of the economic literature on law enforcement as it applies to the enforcement of social regulation of firms. In part II, I analyze the enforcement policy of fire safety in horeca establishments empirically.

1.5.1 *The structure of the theoretical part*

The main question of this thesis concerns the effective and efficient enforcement of fire safety in horeca establishments. Before considering enforcement actions, the question is why these actions are necessary at all. What actually is the problem that enforcement has to solve? Can't we expect proprietors to comply with the regulations by themselves? Can't we expect that the market will solve the problem of fire accidents? In chapter 2 I address these questions. I first describe the accident problem. I explain why I focus on the analysis of the incentive to take precautions and do not concern myself with compensation. I provide arguments why we can not expect that proprietors choose compliance voluntarily or that they are forced to comply by their customers. Therefore we have to rely on regulation of the market, and as a consequence, we have to consider the optimal enforcement policy for this regulation. A first alternative is private enforcement through liability. Section 2.3 provides a summary of the conditions for effective and efficient private enforcement through liability rules. As such, chapter 2 does not provide new analyses but relies on the standard literature available in textbooks and tries to relate this to the case-study of fire safety in horeca establishments. This is useful for the empirical analysis of private enforcement in part II.

The remainder of the theoretical part is devoted to public law enforcement (both criminal and administrative). I study the question which public enforcement policy provides an adequate incentive to take precautions. In particular, I consider the difference between a compliance and a deterrence strategy. My goal is to show that the economic literature has more to offer than simply recommending sufficiently severe sanctions, aimed at deterrence, and that it is able to provide the conditions under which a compliance or deterrence strategy would be the optimal enforcement policy. In order to be able to participate in this debate, I first survey the literature, which enables me to reflect on the discussion about the enforcement strategy later on.

The analysis starts from the standard economic model of law enforcement, as founded by Nobel Laureate Becker (1968) and further extended by a number of scholars. This model implies: (1) a higher expected sanction decreases the level of non-compliance, (2) costly non-monetary sanctions should only be used when costless monetary sanctions have already been used to the maximum, (3) decreasing the probability of detection and increasing the magnitude of the sanction might ensure deterrence but save enforcement costs, hence it is socially desirable. However, important questions remain, in particular with regard to the empirical challenge resulting from the fact that these results are not commonly observed in practice. The analysis evaluates the extent to which the trade-off between detection and sanctions is indeed the optimal enforcement policy. Is it efficient to let enforcement authorities punish all violations with relatively severe, extreme sanctions? Or are *compliance strategies* more effective than *deterrence* strategies? What is the level of compliance that the enforcement authority is trying to achieve? To answer these questions I undertake the following steps.

First of all, it is necessary to study the enforcement problem in more detail. In chapter 3 I consider how the efficient enforcement policy depends on the information available to the enforcement authorities, on the apprehension technology and on the nature of the sanction. This is a review of the most conventional economic literature on enforcement following Becker (1968). I show that the trade-off between the probability of detection and the magnitude of the sanction is only possible if the level of detection is determined for every single offense separately. Otherwise, the possibilities of lowering the probability of detection are limited and extreme sanctions must be reserved for the most severe offenses; while minor offenses should be punished mildly, or even not enforced at all.

Thereafter, in chapter 4, I extend the analysis to offenses that can be committed more than once. Will compliance be continued? Two important conclusions follow. First of all, it may sometimes be optimal to punish repeat offenders more severely. In that case, the possibilities of trading-off sanction and detection are limited for earlier offenses. Secondly, the objective of enforcement may include remediation, which might require sufficiently high probabilities of detection. Remediation might require other enforcement actions than fines or imprisonment. Moreover, I consider how enforcement is affected when offenses occur both accidentally and deliberately. Generally, it is optimal for firms to report their (accidental) violations themselves, which requires non-extreme penalties.

At this stage in the discussion, it has become clear that it is often impossible to induce full compliance with the standards. Generally, there will be a certain degree of underdeterrence. Chapter 5 shows that if underdeterrence is rather severe, it might be optimal to focus or target enforcement instead of uniformly enforcing a standard. The best targeting strategy – one that maximizes the partial compliance rate – is one in which enforcement depends on past compliance (for example through a warning), extending the previous chapter's discussion on repeated offenses.

Chapter 6 focuses on the influence of the standards on enforcement. This influence can be twofold. First, some uncertainty about the application of the rules to a specific offense may be unavoidable. This in turn might require that the enforcement authority follows an enforcement process in which this application is made clear and/or bargains with the firm about compliance, which might thwart an enforcement policy such as proposed by Becker (1968). Secondly, in cases of uncertainty about the norm and its application, firms may challenge an enforcement authority's decisions, thereby hampering the effectiveness of enforcement. Because higher sanctions provide higher incentives to contest enforcement, it may be optimal to impose non-extreme sanctions. Flexible compliance and enforcement might be required.

Finally, it is important to consider that firms are not only motivated by pecuniary costs and benefits, but also by non-pecuniary ones. In chapter 7, I no longer assume that a firm's decisions only follow from the (direct) cost of compliance and the costs of enforcement actions, but also from its social environment and/or its moral intentions to comply. Compliance might evolve voluntarily or spontaneously. This may yield complicated interactions with public enforcement.

Once these points have been examined, it is possible to return to the debate about the deterrent effect of enforcement in chapter 8. In particular, I discuss why enforcement agencies should adopt a *deterrence* or a *compliance* strategy of enforcement.

Chapter 9 focuses on the structure of incentives for enforcement officials. This is important for two reasons. First, the fact that enforcement officials are not necessarily maximizing social welfare, limits the effectiveness and efficiency of the enforcement policies described before and may affect the optimal enforcement policy. Secondly, in particular, incentives are of interest for the discussion on the optimal enforcement strategy because they explain why enforcement officials might choose an inefficient enforcement strategy. On the other hand, incentives can be structured. The analysis explains that creating incentives for one-sided enforcement objectives and strategies (for instance strict punishment of all violations) can be more welfare-enhancing than the discretion to act in society's best interest.

Chapter 10 concludes the theoretical part by summarizing the conditions for the optimal enforcement policy. In discussing the optimal enforcement policy for a specific problem, it should be verified whether these conditions are satisfied. There are several important elements of the enforcement policy. First, I describe the conditions for the optimal timing of enforcement (ex ante or ex post). Secondly, I describe whether public law enforcement should be through criminal or through administrative law. Thirdly, I describe the factors that determine the choice between private and public enforcement. I also investigate the joint use of private and public enforcement. I provide the conditions for the use of a deterrence or compliance strategy, related to the objective of the enforcement actions. I conclude by summarizing the conditions for an effective and efficient use of private, administrative and criminal enforcement and discussing their relevance for the case of fire safety in horeca establishments.

1.5.2 *The structure of the empirical analysis*

Further conclusions can only be drawn once the analysis is filled in with specific, empirical data. In the second part, I apply the analysis to the enforcement policy for fire safety regulation in horeca establishments in the Netherlands. I evaluate the currently observed enforcement policies, in particular the administrative enforcement efforts following the Volendam disaster, and try to determine which enforcement policy is effective and efficient.

I begin by analyzing the policy and enforcement actions that have been adopted following the Volendam disaster. First, in chapter 11, I provide a more detailed description of the case-study of the enforcement of fire safety in horeca establishments. I will discuss the context and the recent pressure to professionalize enforcement.

Secondly, in chapter 12, I describe the current administrative law enforcement policies in 13 municipalities. I describe the interviews I carried out to obtain information on these policies. I interpret these policies in light of the economic literature described in part I. Are these policies effective? Particular attention is given to the evaluation of the use of a compliance strategy.

The third step, in chapter 13, is to question whether the enforcement actions are efficient. I conduct a cost-benefit analysis to answer the question whether the increase in administrative enforcement efforts due to the Volendam disaster has been a socially desirable investment. Chapters 12 and 13 thus analyze the question whether the currently used administrative enforcement policy for enforcing fire safety in the catering industry is effective and efficient.

A further step, in chapter 14, is to consider whether private actions by victims, such as those of the Volendam disaster, may, through filing and settling damage claims, be expected to be (more) effective. Therefore, I study the settlement of the compensation to victims of the Volendam disaster.

The analysis of the actual enforcement actions following the Volendam disaster reveals that important conclusions can be drawn but that many questions remain unanswered. This is due especially to a lack of information on the relationship between enforcement and expected fire damage. Therefore, in chapter 15, I take a different approach. I carry out a simulation of different enforcement policies for a hypothetical but representative Dutch municipality. I no longer examine only currently observed policies, but examine which enforcement policy is most welfare-enhancing.

Finally, chapter 16 concludes. After summarizing the most important findings, I try to answer the main question. Which policy is recommended for the enforcement of fire safety in horeca establishments?

Part I

Theory

In part I of this thesis I discuss the economic literature on the optimal solution to the enforcement problem of fire safety regulations in horeca establishments. I therefore first (chapter 2) generalize the problem of fire safety to a general accident problem to fit the model from the literature. I provide the socially optimal solution to the accident problem and discuss why this solution requires enforcement. Finally, in chapter 2, I examine the conditions under which private enforcement by victims (through the liability system) is efficient. Next I examine public law enforcement (chapters 3 to 9) according to the steps described in the Introduction (section 1.5.1). Most attention is given to act-based enforcement, particularly through administrative law enforcement. Attention is especially drawn to the question how a system of advice, persuasion and warnings (so-called compliance strategies) may be beneficial from an economic perspective. Chapter 10 concludes part I by discussing the optimal structure of law enforcement.

2 The social problem and the optimal solution: Why enforcement is needed

This chapter discusses why enforcement of fire safety precautions is necessary. In section 2.1 I first describe the general accident problem examined in part I of this thesis. This leads to an analysis of the social problem and the optimal solution of the accident problem. In section 2.2, I discuss why this socially optimal solution cannot be obtained without enforcement of the rules. This conclusion builds a bridge to the study of enforcement methods. In section 2.3 I examine whether private enforcement through liability rules solves the accident problem. The next chapters are concerned with public enforcement policies.

2.1 The accident problem

The social objective is the maximization of total well-being in society, where social well-being depends only on the well-being of individuals.[21] The optimal solution to the problem of fire safety in horeca establishments can thus be found by studying the consequences of more or less safety for the well-being of horeca proprietors and visitors. The fire safety problem can be formulated to a general version of the accident problem that is studied in the literature: a firm, also called the injurer, can take precautions to reduce expected harm to some unknown potential victims. These precautions prevent accidents from occurring and/or limit the harm when it does. Of course, taking these precautions and complying with the regulations is costly for the firm.[22]

2.1.1 *The utility of victims and injurers*

The expected harm to victims

When victims are more or less severely injured or even killed in an accident, the consequences for them are threefold. First, a victim being injured or killed leads to monetary costs from medical expenses, costs of decease, and lost earnings from labor participation. These costs are born either by the victim himself and/or his relatives, or by his insurance company.

Secondly, an injury or fatality leads to pain and suffering, to forgone joy of life, by both the victim and his relatives. This non-monetary damage is

---

21 Kaplow and Shavell (2001).

22 A formal presentation can be found in for instance Shavell (1987).

characterized by its irreplaceability as this pain and suffering can never be recovered by receiving similar 'goods' or money, although money can somewhat soften the pain (see Cook and Graham, 1977; Arlen 2000).

Besides these direct damages, an injury or fatality may also influence the value of money, or: the marginal utility of income, for the victim or his relatives (Friedman, 1982, 2000; Viscusi, 1997). Particularly in the case of permanent injuries, the value of money may decrease after an accident, because due to the injuries the victim can derive less utility from his income. Someone who is paralyzed or in a coma, cannot enjoy holidays or visiting a restaurant. In the most extreme case, money has no value for a dead person (apart from the value of bequests).[23]

Besides the expected monetary and non-monetary losses, the utility of a victim depends on his initial level of wealth. A potential victim is generally risk-averse so that his utility is increasing in his level of wealth, but at a decreasing rate. In addition, (potential) victims bear the costs of enforcement, either as actual victims for private enforcement or as tax-payers for public enforcement. On the other hand, they may receive part of the monetary value of the sanction that is imposed on the injurer, in the form of a fine or damage compensation.

What firms care about

The firm must decide on the level of precautions intended to limit the expected harm from accidents. He does so by comparing the relevant costs and benefits. Of course, these involve the (direct) costs of taking precautions (*compliance costs*), such as the time or money spent on taking precautions, or the extra profit that can be made from violating the rules. The firm also considers the *sanctioning risk* that follows from the enforcement activities of the government or the victims. These may lead to costs such as damage compensation, a formal sanction (the payment of a fine or the costs of a (temporary) closing of the establishment), and/or the costs in terms of time and money (paperwork) spent on inspections or on objection and appeal procedures. These costs should be discounted by the probability that a firm has to make these costs. The product of this probability and the magnitude of the 'sanction' is labeled as the sanctioning risk. So the sanctioning risk reflects the expected costs for the firm in case of non-compliance due to enforcement activities. This sanctioning risk depends on the level of precautions taken by the injurer. If the injurer takes more precautions, it is more likely to escape sanctions.

Large firms may be neutral to the risks they face. Since they have large assets they can bear large losses. However, small firms, including most horeca proprietors, are probably risk-averse. Firms face two types of uncertainty which are relevant if they are risk-averse. First of all, it is uncertain whether an accident will occur. Secondly, it is, in general, uncertain whether a sanction will be imposed.

Of course, the injurer such as a proprietor may not only be motivated by

23 Empirical evidence is presented in Viscusi and Evans (1990) and Sloan et al. (1998).

the direct costs of taking precautions, but also by other *compliance-enhancing concerns*. There may be, for instance, *personal concerns* that reflect that taking precautions more or less directly increases the benefits from exploitation of the establishment. These benefits include the reduced probability that the proprietor himself or his establishment will be harmed in a fire accident, the reward for taking precautions offered by the consumers, or a benevolent status with government agencies which might be helpful in the context of other activities. *Social concerns* reflect the fact that the injurer may be concerned about (dis)approval of his behavior by others for other than economic motives. The injurer may be concerned about the evaluation of their behavior by friends or relatives, neighbors, or colleagues. *Moral concerns* reflect independent, intrinsic motivations for taking precautions. The proprietor may be concerned with the safety of his customers, because of (altruistic) regard for other people. Or he may feel the need to obey governmental prescriptions as a civic or moral duty.

A comment
The description above of the accident problem contains some simplifications that are worth noting. I consider so-called *unilateral* accidents (Shavell, 1987) in which only the injurer can take precautions to prevent and limit expected harm. The role of victims is neglected as they are no part of the described regulations.[24] Furthermore, there is no concern for the activity level. This is justified because the level of activity may be part of the standard, as is for example the number of visitors allowed. Moreover, the level of activity is not a social problem if the activity is always desirable given that efficient precautions are taken (the benefits of the activity are constant per period). The production level will be efficiently determined by the market. I assume that if the proprietor complies, exploiting the establishment is privately and socially beneficial. The visitor is always able to enjoy the benefits of a horeca visit.

2.1.2 *The socially optimal solution to the accident problem*

The socially optimal solution to the fire safety problem should satisfy two requirements[25]:

1. *Optimal risk-spreading*: The wealth of a potential victim should be optimally distributed between the "no-accident state" and the "post-accident state". This objective is labelled *compensation*.
2. *Optimal precautions*: Injurers should choose a socially optimal level of precautions. This objective is labelled *deterrence*.

24 Although the way visitors deal with candles, ashtrays or (as in the Volendam disaster) sparklers do affect harm, it can be argued that this behavior is also the responsibility of the proprietor as he and his personnel are able to control the extent to which it is possible or allowed.

25 The analysis in this section is described in Suurmond and Van Velthoven (2005*), particularly in relation to the optimal magnitude of compensation damages.

Optimal compensation and the demand for insurance

The well-being of a potential victim depends on whether or not there is an accident.[26] The maximization of someone's well-being, and hence that of society, implies that his wealth should be optimally spread between the "no-accident state" and the "post-accident state". This implies that compensation for losses (in the "post-accident state") should be at the optimal level.[27] This objective is therefore labeled *compensation* and is defined as the demand for insurance, as it equals the level of monetary compensation that a victim wants to receive in case of an accident. Remember that monetary losses are replaceable, so monetary compensation allows replacing these losses. Non-monetary losses are irreplaceable, so compensation is only possible in the form of monetary benefits with which other goods or services can be bought, through which the pain may be softened.

Risk spreading is optimal if an individual does not want to transfer money from the state of the world in which no accidents happen, into the other state in which accidents do happen, or vice versa.[28] Transferring money to the other state is not worthwhile if receiving additional money has the exact same value in both states of the world, i.e. if the individual can not make himself better off by giving up €1 in one state of the world to receive it in the other. If an accident does not affect the value of money, risk-averse victims will always want to have the same level of wealth and they will want to eliminate all uncertainty about their level of wealth. Hence, victims will want to insure themselves against all monetary losses. If, however, money is less valuable to victims after an accident than before, they will want to have less money in case an accident occurs than in case no accident occurs. Money that is given up when no accident occurs is relatively expensive. The individual would have to give up pleasant activities (such as holidays or visiting a restaurant) in good health, while being unable to undertake these activities when in bad health. Therefore he will want to insure himself against less than the monetary losses. If the monetary losses are relatively small, an individual might even want to obtain 'negative' insurance, i.e. he might wish to buy an insurance contract that pays out in case no accident occurs.[29]

---

26 The same argument applies to the risk spreading of the injurer when he faces uncertainties like sanctions or compensation damages. In the next chapters this problem is not further dealt with, except for the consequences of insurance for the incentive to take precautions (section 2.3.2).

27 See for instance Shavell (1982, 1987, 2004), Friedman (1982, 2000), Arlen (2000), Schäfer and Ott (2004). Note that the objective of compensation does not imply concern about the income distribution (Shavell, 1987). We are not considering whether wealth is optimally distributed between the injurer and the victim. However, we are considering whether the wealth of the victim (or the injurer) is optimally distributed between his possible states of the world.

28 See the aforementioned references.

29 Such insurance contracts are problematic as individuals would have to pay once an accident occurs, creating possible credit constraints and potentially high administrative costs because of the frequency of pay-out and the lack of incentives to report accidents (Friedman, 1982, p. 91; Arlen, 2000, p. 700). Negative insurance is not observed in reality.

The potential victim would never want to insure himself for the non-monetary loss of pain and suffering, as this loss is irreplaceable. Suppose someone might loose his child in an accident, which occurs with a probability of 1%. There are no monetary losses. He can buy an insurance contract that will pay out €100,000 in case of an accident. Without insurance, the individual will have a *certain* wealth of for instance €25,000 and a probability of 1% of suffering the severe pain of losing his child. If he insures himself, he will have a wealth of €25,000 - €1,000 (the actuarially fair insurance premium) = €24,000 if no accident takes place, and €25,000 - €1,000 + €100,000 = €124,000 if there is an accident, on average €25,000. However, this does not prevent the possible loss of his child. So the individual faces an *uncertain* expected wealth of €25,000 as well as a probability of 1% of the severe pain and suffering of losing his child. For a risk-averse individual this can never be beneficial. If all other things remain equal, he would want to have certainty about his wealth.

That potential victims are left with uncertainty about their level of wealth as a result of a potentially severe monetary loss is thus an important element of the accident problem. However, as Shavell and others have argued, it is not something we should worry about when analyzing enforcement problems.[30] The reason is that an individual can provide himself with the desired level of compensation through buying insurance. If insurance is freely available on the private market, the optimal level of compensation can always be obtained. Insurance is available on the private market if (among others) the following conditions are satisfied:

- Administrative costs are small enough to prevent refraining from risk-spreading, and insurance premiums are sold in a competitive market, hence actuarially fair.
- The insurance market is not characterized by adverse selection, i.e. insurance companies are as informed about risks as the insured parties. For fire safety in horeca establishments, this seems to be a reasonable assumption.
- The insurance market is not characterized by moral hazard, i.e. insured parties do not take fewer safety precautions once insured. For victims, this is not a problem as they are assumed to be unable to influence expected damage.[31]

If these conditions are not satisfied, an individual obtains no compensation or a suboptimal level of compensation in the private insurance market. However, even if this is the case, compensation should not be the objective of regulation and enforcement, as other government actions might be more effective or efficient in providing compensation, such as regulation of the insurance market or providing public compensation by means of the social security and tax system, public health care, or a compensation funds.

Therefore, I focus on the importance of enforcement in creating deterrence, the second element of the accident problem.

30 See for instance Shavell (1987, 2004, 1984a), Friedman (2000), Schäfer and Ott (2004) and Arlen (2000).

31 For injurers this may prove to be a severe problem, as discussed in section 2.3.2.

Efficient precautions
The second problem is that the level of precautions taken by the firm should be efficient. It is not efficient to prevent harm at all costs but to balance the costs of taking precautions against the reduction in expected harm, both monetary and non-monetary. As a consequence, it is not necessarily efficient to fully comply with the regulations. Spending €1 more on precautions is only efficient if it reduces the expected harm by more than €1. The efficient level of precautions is obtained once the marginal costs of taking precautions equal the marginal benefits of taking precautions.[32] Therefore we have to compare the costs of precautions with the costs of expected harm.

The monetary losses can be estimated relatively easily, based on the costs of replacement or recovery. Market prices are often available. For non-monetary losses, there are no markets, but they can be based on the ex-ante willingness of potential victims to pay for higher precautions, or on the willingness to accept higher accident probabilities.[33] With such an approach, we are not estimating the value of an identifiable life, but the so-called *statistical value of life* or injury. This value estimates the amount of money and time society wants to spend on preventing accidents.[34] Suppose the simple case in which parents can lose their child in an accident without any monetary losses and without any effect on the value of money. Suppose that parents want to pay €100 maximally for a decline in the probability of a fatal accident of 0.0001 (1 in 10,000). Then the statistical value of the life of their child can be estimated to equal €100 / 0.0001 = €1 million.

The willingness to pay can be derived from common implicit or explicit decisions involving accident risks. For example, some employees are willing to work in dangerous jobs (such as maintenance of roads and railways) in exchange for higher salaries. In buying a car, one balances the benefits of car safety against the price and the looks and status of the car. In driving a car, one balances the benefits of mobility with the possibility of an accident. From such choices, we can derive how reducing (the probability of) non-monetary losses is valued by potential victims.

2.2 Voluntary compliance will be insufficient

This section discusses that the efficient level of precautions is not automatically obtained. First, the incentive for the proprietor to take precautions voluntarily, because of personal concerns, or because of social and moral concerns, is insufficient. Secondly, there is no market for precautions, where victims can buy deterrence.

---

32 See Shavell (1987) for a formal presentation.
33 E.g. Shavell (1987).
34 See Viscusi (1993 and 1998) on valuing life and risks to life.

2.2.1 The proprietor's personal interest

Compliance with the fire safety regulation, or more generally: choosing precautions to prevent fire damage is not always conflicting with the proprietor's payoff. A clear example is the maximum number of visitors allowed in case of fire safety in the horeca. For many establishments, the optimal number of visitors for a profit maximizing proprietor is lower than this maximum. When there are too many people present, they cannot find their way to the bar and hence the proprietor's turnover will be low. However, this argument does not apply to the majority of the regulations. Keeping emergency exits free or guaranteeing that escape-route signs burn are costly for the proprietor because money is paid on equipment, time is lost on maintenance or his establishment is less burglar-proof.

Similarly, a proprietor is inclined to choose some precautions because these will reduce the probability that he, his relatives or his personnel are hurt in a fire accident, that his property is destroyed or his establishment is closed. The damage to the proprietor himself is part of his costs and he will hence take them into account. However, from a social point of view, he should not only try to prevent damage to himself, but also to the visitors. The regulations are precisely intended to focus the attention of the proprietor to the interests of the visitors, because he is not naturally doing so.

In addition, the insurance company of the proprietor might stimulate precautions, but is primarily interested in limiting the insured damage. This includes the monetary loss of the proprietor, but not, or not necessarily, the damage to visitors. Moreover, the efforts of the insurance company to induce the proprietor to take precautions are costly. Therefore, insurance companies are helpful in improving the structural state of the building and its technical devices, but they have difficulty to control the daily behavior of the proprietor, like keeping emergency exits free, being attentive, etc.

Another problem arises if proprietors do not have sufficient information about the expected damage and its relationship with compliance costs. In such cases, even if they are somehow concerned with providing precautions, they simply lack the information required to do so. In general, if a proprietor is concerned about the expected damage, he also has a perfect incentive to obtain the optimal amount of information about the expected damage, and no failures would occur. This however may not be the case if the costs of obtaining information are borne by an individual proprietor, but the benefits are publicly available. Since the information on expected fire damage is quite technical and is learned in technical laboratories, the costs of acquiring information are characterized by decreasing returns to scale. This causes a threat of free-riding by proprietors, so that they obtain insufficient information about the possibilities and necessities of reducing expected fire damage, which may in turn result in a suboptimal level of precautions. Moreover, cognitive failures may lead the proprietor to underestimate the harm, and hence the importance of taking precautions.

2.2.2 *Social or moral concerns*

As mentioned in the previous section, the proprietor might care about his reputation and/or might have an intrinsic motivation to reduce harm to his customers. This will induce the proprietor to take some precautions. However, these social and moral concerns are unlikely to induce the proprietor to take efficient precautions, especially when there are no regulatory or enforcement activities. A proprietor will, in general, not be fully altruistic. Moreover, the importance of reputational concerns and intrinsic motivation is likely to vary among proprietors. While some will be motivated to take precautions, others are not. There are always some 'bad apples' who will not voluntarily comply with the regulations. Therefore, precautions are not first-best efficient, although they are larger than zero.

Furthermore, without enforcement activity, social or moral concerns may be weak.[35] For example, reputational concerns are only relevant if non-compliance is revealed. This could be carried out by for example the branch organization. However, this is not the primary goal of this organization, so it is unlikely that they will exert sufficient effort in publicizing non-compliance. What's more, no initiatives of such activities are observed. Visitors are also unable to detect (non)compliance and punish or reward the proprietor accordingly (this argument will be further developed below). Therefore, without enforcement the detection of non-compliance is very infrequent and the proprietor does not have to worry about his reputation.[36] Only if an accident with injuries occurs, this is likely to affect the proprietor's reputation.

Similarly, moral concerns may only be important if it is clear that something is forbidden and if the government propagates that it is important to follow the rules. If there is no enforcement, this may signal to the proprietor that compliance with the regulation is not important. Hence, the proprietor will not be motivated to comply.

2.2.3 *The problem of transaction costs and limited information of visitors*

So, in the absence of any laws or regulations, or more specifically in the absence of any enforcement activity of these laws or regulations, the proprietor does not efficiently consider the expected damage to visitors when deciding about the level of precautions. It can even be argued that, in the complete absence of enforcement, he will not choose any precautions to prevent harm to the visitors. His personal interests concern his personal damage. Reputational losses are absent. And intrinsic motivation to comply will break down.

---

35 There is some support for this argument in the literature, for example Axelrod (1984) and Sugden (1989). The interaction between enforcement and social or moral concerns is discussed in chapter 7.

36 See for a summary of the general argument Posner and Rasmusen (1999, p.380).

In economics, this is to say that there is a problem of *externalities*. The injurer, here the proprietor, does not consider the negative consequences of his behavior to another party, here the harm to visitors. Coase (1960) demonstrated that nonetheless efficient precautions will result if transaction costs are sufficiently low. If transaction costs are sufficiently low, visitors and proprietors are able to negotiate about the level of precautions. If there is no regulation, or if the regulation is weak (i.e. the proprietor has the right to cause damage), well-informed visitors will pay the proprietor to choose the efficient level of precautions. They are willing to compensate the costs of precautions as long as these costs are smaller than the damage these precautions prevent. On the other hand, if visitors have the right to be refrained from any harm from fire, they will sell this right for a price that compensates sufficiently for the damage. The proprietor is willing to pay this price as long as it is cheaper than taking precautions. So, in both cases the costs of precautions and the expected damage are efficiently balanced. However, transaction costs in the fire damage problem are too large for such negotiations. There are too many potential victims, creating high costs and also free-rider problems.

However, even if transaction costs are high, an efficient outcome would result if visitors were sufficiently aware of the risks involved in a horeca visit. The relationship between the proprietor and the victim is a market relationship between producer and consumer. Competition in the market would guarantee that customers only visit establishments that take optimal precautions. Proprietors that do not take optimal precautions would go bankrupt, effectively forcing them to take optimal precautions (e.g. Shavell, 1987). However, visitors do not have sufficient information about expected fire damage. This information is of an expert, technical nature and it is not freely available. Furthermore, fire damage does not occur frequently, so learning is difficult, and the expected damage varies from one establishment to the next. As discussed, human beings have much trouble estimating low probabilities and high damages. The impression is (in the eyes of fire experts) that people underestimate the expected fire damage in horeca establishments. They therefore cannot force the horeca proprietors to take an optimal level of fire safety precautions. Moreover, given the underestimation of fire damage and the free-rider problem, visitors do not have an optimal incentive to obtain information about fire damage.

2.3 Liability for damages

The enforcement and information costs of property rights are too high to be an effective and efficient alternative. The next alternative is to induce precautions by enforcing liability rules. Liability rules allow the victim, ex post, to claim damages. For that, the victim has to show that the injurer behaved negligently by taking insufficient precautions. Whether precautions are insufficient can be determined in different ways, such as whether it violates the victim's rights or is improper social conduct because of a rule of unwritten law. But also an act or omission violating a statutory duty is deemed unlawful. There are also cases of strict liability where, according to law or common opinion, the injurer is liable for damages irrespective of the precautions he has taken.

Liability rules provide the injurer with an incentive to take into account any damages he may cause, so that he will try to prevent such damage against reasonable costs. However, private enforcement by victims is only effective if injurers are really forced to internalize the harm they inflict. If the injurer can somehow escape liability, fully or partially, liability for damages does not provide the socially optimal incentive to take precautions. Let me briefly discuss the conditions under which liability leads to the efficient level of precautions.[37]

2.3.1 *The incentive to take precautions and standard-setting*

Under the negligence rule, the injurer only has to pay damages if the level of precautions that were taken is below the level of due care determined by the court. Therefore it is important what courts consider to be due care.[38]

Efficient standard-setting

Suppose that courts consider due care to equal the efficient level of precautions because they require that benefits and costs of taking precautions are balanced. If the magnitude of liability is large enough, this will induce the proprietor to choose this level of precautions. Choosing (little) less precautions will render him liable, paying damages. Choosing higher precautions only increases his costs of taking precautions.

Inefficient outcomes will result if courts have difficulty with determining the socially optimal level of precautions. If courts miscalculate the efficient level of care or are unwilling to enforce this level, it will usually still be optimal for the firm to choose the level of due care and, hence, to choose a suboptimal level of precautions. By taking this level of care, the firm completely avoids the costs of being held liable. If standard setting by the courts is systematically too low, firms can escape liability by choosing less than efficient precautions. If standard setting is too high, they take excessive precautions.[39]

---

37 The analysis in this section only covers the major results of the functioning of liability rules. For more detailed analyses I refer to the literature. There are several good literature overviews available. See for example Shavell (1987, 2004), Cooter and Rubinfeld (1989) and Schäfer and Ott (2004). An extensive (economic) review of Dutch tort law can be found in Visscher (2005*).

38 The economic theories of liability are introduced in Shavell (1987, 2004) and Schäfer and Ott (2004).

39 If the standard becomes very high, the injurer is better off being negligent, taking efficient precautions and paying the associated damages. See Ott and Schäfer (1997) for a somewhat different way of determining negligence, which requires less information by courts.

Uncertainty about the level of due care
Another problem is that courts might set due care at the efficient level of precautions, but do so with a margin of error, so that injurers are uncertain how their behavior will be evaluated by the court.[40] Suppose that the injurer is not certain about the due standard, but only knows the distribution of the standard, for instance because the predictive power of the courts' rulings (e.g. reasonability and fairness) is imperfect. For example, it may be uncertain for a proprietor whether the court values the removal of decorations as sufficient fire precautions, because it is uncertain how the court will assess the costs or the effectiveness of this precaution. In deciding about taking precautions, the proprietor only knows that such precautions will increase the probability that he escapes liability, instead of being certain about it. Such uncertainty has two opposing effects. On the one hand, the injurer may increase the level of precautions in order to reduce the probability that he will be found negligent. On the other hand, the injurer may choose a lower level of precautions because the benefits of taking more precautions are smaller as the precautions only reduce the probability of being found negligent. Which of these effects dominates is undetermined. Hence, uncertainty about the level of due care might create both under- and overdeterrence.[41] Underdeterrence results if the uncertainty about due care is very high. Because the injurer will be found liable over a wide range of precautions, a higher level of precautions will only result in a marginally reduced likelihood of being held liable.

Uncertainty about actual care
A second type of uncertainty is that courts encounter difficulties in verifying the facts. It might be uncertain whether or not a court will be able to observe the precautions taken by an injurer. For example, after a fire accident, a proprietor may be unable to demonstrate the number of visitors that were present or whether escape-route marks were visible. This may lead to two possible errors. An injurer may be found negligent even though he took sufficient care. Or he may escape liability despite having taken insufficient care. When courts misperceive facts in an unpredictable way, the injurer may exert either too much or too little care according to the same logic as above. He may be induced to take too few precautions because there is a probability that courts will find this level of precautions sufficient. However, he also has a high incentive to take too many precautions because he wants to increase the probability of escaping liability.

Strict liability
Under strict liability, the injurer always has to pay damages, irrespective of

40 See Miceli (1997, pp.44-59) for a general overview and analysis of the relevant uncertainties.

41 Shavell (1987, p.80 and 94-96) argues that in most relevant cases the injurer will take more than due care. See also Kolstad et al. (1990).

his level of precautions. Therefore the abovementioned problems of standard-setting are absent. Under strict liability the injurer balances the costs of precautions and the magnitude of liability. If the injurer must always pay damages in the event of an accident and it must pay in full the harm done, he will be forced to balance the costs of taking precautions and the expected damage, which will lead to the efficient level of precautions. The expected damage is internalized in the decision of the injurer and he will weigh social costs and benefits efficiently.[42]

2.3.2 *Risk aversion and insurance*

If the injurer is risk-averse, he will wish to insure himself against the risk of having to pay damages to victims (liability insurance).[43]

Strict liability and cost-free observing of precautions
Under strict liability, the injurer always has to pay damages; hence he will want full insurance against these losses. If the injurer is fully insured, he will not directly benefit from taking precautions (the *moral hazard* problem). Taking precautions will reduce the expected magnitude of liability, but this will benefit the insurance company, while the injurer himself bears the costs of taking precautions. Suppose insurance companies are able to observe the level of precautions taken by the injurer at no cost. For instance, insurers can easily observe whether a proprietor has installed and maintained fire extinguishers. They can, therefore, require that fire extinguishers be installed before an insurance contract is accepted. Alternatively, insurers can contract that no benefits are paid if it appears that the proprietor does not possess or maintain fire extinguishers. This is not necessarily an expensive procedure, especially if the proprietor is required to submit maintenance certificates.

If it is possible for insurance companies to observe the levels of precautions without costs, they can fully control the level of precautions taken. The level of precautions will directly affect the insurance premium and/or insurance coverage. Therefore both the injurer and the insurance company will have an incentive to minimize the costs of precautions plus the expected magnitude of liability. Therefore, if the magnitude of liability includes all dam-

42 In the next subsections the analysis of several complications is always developed by considering first strict liability and by arguing thereafter how it will apply under negligence. This line of reasoning is common in the literature because the analysis of strict liability is more straightforward, thus avoiding different complications to happen simultaneously.

43 This section is largely based on Shavell (1987), Schäfer and Ott (2004), Shavell (1979, 1980, 1982) and Spence (1977). The most readable and accessible introduction is (as always) found in Shavell (2004, part II). This section supposes that the injurer is only concerned with wealth, that insurance premiums are actuarially fair (perfect competition), and that there are no administrative costs of insurance.

ages, the injurer internalizes the full degree of expected harm and will choose the efficient level of precautions.

Strict liability and costly observing of precautions

In some cases, insurance companies have no opportunity to monitor the injurer's behavior at all. For instance, many aspects of fire regulations include rules about the proprietor's daily behavior, such as opening and keeping free the emergency exits and escape routes, limiting the number of people admitted, and more generally 'being attentive'. It may be difficult for insurance companies to control this behavior. In this situation, it is optimal for the injurer to partially insure himself. For the part of the damages for which he is insured, he has no incentive to prevent harm. For the part of the damages for which the injurer is not insured, he has an appropriate incentive to take precautions. On balance, too few precautions result.

Usually, insurance companies are able to observe precautions, but only at some cost. For example, an insurance company can invest in costly ex-ante inspections of horeca establishments to control whether or not the establishments have been furnished in a fire-proof manner. Ex-post, i.e. after an accident has occurred and a claim has been filed, the insurance company can invest in (technical) investigations into the cause of the fire and the amount of damage, to verify whether or not the proprietor had taken appropriate care. If not, they can refuse to pay in full. If the costs of observing precautions are not too high, observation is socially beneficial, because both deterrence and compensation are more efficient under observation of precautions. In general, investigation ex-post is less expensive because it needs only be done if an accident has occurred (and resulted in damage). However, the costs of such an investigation may be higher than a (quick) inspection ex-ante, or the information gathered ex-ante may be of higher quality.[44]

More generally, the choice is not between no observation and perfectly accurate (although costly) observation, but the accuracy of monitoring depends on the amount of money insurance companies invest in monitoring precautions. The more an insurance company invests in monitoring, the more accurate its observations will be and the more possibilities there are to make the insurance premium and/or benefits conditional on the level of precautions. The injurers will then choose to partly insure themselves. Whether or not the injurer takes too many or too few precautions depends on whether or not the insurance company under- or overestimates the necessary precautions.

So, it is possible to obtain a reasonable outcome under strict liability. If insurance companies can observe the level of precautions without incurring costs, optimal precautions will result. If insurance companies have to spend money on monitoring, the outcome will be 'reasonable', but not first-best. Monitoring costs directly decrease social welfare. Indirectly, the injurer will

44 Shavell (1979).

choose partial insurance[45] and his level of precautions will be higher than under no enforcement, but less than the first-best level of precautions.[46]

Negligence and risk-aversion
If a risk-averse injurer is non-negligent, he will not have to pay any damages, and he will therefore not want to buy any insurance. If he is negligent he will face the risk of being held liable and he will want to insure himself against this risk. If courts enforce the socially optimal level of precautions, it will always be beneficial for the injurer to choose this level of precautions, and hence not to insure himself. The analysis becomes more complicated if the negligence rule does not function perfectly. If the injurer is sometimes found liable, he will have an incentive to insure himself. The results from strict liability apply to the situations in which he may be found liable. Moreover, a risk averse injurer might choose higher levels of precautions to reduce the uncertainty of being found liable when courts set standards of negligence with error.

2.3.3 *The optimal magnitude of liability and non-monetary damages*

The next step in the analysis is to examine the optimal amount of damages, especially as relating to non-monetary losses. Moreover, because under liability victims actually receive compensation, the effects of liability on the allocation of risk cannot be ignored.[47]

Strict liability
Under strict liability, a socially optimal solution to the accident problem can generally not be achieved, because the optimal award for compensation diverges from the optimal award for deterrence. The optimal compensation

45 Partial insurance implies that the injurer does not obtain a first-best level of risk-spreading. This highlights the condition of the absence of moral hazard for obtaining optimal risk spreading through private insurance.

46 I have assumed that injurers are identical. This analysis is relevant either when differences between individuals are unimportant or when these differences are somehow recognizable. If this is not the case and if there is asymmetric information between the injurer and the insurance company about the (individual) expected damage, there might be a problem of *adverse selection*. Those injurers that cause relatively limited expected damage might choose not to insure themselves against the relatively high 'average' premium. The insurance company can try to differentiate between injurers by offering different amounts of coverage. Due to the possibility of ex-ante monitoring, the existence of credible signals of the risk level, the availability of differentiating insurance contracts and the fact that injurers themselves have only limited information about expected harm, I expect that the problem of adverse selection is not severe in the case of liability of proprietors for fire accidents in the horeca. The problem of adverse selection can also be solved by making insurance obligatory.

47 See for example Shavell (1987, pp. 186-235), Arlen (2000), Friedman (2000; pp. 95-102) and Calfee and Rubin (1992). This problem is analyzed in relation to the Dutch debate on hedonic damages in Suurmond and Van Velthoven (2005*).

for a victim depends on the monetary losses and on the extent to which an accident affects the value of wealth for the victim. As discussed in section 2.1.2, the optimal amount of compensation (insurance) is not expected to exceed the monetary losses. Therefore, in order to provide the victim with optimal compensation, the magnitude of the liability should equal the monetary losses or be lower.

However, in order to induce the injurer to take the efficient level of precautions, a damages award is required that exceeds the monetary losses. We also want to prevent the non-monetary losses of the pain and suffering of losing relatives and friends, of becoming disabled, etc. as these also constitute a decrease in welfare. It is desirable to prevent the accidental killing of teenagers in a fire, even if it does not involve any monetary loss and parents do not want to insure themselves for this non-monetary loss. To induce the injurer to spend costs on taking precautions to reduce these non-monetary losses, the magnitude of liability should include both monetary and non-monetary losses. The optimal magnitude for non-monetary damages can be based on the ex-ante willingness of potential victims to pay for higher precautions (section 2.1.2). If the injurer has to pay damages based on this willingness, he will take precautions in order to balance the social costs and benefits of taking precautions. Consider the simple example of section 2.1.2 in which parents can lose their child in an accident. Suppose these parents want to pay €100 maximally for a decline in the probability of a fatal accident of 0.0001 (1 in 10,000), so that the statistical value of the life of their child can be estimated to equal €1 million. The rule that potential injurers have to pay this €1 million for each child they accidentally kill will be efficient. Taking precautions to prevent accidents completely will be beneficial for the injurer if it costs less than €100, because it will reduce their expected liability payment by 0.0001 * €1 million = €100 – precisely the maximum amount the parents want to pay themselves.

Thus, it is clear that an efficient solution is impossible without additional arrangements, because the optimal awards for compensation and deterrence diverge. Monetary losses should be awarded for both compensation and deterrence. For deterrence, injurers should also be held liable for non-monetary losses. But awarding non-monetary damages leads to undesirable compensation of victims. Of course the victim will not oppose higher damages. But the (potential) victim could be better off if the expected money is available ex-ante, before an accident occurs. Optimal deterrence and optimal compensation cannot be achieved simultaneously.

One solution is 'negative', countervailing insurance contracts.[48] Then the damages award can be used to induce the injurer to take the efficient level of precautions, and the victim can use insurance to provide himself with optimal compensation. The victim will want to conclude an insurance contract in which he agrees to give away any damages above the optimal compensation amount, in exchange for a payment of the expected value of that amount

48 Friedman (1982) and Arlen (2000).

irrespective of the occurrence of an accident. This will remove the uncertainty about his wealth. There are two important problems with such contracts. First, the administrative costs of such contracts are relatively large. Pay-out is very frequent. For example, the victim receives €1 every month in exchange for paying the insurance company €1 million in the very unlikely event that he is injured in a fire (and receives such an amount of money from the liable proprietor). Such contracts also require that search and transaction costs be sufficiently low. Secondly, the incentive for victims to report damages is small as they have to pay in case of damages. Therefore, insurance companies have to spend large amounts of money on finding actual victims and forcing them to pay. This problem is mitigated if insurance companies arrange the monetary damage compensation of the victims and if victims actually sell their right to non-monetary damages to their insurance companies.

An alternative system involves a damages award that is given to victims in order to provide them with optimal compensation and a fine which is paid to the state in order to induce the injurer to take efficient precautions (Shavell, 1987). The state can use the benefits so that they become available to potential victims ex-ante (like a tax reduction or subsidy fund). However, such a system lacks the incentive for victims to report and obtain the damages award as described above.

In short, in the case of strict liability, a socially optimal solution is only achieved either if insurance markets work perfectly, so that victims can always acquire the optimal level of compensation, or if damages for compensation and deterrence are decoupled. Otherwise, we have to balance the award that is optimal for deterrence with the award that is optimal for compensation.

Negligence

If liability is based on negligence, it is possible to provide both optimal compensation and optimal deterrence even if 'negative insurance' is impossible. As discussed above, if due care is properly set and the magnitude of liability involves both monetary and non-monetary losses, this will induce the efficient level of precautions. However, given that due care is taken, no damages will be paid and hence over-compensation does not occur. Victims will receive no compensation from the injurer but can obtain compensation by insuring themselves.

2.3.4 *Limited assets*

Under strict liability, the firm will choose an inefficient level of precautions if its assets are insufficient to pay the damages to which it is convicted (the so-called judgment proof problem).[49] If the magnitude of liability exceeds an injurer's assets, the assets of the liable injurer will be completely seized and, in the case of firms, the injurer will go bankrupt. The injurer only considers to which extent

49 E.g. Shavell (1987, 2004), Schäfer and Ott (2004).

taking precautions reduces the probability that he will lose his assets in paying damages. Imposing liability above his assets will not further increase the incentive to take precautions.[50] Four remarks are relevant in this context:

1. Under the negligence rule the judgement proof problem is weaker because taking due care entirely relieves the injurer from liability and thus from possible bankruptcy. This advantage may be large enough for the injurer to choose due care, even if the damages for negligence exceed assets.[51]
2. If the costs of taking precautions are monetary, they reduce the level of assets available for paying damages. Therefore (Miceli and Segerson, 2003), under strict liability, if the assets are only slightly smaller than the damages the injurer has to pay, it may be beneficial for the injurer to increase his precautions above the efficient level, in order to lower the probability of being liable for damages and losing all assets (and going bankrupt).
3. A possible solution to the judgement proof problem is to make full insurance for damages compulsory. That garantuees that the injurer, or rather his insurance company, can always pay the required level of damages. However, whether the injurer will take efficient precautions depends on the possibility for insurance companies to monitor precautions.[52]
4. Having to pay damages might create an incentive to lower or conceal the assets invested in a firm in order to avoid damages, for example by dividing assets over multiple firms, holdings or establishments. This has two important consequences (Ringleb and Wiggins, 1990; Boyd and Ingeberman, 1999).[53] First, if capital investment is determined partly by the need to limit liability, production costs will not be minimized. Secondly, wealth reduction reduces the incentive to invest in safety precautions.

2.3.5 *Uncertainty about expected harm*

If the injurer is not well informed about expected damage, imposing liability may not induce him to take efficient precautions.[54] Moreover, the firm might be uncertain as to *how* to exercise care. Especially in technically and legally complicated settings, firms may not end up taking precautions until they determine how precautions can and should be applied, by for instance

50 See however the comment by Innes (1999a).

51 Cooter (1984) discusses the different incentives following from negligence and strict liability.

52 Note that if injurers have insufficient assets, it may be more difficult (or: costly) for insurance companies to monitor the levels of precaution that are taken. One way to monitor precautions is to contract the condition that the insurance company can recoup the damages from the injurer if he has taken insufficient precautions. However, if the injurer has insufficient assets he will not care (or care less) about such threats since he would not be able to pay anyway. Therefore it is more costly for insurance companies to handle the moral hazard problem.

53 See also Mason (2004).

54 See Miceli (1997, pp. 44-59) for a discussion.

hiring an expert for advise, asking colleagues, investing in R&D etc. (Pfaff and Sanchirico, 2000).

Under strict liability the injurer internalizes all social costs of accidents. Because the injurer is liable regardless of the level of precautions taken or of his level of information, the injurer will make the socially optimal decisions about both information acquisition and precautions. The injurer will minimize the harm he causes, the costs of obtaining information and the costs of taking precautions, i.e. the social costs. Under a negligence rule, an efficient result is only obtained if the optimal acquisition of information is incorporated into the level of due care. This includes situations in which the injurer is held liable if he failed either to choose efficient precautions *or* to acquire information when it was efficient to do so; or situations in which the injurer is held liable if he failed to take efficient precautions assuming optimal information acquisition (the injurer "knows or should have known" the facts).[55]

Therefore, if the liability rules are determined efficiently, the injurer can not claim to be uninformed about effective precautions. Being held liable gives the injurer the optimal incentive to be well-informed concerning precautions. However, problems might arise if the acquisition of information has characteristics of a public good, so that it creates free-rider problems. The analysis above assumes that the costs and benefits of information are private, only for the firm itself. If the information about precautions and expected harm is quite technical, expert information that has non-rival benefits, the injurer may not have the optimal incentive to obtain information. This problem will be further discussed in section 6.1.[56]

2.3.6 *Vicarious liability*

If liability concerns a firm as the injurer, another remark is relevant. So far, I have assumed that the firm acts as a single person. However, 'firms' do not act, their agents do. Offenses are committed by individuals acting on behalf of the firm.[57] Often, an employer is (strictly) liable for any misconduct of his employees committed in furtherance of their employment, so-called 'vicarious liability', or the principle of *respondeat superior*. This subsection shortly

55 These results are extensively described in Shavell (1992a).

56 Pfaff and Sanchirico (2000) point at a second problem associated with obtaining information. Obtaining information might increase the probability that a firm is found liable, because this information may be used against them by governments, victims or other adverse parties in prosecution or a tort case. By investigating their own compliance status and the possibilities of taking precautions, firms are effectively aiding their own prosecution or adverse suit. Therefore the firm's incentive to obtain information may be too weak.

57 Kramer and Sykes (1987), Kraakman (2000), Alexander (2004), Faure and Visser (2004).

points at the consequences of different individuals who are responsible for precautions.[58]

In general, we can distinguish between three different types of individuals related to the firm. The first type covers lower-level managers or employees who take day-to-day decisions that affect the level of precautions and hence expected harm. For example, a waiter in a café or restaurant may decide to place a beer-crate in front of an emergency exit. The second type consists of upper managers who set overall safety policies and make decisions about safety investments. These people also decide on the production level. Finally, there are the shareholders, or the owners, of the company. They receive the profits made by the firm, which depend on the costs of taking precautions and on sanctions. The party bearing the costs and consequences of taking precautions (either monetary or non-monetary) may be different from the one deciding about taking precautions. The important question is how the enforcement approach should be structured to ensure that all parties face efficient incentives to reduce the expected harm (Segerson and Tietenberg, 1992).

Generally, shareholders have some control over the firm because they are able to designate incentives for the upper managers, to ensure that the firm is trying to maximize profit. It is generally not efficient to extend the liability of the firm to its shareholders (Shavell, 1987). In small firms, the shareholders (one or two) are usually the managers themselves and they will have invested all their assets in the firm. In other cases, the shareholders are not clearly able to observe and control the behavior of individual actions by employees in such a way as to reduce accidents, because they lack the expertise to do so and because of the free-rider problem arising if every shareholder only has a limited share. Liability of the firm implies the liability imposed on the assets of the firm with which the upper managers are concerned. The most relevant agency problem is the relationship between the lower level employee and the upper manager who represents the firm.

There are several situations in which corporate and/or vicarious liability suffices to induce precautions, even if employees are responsible for precautions. If there are no transaction costs, it is not important who is sanctioned for an offense as the sanctions are freely transferable. This is a direct application of the Coase theorem. If the firm is held liable for the actions of its employees, it will charge the damages or fine it paid to the liable employee. Vicarious liability may thus be efficient if the firm is either able to monitor the actions of its employees at no cost (ex-ante) or if the wage rate can be made dependent on sanctions (ex-post). However, in many cases such shifting of sanctions is

58 More extensive analyses, which underlie this subsection, can be found in Segerson and Tietenberg (1992), Kraakman (2000), Shavell (1987) and Polinsky and Shavell (1993). See also Lott (2000), Alexander (2004) and Faure and Visser (2004). This section does not intend to provide a full analysis of vicarious liability, but rather to explain which principal-agency problems might be present in organizations more complex than the simple hierarchies I analyze. The remarks apply to both civil and criminal liability.

imperfect. Firms may not be able to vary wages directly or indirectly with the actions of its employees or with the size of the penalties it is issued. At least in the short run, wages are set by contract, and transaction costs prohibit continually bargaining over the contract. Or the wages might be governed by a process of collective bargaining with unions. The firm might have difficulty observing the daily behavior of their employees. The consequences of actions might only become apparent in the long run or the responsibility of employees might be diffused. Hence, if only the firm is held liable for all the actions of its employees, insufficient precautions will result, and welfare may be improved by making employees personally liable for their actions. In addition, the state might be able to control the employees' behavior more effectively or at lower costs, because the state can impose imprisonment as an alternative or supplement to the fine.

On the other hand, vicarious liability might improve deterrence if the individual employee is unable to pay for the harm done (while the firm is not), if it is hard for victims to identify the responsible employee (while the firm is not) or if upper managers are able to reduce accident risks by reorganizing the production process (which is hard to judge by courts).

For fire safety in horeca establishments I assume that the proprietor is both the shareholder and the manager of the horeca firm and that he is able to control the behavior of his personnel. Therefore, it suffices to provide the proprietor with an efficient incentive to take precautions, and I simply describe the firm as the injurer of an accident. If, however, these assumptions cannot be made, solving the accident problem would be more complicated. If the firm is unable to observe and control employees' precautions, corporate and/or vicarious liability will not lead to harm reduction. If the upper managers and shareholders are different persons, shareholders face the problem of providing the managers with incentives to maximize the firm's profit.

2.3.7 *Litigation*

Liability rules are only an effective mechanism in inducing injurers to take precautions if victims are actually able to enforce a claim for damages. Otherwise, the injurer will not be inclined to prevent accidents. However, the litigation process is not without problems because the court system is costly, both privately and socially.[59]

The decision to bring suit and deterrence

A victim will bring suit if the amount he can claim from the injurer (the loss suffered) exceeds the costs of a claim. Therefore, a victim is less likely to claim damages if his costs of litigation are high, if his estimate of winning a trial is

59 The results of this subsection can be found in Miceli (1997) and Shavell (1987 and 2004). See also Cooter and Rubinfeld (1989).

low, or if the expected magnitude of the award is small. These conditions are more likely to be satisfied if harm is widely dispersed over a large number of victims, if the harm from an activity becomes apparent after a long time, if the injurer is untraceable, or if it is difficult to find the causal relationship between act and harm.

If, due to the costs of litigation, the victim might not always bring suit, this implies that the injurer is not always forced to pay damages. Or, even if the victim brings suit, but the costs of litigation are high, the victim might accept a relatively low settlement amount, and hence the injurer will not be forced to face the full harm he caused. This has a negative effect on the incentive to take precautions. On the other hand, there is also no danger that injurers might choose excessive precautions in order to reduce the probability that they will incorrectly be held negligent (section 2.3.1). If the proprietor behaved non-negligently, a victim's failure to bring suit is not a problem.

The efficient levels of precautions and of litigation

If accidents lead not only to losses for the victims, but also to litigation costs, it means accidents are socially more expensive, so precautions should be higher. Generally, however, an injurer only considers the extent to which taking precautions reduces expected damages plus his own litigation costs, instead of considering the full social costs of litigation. Victims do not naturally make efficient use of litigation either. From a social point of view litigation is desirable if the costs of litigation are lower than the deterrence that is gained from the procedure, i.e. the reduction in expected harm plus compliance costs. However, victims make a different consideration: they tend to balance their own costs of litigation with the private benefits obtained from the wealth transfer. As a result, a victim may choose for too many or too few lawsuits.

Litigation costs may differ under strict liability and under negligence. Negligence can induce optimal precautions while avoiding a lawsuit, whereas strict liability requires a lawsuit plus its costs for injurers to take precautions. However, given a claim negligence is costlier because the courts have the additional tasks of determining the due standard and actual care.

Settlement

Settling the damages claim allows parties to save on trial costs. There are several reasons why parties might fail to reach an agreement (Kobayashi and Parker, 2000). First, a settlement will not be reached if parties over-estimate their chance of trial victory, making them unwilling to accept a settlement. Secondly, the existence of external effects may cause trials, such as the precedential and preclusive effect of the underlying case, which will determine the parties' payoff in similar future cases. Thirdly, a settlement may not be reached if information about the expected trial outcome differs among parties. Information problems can be solved through costly trials. Costly trials are a signal of the strength of the case, which improves the bargaining position. Finally, agency costs, such as costs of hiring a lawyer, can give rise to the occurrence of trials. For example, an attorney may receive a higher compensation in a trial because that requires more hours of work.

This shows that litigation is a complex process. The existence of a settlement, or bargaining, range is a necessary, but not a sufficient, condition for settlements. The outcome of the litigation process also depends on the amount of money and time parties spend on it. Generally, the private and social incentive to settle and the incentive to spend money on litigation diverge.[60]

Legal aid

If the legal costs are too high in relation to the expected judgment, victims will not bring suit even if it is socially desirable to do. Moreover, even if a victim estimates that the expected judgment exceeds his legal costs, he will have to advance the legal costs. If the victim has limited (liquid) assets or a low income, this might be a problem. If these credit constraints are sufficiently high, the victim will be unable to bring suit. A system of public legal aid may overcome credit constraints, thereby enhancing deterrence and justifying the use of public resources – although legal aid also has some negative effects on litigation.

Credit constraints are absent under contingent fees or under legal cost insurance. Under these systems the outcome depends on who has control over the claim and the settlement decision. Accident or liability insurance may also imply that the claim is no longer in the hands of the parties themselves but in those of the insurance companies. The interests of the insurance company or attorney are not necessarily in line with those of the parties themselves.

Besides financial buffers, claiming damages also requires social and communicative skills that will enable the victim to know how to claim damages and how to bargain for the highest possible settlement amount.

Standardization

Uncertainty concerning the outcome of a trial may be caused by differences among judges or districts, by uncertain or open-ended norms, by uncertainty about the way in which harm will be compensated, etc. Such uncertainty may create under- or overdeterrence (section 2.3.1), it may induce victims to inefficiently refrain from litigation and it may prevent parties from settling the claim.

This uncertainty can be reduced by standardizing or harmonizing litigation outcomes. One of the clearest examples is the computation of damages. It might be an unduly hard task for the victim to show his precise level of losses, especially when these losses are still unrealized future events (such as the loss of earning capacity) or difficult to measure (such as non-monetary

60 An overview of the relevant literature (and further references) on this bargaining process can be found in Cooter and Rubinfeld (1989), Kennan and Wilson (1993) and Daughety (2000).

damages). It might be efficient to introduce a classification of losses paired with appropriate compensation amounts, so that the victim will only have to prove to which class he belongs. This might save the time and costs spent on a trial (such as hiring medical experts), or even induce settlement by reducing the uncertainty and the possibility of disagreement. If the classification is done properly, so that it approximates a victim's average losses, deterrence is not lost: injurers will choose their precautions based on expected, rather than actual damages.[61]

Joint litigation and class actions

In the case of severe accidents, multiple victims suffer damage. In such cases, there are some ways to save litigation costs.

First, courts have the possibility to route similar cases to a single court for coordinated management. Moreover, it is possible to unite similar questions into one case. For example, victims can file a case to determine whether or not the injurer was negligent. If he was, individual victims can try to claim their damages from the injurer in individual procedures. Secondly, an interest group can represent the plaintiffs, and negotiate on behalf of all victims, so that victims settle their case collectively. This might prove difficult when many victims are unknown or widely spread geographically, making it impossible, or at least very costly, to unite all claims into one settlement. The most far-reaching system is the *class action*. The class action implies that a victim, or an injurer, can ask the court to bind all persons with related claims (a class) to the settlement or trial outcome, even if these persons are no parties at the proceedings (Silver, 2000).

The main advantages of class actions for victims are that litigation costs are saved because common questions about facts or law have to be decided only once. So long as the benefits of litigation are non-rival, as is the case with document preparation, expert witness testimony, computer simulations, etc., costs are decreased as a result of being shared. This is also socially worthwhile. Especially when the damage is spread among many victims, so that an individual trial might not be worthwhile for a victim, class actions may increase deterrence. The benefits of class actions for the defendant lie in the certainty he receives about the (financial) consequences of a settlement. He is sure that no more litigation actions will follow and that the case is closed. This certainty might also be socially beneficial.[62]

61 The only disadvantage is that compensation only approximates first-best, because a victim is seen as a member of a class rather than an individual receiving individual compensation. In general, the savings in litigation costs are expected to exceed the (minor) loss in optimal compensation.

62 Class actions, however, have also disadvantages. A more extensive summary and discussion of the system and its (dis)advantages can be found in Silver (2000) and Miller (1997).

2.3.8 *Conclusion on the efficiency and effectiveness of liability*

This section investigates the extent to which private enforcement by victims can result in efficient precautions. Generally, the injurer is induced to take efficient precautions if the expected magnitude of the liability equals the expected harm. This induces him to balance the social costs and benefits of accidents. This in turn leads to several requirements, which have been discussed: (1) Courts should enforce socially optimal levels of precaution in a predictable way; (2) If necessary, insurance companies should be able to control moral hazard problems; (3) The magnitude of liability should involve full awards for both monetary and non-monetary losses; (4) The injurer should be able to pay all damages; (5) The injurer should know or should ought to know how to prevent liability; (6) (a) The victim should be willing and able to file a claim, (b) The injurer should be forced to offer compensation not only for harm done, but also for all other accident-related costs, such as litigation costs, (c) Litigation costs harm social welfare and should therefore be minimized given the incentive to take precautions.

If these conditions are not satisfied, the injurer will choose less than efficient precautions. In order to maintain efficient precautions, the incentive to take precautions must be adjusted, for instance through punitive damages, obligatory insurance, public legal aid, public apprehension and accident investigation. For the case of fire safety in horeca establishments it is not clear whether or not all requirements are satisfied. For that an empirical analysis of actual data is required, for instance to determine whether litigation costs are a barrier, whether proprietors are able to pay for damages, what the actual amount of damage compensation is, etc. For now, it suffices to conclude that private enforcement by liability rules might be a relevant method to obtain efficient precautions, but that there are also sufficient reasons to believe that it might fail. Then, public regulation and enforcement is an alternative method of creating deterrence.

2.4 Conclusion: The need for public enforcement

This chapter starts with identifying the relevant accident problem. The problem is that the proprietor must be provided with an incentive to take precautions. I discussed several reasons why the proprietor will choose an inefficient level of precautions, or even no precautions at all, when nothing else is regulated or enforced. Contracting between proprietors and visitors will fail because of the high transaction costs. Therefore, the externality problem should be solved in a different way, by making the injurer pay a price for the harm he inflicts on others. If the injurer fully internalizes all harm he inflicts, he will have an efficient incentive to take precautions. He will be forced to balance the costs of precautions and the benefits of harm reduction. The externality problem will thus be solved. For instance, victims can force the injurer to pay compensation for the harm done. Thus the enforcement of liability rules by actual victims induces the proprietor to take precautions. But this will only result in the desired level of precautions if several important conditions are satisfied. Hence, it is necessary to consider another alternative. This alternative is the introduction of public regulation, or more precisely, of public enforcement of the regulation. The next chapters consider whether the imposition of fines, imprisonment and other public sanctions on violation of the regulation, is efficient.

Given that firms do not voluntarily comply with the regulation, maximizing social welfare implies a third requirement next to compensation and deterrence: the minimization of enforcement costs.[63] Before, I argued that the efficient level of precautions is determined by a balance of the costs of compliance and the reduction in expected harm. Given the need for enforcement activities, this balance should be adjusted. The enforcement costs of obtaining a higher level of precautions should be weighed against the balance of compliance costs and expected damage. Even if the reduction in expected damage resulting from taking more precautions is higher that the compliance costs, it is not necessarily efficient to obtain that higher level, due to the costs of enforcement. If there are (unavoidable) costs of enforcement, it is efficient to enforce a level of precautions that is lower than the first-best.

63 Calabresi (1970).

3 Introduction to public law enforcement: detection and sanctions

This chapter introduces the economic analysis of public law enforcement. I examine the conventional literature on efficient enforcement policies. The following chapters further investigate optimal public enforcement actions involving more complex problems. This review provides the arguments for chapter 8 on the distinction between a deterrence and a compliance strategy.

In this chapter I analyze the optimal probability of detection and the optimal magnitude of the sanction. In addition, the optimal nature of detection and of sanctions is examined. These depend on the information that is available to the enforcement authority: does it know whether a violation is desirable and which sanction will deter? The available information also determines the nature of detection. Detection might be perfectly certain or the enforcement authority has to invest in detection. Such investments might depend on information following reports of victims or following self-reports of firms.

First of all, I start with a description of the enforcement problem that is analyzed. In section 3.2, I consider how the efficient enforcement policy depends on the information available to the enforcement authorities, on the nature of detection (apprehension technology) and on the nature of the sanction. In section 3.3 I discuss how the probability of detection might depend on the compliance rate. In section 3.4 I consider detection by means of self-reporting.

3.1 The model of public enforcement

3.1.1 *The description of the enforcement policy*

The enforcement problem examined here is as follows. A risk-neutral firm has to decide about its level of precautions, where some standard level of precautions is prescribed (in some law, regulation, license etc.). Taking precautions reduces the expected harm from its business. Compliance with the standard is costly for the firm. A public enforcement authority is able to enforce compliance by imposing sanctions. There is a probability of inspection of p. If there is an inspection, the enforcement authority perfectly detects the level of compliance. A level of compliance lower than the standard may be sanctioned with either a monetary or a non-monetary sanction or a combination of these, denoted f and s respectively with s as the personal disutility of an imprisonment time t, $s= s(t)$.[64] The magnitude of the sanction is limited to some upper bound, f^{max} and s^{max} respectively. The fine is limited by the assets of the firm. Non-monetary sanctions are limited by for instance life expectancy (age), time horizon etc. Sanctions are also constrained by the unwillingness of society, courts and prosecutors to impose large sanctions. The limits to sanctions such as the withdrawal of licenses and imprisonment depend on the gain lost once the activity can no longer be pursued.

We can distinguish between two cases. The simplest case involves a firm which has to make a binary choice about either taking no precautions, which causes harm h but costs nothing, or choosing the standard, which leads to costs of compliance c but prevents harm. This is similar to choosing an activity that is forbidden. Examples include locking an emergency exit, failing to check fire extinguishers etc. This is therefore referred to as the binary case or single act model. A more complicated case is the one in which the firm has to choose a level of precautions x that costs $c(x)$ with $c'(x) > 0$, and leads to harm $h(x)$ with $h'(x) < 0$ (which is similar to choosing one of multiple acts). Examples include overcrowding in horeca establishments, incorrectly hanging decorations in a café, or making a decision regarding several precautions simultaneously, such as only locking an emergency exit versus locking an emergency exit and choosing dangerous decorations. In these decisions, the issue of *marginal deterrence* (Stigler, 1970) is important, i.e. individuals who are not deterred must be induced to choose less harmful offenses.[65]

Assume that imposing a fine does not lead to social costs. Fines merely involve a transfer of wealth. It is costly for the penalized individual, but the benefits from the fines can be used in socially beneficial projects or tax reductions. For simplicity's sake, I refrain from including the administrative costs of collecting fines as well as costs of prosecution. With imprisonment, important social time is lost and resources are spent on prisons, guards, parole officers etc., which costs are denoted as $k(t)$.[66] Until chapter 6, I assume that courts make perfect decisions, i.e. they convict firms if and only if they are guilty given the rules and actual behavior and according to the enforcement policy. In cases of conviction, sanctions are always carried out.

---

64 'The firm' cannot be sent to prison, only the responsible manager or employee can. I assume that the proprietor is equal to the firm and has full control over its employees. He can be imprisoned.

65 I do not distinguish between harm-based enforcement (such as investigating a fire accident) and act-based enforcement (such as taking fire prevention measures). The analysis here starts with h as expected harm including both accidental and certain harm. Chapter 11 discusses whether harm-based or act-based enforcement is optimal.

66 Of course, in real life, prosecution and enforcing the payment of fines is costly. The reader can infer the influence of these costs from the results of imposing costly non-monetary sanctions.

3.1.2 *The trade-off between sanctions and detection*

The (modern) economic approach to law enforcement is based on the seminal article by Becker (1968). According to this analysis of the binary case, a rational and risk-neutral firm will comply with the rules if the costs of compliance exceed those of violation. An expected sanction equal to or larger than the costs of compliance will deter the firm from violating the standard. Under a system of fines, it is optimal to use extreme sanctions and low probabilities of apprehension. If inspection is costly and imposing a sanction is not, enforcement costs are minimized by a probability of inspection arbitrarily close to zero and a maximal sanction sufficient to deter. This conclusion still holds, according to Becker, if punishment is costly, as with imprisonment. If all activities are undesirable, the threat of severe imprisonment suffices to deter, and hence costly imprisonment will never have to be carried out. Even if not all activities are undesirable and deterred, the probability of detection can be decreased and the prison time increased, so that the expected time in prison remains constant. The same level of deterrence is achieved by lower costs of detection and the same level of costs of imprisonment. This has a beneficial effect on social welfare.

In sum: (1) a higher expected sanction decreases the level of non-compliance, (2) costly non-monetary sanctions should only be used when costless monetary sanctions have already been used to the maximum, (3) decreasing the probability of detection and increasing the magnitude of the sanction might ensure deterrence but save enforcement costs, hence it is socially desirable. Therefore, the sanctions should be at the maximum level with a probability of detection that suffices to deter.

3.1.3 *Risk attitude and the costs of imprisonment*

I will generally adhere to the assumption of a risk neutral firm. However, it is useful to shortly describe how the risk attitude may influence the optimal enforcement policy. Because of the probability of detection, non-compliance is a risky choice (Becker, 1968). Risk aversion with respect to imprisonment refers to cases of increasing marginal disutility of imprisonment, so that four years in prison is more than twice as bad as two years. A risk preference for imprisonment implies that the marginal disutility of imprisonment decreases, so that serving four years in prison once is valued less badly than serving two years in prison twice.

If the firm is risk-averse with respect to wealth or imprisonment, a smaller probability of detection increases the risk costs, and hence increases deterrence. Therefore a lower expected sanction will be sufficient to deter non-compliance. The benefits from a decrease in the probability of detection and a proportional increase in the magnitude of the sanction are higher. However, if there is a risk preference for wealth or imprisonment, a decrease in the probability of detection increases the risk of committing a violation, and hence makes non-compliance more attractive to the firm. This means that the sanction should increase more than proportionally in order to retain deterrence. This is no problem under costless monetary sanctions. However, under imprisonment, this implies that the expected man-years in prison increase, so that it is not necessarily beneficial to impose extreme sanctions. That might lead to large costs for carrying out imprisonment. It might therefore be desirable to choose for a higher probability of detection and a smaller sanction.[67]

The results depend on the marginal costs of imprisonment. The social costs of carrying out imprisonment consist of the private disutility of imprisonment plus the public costs of imprisonment. If the marginal social costs of imprisonment are constant (i.e. $k''(t) = 0$) and if there are no fixed costs of imprisonment, imprisoning a single man for four years is as costly as imprisoning two men for two years. For example, a decrease in the probability of detection from 10% to 5% and a corresponding increase in the imprisonment sanction from two to four years will both lead to an expected imprisonment sanction of 0.2 years. Unless the firm is a risk lover, deterrence remains at a constant level, the expected man-years in prison and hence the costs of imprisonment are constant, but the costs of detection decrease. This has a beneficial effect on social welfare. If the marginal costs are decreasing (i.e. $k''(t) < 0$) and/or there are fixed imprisonment costs, it will be less costly to imprison a single man for four years than to imprison two men for two years. On the other hand, if the marginal social costs are increasing (i.e. $k''(t) > 0$), it is less costly to imprison two men for two years. In that case, even if the firm is risk neutral, trading-off the probability of detection and the magnitude of the sanction might increase punishment costs. This increase should be balanced with the decreased costs of detection.

3.1.4 *Subjective and objective sanctioning risk*

The firm bases his decision about compliance on its subjective perception of the probability of detection and conviction, and of the sanction. This might be different from the objective, actual sanctioning risk implemented by the enforcement authority. However, in general, the objective and subjective sanction risk are related. The subjective value of the probability of detection and the magnitude of the sanction depends on the information that the potential offenders possess, which in turn depends on, for instance, the visibility and predictability of enforcement. If there is uncertainty concerning the probability of detection or the magnitude of the sanction, firms will update their belief about the expected sanction if they observe enforcement actions against themselves or others (see section 4.1). Generally, it is not possible to try to

67 If not all firms can or should be deterred, the realized risk costs or benefits should be incorporated into social welfare (Polinsky and Shavell, 1979).

only threaten with high probabilities and sanctions – in the hope that people will believe it and comply – without actually imposing them. Such a threat would have no credibility, at least in the long run.[68] If there is at least some positive relationship between the objective, actual sanctioning risk and the subjective, perceived sanctioning risk, a higher sanction or higher probability of detection will still result in increased compliance, even though effectiveness may be decreased.

A step further is to argue that firms are not only imperfectly informed about the expected enforcement policy, but are also not perfectly able to update and process all available information. There are important limits on human cognition (section 1.4.2). Jolls (2004) has sketched some important implications of *bounded rationality* for public law enforcement. She mentions three important aspects of bounded rational behavior which affect the potential offenders' comparison of the benefits and costs of committing offenses:

1. *Optimism bias*. Individuals make judgment errors because they underestimate the probability that a negative event will happen to them rather than to others. This implies for instance that potential offenders might think that the probability of arrest is lower for them than for others. Optimism bias might also involve an underestimation of the probability of detection relative to the actual probability.
2. *Availability*. The ease with which a given event comes to an individual's mind affects the probability estimates. The estimated probability of arrest depends on the salience and vividness of observed arrests. This implies that it is important *how* arrests takes place. If it is carried out in such a way that everyone knows about it, this will increase others' estimate of the possibility of arrest. Furthermore, the estimated probability of arrest depends on the observed number of arrests. Possibly, if this number is below a critical treshold, the probability of arrest simply disappears from the radar screen of potential offenders.
3. *Prospect theory*. The main point here is that people evaluate outcomes based on the change they represent from an initial reference point rather than based on the nature of the outcome itself. This implies, for example, that people exaggerate the difference between a small probability of arrest and a zero probability of arrest. Another implication is that changes far from the reference point matter relatively little. For example, the deterrent effect of increasing the magnitude of the penalty will have a strongly diminishing effect. The difference between a fine of €10,000 or €11,000 is much less than the difference between a fine of €100 or €200, although in the first case the fine increases with €1000 while in the second with only €100.

A different interpretation of this status quo bias is that the value function of individuals is concave for gains, but convex for losses, i.e. they are risk-averse with respect to gains, but risk-seeking with respect to losses. Evaluation of changes relative to an initial reference point follows from an "endowment

68 See for example Shavell (2004, p.481 including references).

effect", i.e. the fact that the possession of a good implies some entitlement to that good.[69]

In the analysis hereafter I will ignore these complications and assume that the firm makes a fully rational decision about compliance.

3.1.5 *Detection technology*

There are two basic forms of detecting non-compliance. Apprehending offenders can be in a *specific* or *general* manner.[70] Specific enforcement, also called investigation or inspection, implies that expenditures on detection are allocated (ex-post) to tracing the offender of a specific, known violation, such as when investigating a fire accident or the illegal dumping of waste. Under specific enforcement, resources for detection can be conditioned on the information about the severity of the offense (for example from reports by victims). General enforcement, also called surveillance or monitoring, implies that expenditures are allocated as a fixed budget ex-ante to detect offenders and prevent several types of offenses, for instance surveillance police officers, flash guns and regular inspections by fire prevention and environmental protection officers. The main characteristic of monitoring is that it cannot be conditioned on the severity of an offense, because it is as yet unknown whether an offense has occurred. The enforcement authority must therefore spend resources before receiving information about the offense. Hence, a general probability of detection applies to the same type of offenses regardless of the harm done.

3.2 Basic efficient enforcement policy

This section extends the basic steps of Becker (1968) to verify whether the economic analysis predicts high sanctions and low probabilities of detection.[71] The efficient enforcement policy is identified in steps. First, I assume that enforcement is certain because the offender is publicly known. This is sometimes a realistic assumption, for example in case of ex-post enforcement after an accident. Moreover, this case shows the optimal expected sanctions when costs of detection are absent. Secondly, I consider enforcement through monitoring where the probability of detection is fixed but less than one. Thirdly,

69 It is far beyond the scope of this thesis to discuss prospect theory. See Curran (2000) for a very brief introduction into ideas put forward by Kahneman and Tversky (for example 1979, 1981, 1992).

70 See Mookherjee and Png (1992) and Shavell (1991, 1992b).

71 I provide only the most basic results, because several extensive literature reviews are available. See for example Polinsky and Shavell (2000a and 2000b). Shavell (2004, part V) provides the most extensive, recent and accessible theory of optimal enforcement. These references also provide or refer to formal analyses.

I consider enforcement through investigation where detection can be conditioned on the harm done. I then discuss the joint use of monitoring and investigation. Finally, I discuss the problem raised by the fact that the probability of detection (or sanctions) depends on the number of offenses.[72]

3.2.1 *Certain enforcement*

Let us assume first of all that all offenses are perfectly detected. The enforcement authority can observe that an accident has occurred at no costs. Furthermore, consider the binary case in which a firm must decide whether or not to comply with a standard. Under certain enforcement, the authorities know who committed an offense without allocating resources to detection. Without costs of detection, it is publicly known who caused damage. Hence, the important question is the optimal magnitude of the sanction.

Perfect information

If the enforcement authorities are also perfectly aware of the individual harm done and the costs of compliance, they know perfectly whether a violation is (in)efficient. Should they also know the magnitude of the maximal sanction, they will know whether the sanction can be made large enough to deter violation. The enforcement authority can then focus its resources on offenses that are inefficient ($h \geq c$) and can be deterred by making the sanction large enough. A sanction at least equal to the costs of compliance will induce these firms to comply. The costs of imposing sanctions – if present – will never be realized. There are no costs of detection, no costs of imposing sanctions and the optimal level of deterrence is achieved; the best possible result.

Under monetary sanctions, another policy is also optimal. Because imposing sanctions is costless, it is also possible to sanction all offenses with a fine equal to the harm done ($f = h$). The firm will comply if the harm, hence the sanction, is larger than the costs of compliance. This leads to optimal deterrence of offenses as long as the firm is able to pay the fine. If the firm is not able to pay, that should be taken for granted. Again, there are no costs of detection, no costs of imposing sanctions and the optimal level of deterrence is achieved.

If the legal standard is efficient, for example because the standard is set at the individual level in licenses (so that inefficient offenses are exempted), the

72 I do not distinguish between strict and fault-based liability (see Shavell, 2004 for a different approach). One reason is that strict liability may be seen as a form of negligence in which the level of due care equals the legal standard. Moreover, as follows from Shavell (2004), a fault-based system is optimal when authorities have perfect information. If authorities only have information about harm, then only strict liability is available. I discuss the optimal enforcement policy given the available information. Comparing the optimal regimes under different information structures would yield a comparison of fault-based and strict liability.

objective is simply to eliminate all offenses. If, for example, an imprisonment sanction of 10 years suffices to deter all offenses, there is no harm, no costs of detection and no costs of imprisonment, hence the first-best result.

Only harm is known

Now suppose that if an accident occurs, enforcement authorities only observe the damage to victims and property and not the costs of compliance. This implies that the enforcement authorities cannot distinguish the different types of offenses. Offenses that cause the same level of harm will be penalized identically. I argued that under perfect information and monetary sanctions, it is optimal to impose a fine equal to the harm done. Because this does not depend on information about the costs of compliance, the same enforcement policy is still possible and optimal.

The result is different for costly sanctions. Since it is not possible to distinguish between the different types of offenses, enforcement costs will be made for inefficient offenses or for offenses that cannot be deterred due to the maximum imprisonment sanction. Both under- and overdeterrence can be optimal. The efficient enforcement policy is summarized in the following points.

1. The optimal sanction includes the costs of imposing a sanction (Polinsky and Shavell, 1992). Assume for simplicity's sake that these costs, k, are independent of the imprisonment time. Then $\hat{s} = h + k$. The reason is as follows: firms who comply are not prosecuted and punished. The social costs involve the compliance costs c. Firms that violate cause harm and are prosecuted and punished, leading to costs of harm done h and costs of imposing a sanction k. Hence, the socially optimal decision is to comply if the costs of compliance are lower than the costs of harm and punishment, $c \leq h + k$. The optimal sanction equals $\hat{s} = h + k$. Note that this results in overdeterrence because some efficient offenses will be deterred (if $h < c \leq h + k = s$).
2. From the previous point it directly follows that it is optimal not to impose any non-monetary sanction for minor offenses. If the costs of imposition are relatively large and the benefits of deterrence relatively small, it is not worthwhile to spend resources on imprisonment. If authorities choose not to impose any sanction, all firms will violate, leading to costs h. Hence, not imposing imprisonment is beneficial if h is sufficiently small and k or the estimation of c sufficiently large.
3. Under the more realistic assumption that the costs of imprisonment are larger for higher imprisonment sanctions ($k'(t) > 0$), either under- or overdeterrence can be optimal. First, note that it is optimal to impose a sanction lower than $\hat{s} = h + k$. Reducing the sanction has the beneficial effect of reducing the costs of imposing a sanction, so the optimal sanction will be lower than it would have been under fixed costs (Polinsky and Shavell, 1992). Moreover, if these costs increase rather strongly, the optimal sanction may be below harm. Decreasing [increasing] the sanction sacrifices [strengthens] some deterrence leading to more [less] offenses, but reduces

[increases] costs of executing imprisonment for a given offense. Without specifying the cost functions, it is impossible to determine which effect will dominate. Underdeterrence by imposing a sanction lower than h can be favorable because it saves on the costs of imprisonment due to a shorter imprisonment time. Overdeterrence by imposing a sanction higher than h can be favorable because it saves on the costs of imprisonment since fewer offenses are committed, and therefore fewer sanctions imposed.

4. The same applies for the joint use of fines and imprisonment. It is always beneficial to use fines first, because they are costless. Whether it is optimal to supplement fines with imprisonment follows the same lines of reasoning and depends on the benefits of the gained extra deterrence and the realized costs of imprisonment.

Only costs of compliance are known

Suppose that the enforcement authority only observes the costs of compliance. As in the situation where only harm is known, the enforcement authorities will not know whether an offense is (in)efficient and should be deterred. The sanction can be made dependent on the observed level of compliance costs and on the estimated, perceived harm and its distribution. Consider a given expected harm. Low levels of compliance costs should be deterred because compliance costs will probably be lower than the harm and the offense is therefore expected to be inefficient. A sanction equal to at least the compliance costs suffices to deter. If the compliance costs increase, two things will happen. First, in order to deter firms with higher costs of compliance, the sanction will have to rise. Secondly, however, if the costs of compliance become larger, the offense is more likely to be efficient and should not be deterred. Therefore, above a certain threshold, no sanctions are optimal. Hence, there is no monotonic relationship between compliance costs and the sanction.

No information

If the enforcement authority has no information at all about individual compliance costs and harm, but it is able to observe whether a certain standard is violated (because it observes a fire accident or environmental spill), three options are available. First, if the average harm is rather high and average compliance costs rather low, it is optimal to enforce the standard maximally to ensure that everyone complies (if f^{max} or s^{max} are not binding). This will lead to costs of compliance but no harm. Secondly, in the reverse situation, if the average harm is low and average compliance costs high, it is optimal to do nothing, leading to harm but no compliance costs. In all intermediate cases, it is optimal to impose a sanction based on the estimated harm (and adjusted for costs of imposing sanctions as above). Then the firm will comply if its compliance costs are lower than the average harm. Firms with low compliance costs will comply, which is in most cases efficient, and firms with high compliance costs will violate, which is in most cases efficient.

3.2.2 *Monitoring*

Above, detection was certain. Generally, authorities have to spend resources on tracing offenders, either through monitoring or through investigation. This subsection deals with general enforcement (monitoring): the authorities have to spend resources before receiving information about the offense (Mookherjee and Png, 1992).

Under monitoring, detection is costly and applies to all offenses of a certain type regardless of the harm done. Once an offense is detected, the important question is the same as under certain enforcement, namely the optimal magnitude of the sanction dependent on the information that is available after detection. This implies that the results obtained under certain enforcement are applicable, except that the sanction should be adjusted to the fact that the sanction is only imposed with a certain probability.[73] The firm will comply if the expected sanction for non-compliance is higher than the costs of compliance, $c \leq p(f+s)$. The first-best level of compliance is obtained when the expected sanction equals the harm. This forces the firm to balance the costs of compliance with the harm. Hence a fine equal to the harm inflated by the probability of being apprehended, $f = h/p$, implements efficient behavior. Compared to the case of certain enforcement, the magnitude of the sanction should be increased to correct for a lower probability of detection, so that the expected sanction will equal the harm. If the sanction exceeds the maximal available sanction, the firm will not comply, unless the probability of detection is increased.

Let me discuss this more extensively.

Multiple firms and monetary sanctions

Under monitoring authorities do not freely observe the (expected) harm that is inflicted. However, they may learn about the level of harm done, following detection. While detection cannot be made dependent on the level of harm, prosecution and punishment can. Consider the binary choice model with multiple firms who make a similar decision, but differ in the harm they inflict and in their private costs of compliance. This case is extensively discussed in Polinsky and Shavell (1992). See also Shavell (1991).

Let us begin with using only monetary sanctions. The firm will comply if the expected sanction exceeds the costs of compliance: $pf \geq c$. The first-best level of compliance can be obtained when the expected sanction equals the harm. This forces the firm to balance the costs of compliance and the harm. Given the probability of monitoring p, a fine equal to the harm inflated by the

73 From now on I only discuss the most realistic case, i.e. that the enforcement authority only has information about the harm done, and only briefly reflect on cases in which it has perfect information. I ignore cases where the authority only knows compliance costs or only knows that an offense or accident occurred. Results follow from the analysis under certain enforcement and the conclusions about detection as discussed in the main text.

probability that an offense is detected, suffices to implement efficient compliant behavior provided that the firm can pay the fine: $f = h/p$ (hence $pf = h$) if $h/p \leq f^{max}$.

If it is optimal to have some positive monitoring, the probability of detection should be sufficiently large to deter more serious forms of harm. This implies, however, that firms who cause less harm, below the threshold $h = pf^{max}$, are also detected with the same probability and the fine can be adjusted to the harm, $f = h/p$, so that deterrence is first-best. The sanction rises with harm. For firms who cause harm above this threshold the fine is maximal, f^{max}, and there is underdeterrence. Of course, this underdeterrence can be prevented by increasing the probability of detection. However, because detection is costly, it is not beneficial to eliminate all offenses, but to sustain some degree of underdeterrence for the most harmful offenses. The optimal probability of detection follows from equating the marginal benefits of deterring firms who fail to comply because of the maximal available sanction, and the (marginal) costs of increasing monitoring (for all firms). This depends on the specific costs functions, but certainly implies some underdeterrence. It is optimal to have some firms for whom the maximum available sanction is binding.[74] These are the firms that cause relatively much harm.

Multiple firms and non-monetary sanctions

Now suppose that costly non-monetary sanctions are imposed. First, the same conclusion applies as under certain enforcement with respect to the costs of imposition that should be included in the sanction. If the costs of imprisonment k are independent of the imprisonment time, the optimal sanction equals $s = h/p + k$.[75] Generally, the costs of imposing a sanction increase with the imprisonment time ($k'(t) > 0$).

Because the enforcement authority only knows the harm, it is not able to punish only inefficient offenses so that the costs of imprisonment are realized. Given the costs of detection and imprisonment, the optimal enforcement policy is as follows (see for example Shavell, 1991; Polinsky and Shavell, 1984):

---

74 As under certain enforcement, we can identify three types of firms and offenses. (1) Firms for whom compliance is efficient ($h \geq c$) and who can afford to pay the fine ($f = h/p \leq f\,max$). They will efficiently choose to comply. (2) Firms for whom compliance is efficient, but who cannot afford the fine and will not comply ($pf^{max} < c \leq h = pf$). (3) Firms for whom compliance is inefficient ($h < c$) and who will efficiently incur violations either because they cannot pay the fine or because they do pay but still incur the violation. Increasing the probability of detection increases the costs of detection for all firms. It increases deterrence for some type 2 firms, converting them to type 1. Deterrence for types 1 and 3 remains unchanged, as the fine $f = h/p$ is costlessly adjusted in order to prevent overdeterrence. Under monitoring it is optimal to choose the probability of detection in such a way that there will be some firms of type 2. It is only optimal to trade-off the probability of detection and the magnitude of the fine if the maximal fine is not binding for a sufficiently large number of firms.

75 Hence, $ps = h + pk$, or the expected sanction equals the expected costs of harm and sanctions, Polinsky and Shavell (1992).

1. For low levels of harm the optimal sanction is zero. If the harm is low, the benefits of deterrence are also low and it is not worthwhile to spend resources on imprisonment.
2. If harm becomes more important imprisonment becomes socially beneficial. As under monetary sanctions, the expected sanction should rise with harm, by increasing the sanction (the probability is fixed under monitoring).
3. As under certain detection, there can be both over- and underdeterrence for these intermediate levels of harm. A higher sanction implies that more firms will comply, so fewer sanctions have to be carried out. However, the costs of punishment increase for firms who do not comply and are inspected.
4. For serious levels of harm, above a second treshold, the optimal sanction exceeds the maximum imprisonment, hence the firms in question are sanctioned with the maximal sanction and the result is underdeterrence.
5. Compared to the decision under costless monetary sanctions, the optimal probability of detection should also take into account the costs of prosecution and punishment. A higher probability of detection increases the number of imprisonments that must be carried out for non-compliant firms, hence the costs of imprisonment. On the other hand, since more firms comply, fewer firms are found to be in violation and so the costs of imprisonment are decreased. As under point 3, the specific result depends on the costs and distribution functions.
6. When both fines and imprisonment are used, it is optimal to impose a monetary sanction equal to the harm inflated by the probability of detection ($f = h/p$) if the firm can pay this fine, hence if harm is small enough (below the treshold, $f \leq f^{max}$, or $h \leq pf^{max}$). Deterrence for these firms is optimal. Above this treshold, the maximal fine is imposed. In the beginning, optimal imprisonment is zero because the costs of imposing imprisonment are not worth the benefits of higher deterrence, so that some underdeterrence develops. If this underdeterrence becomes severe enough, it becomes optimal to supplement the sanctions with costly imprisonment. For very severe levels of harm, the maximum monetary and non-monetary sanction may be insufficient to deter.

Remarks on monitoring and multiple firms

Before continuing, some general remarks are relevant concerning the results obtained above.

1. It is important to stress that generally, a sanction equal to the harm done (as required under proportionality) is inefficient, especially if imposing a sanction is costly. Under uncertain enforcement, such a sanction will fail to enforce the most efficient level of precautions. Moreover, since firms non-comply, costs are made for prosecution and punishment. Hence, firms fail to comply (no deterrence is obtained), while enforcement costs are made. This is probably the socially worst outcome.

2. If the enforcement authorities learn of the magnitude of the compliance costs c (after detection), they will know whether compliance and violation is efficient. This has an important advantage: the authorities can choose to punish only inefficient violations, which saves the costs of imposing sanctions.[76] It is even possible to not punish firms who cannot be deterred because the maximal available sanction is too low. Moreover, the sanction no longer has to be adjusted to the level of harm in order to induce an efficient balancing of compliance costs and harm. Inefficient offenses can be deterred by $pS \geq c$. Since in the case of inefficient offenses $h \geq c$, this means that a lower expected sanction is sufficient and can be used to save enforcement costs by lowering S or possibly by lowering p (this, however, will affect all offenses).

Monitoring and heterogeneous wealth

In the above analysis, it was assumed that the enforcement authority knows the maximal available sanction (before and after detection) and that this is identical for all firms. The optimal probability of monitoring depends on this maximum. However, generally, the maximal fine or imprisonment differs among individuals and firms. Suppose that the enforcement authority is not informed about this maximum before detection, but only learn of it afterwards (Polinsky and Shavell, 1984). First, consider fines. If firms differ with respect to wealth, it remains optimal to impose a sanction equal to $f = h/p$ as long as a firm can afford this fine. If this maximal fine becomes binding for the firms with the lowest wealth, it is optimal for firms with higher wealth to still have $f = h/p$, a fine that increases with harm. Therefore, firms are punished differently. Because the probability of detection is equal for each firm (independent of wealth), fines for firms with higher wealth should be higher than for low wealth firms. If wealth is binding for low wealth firms, they are underdeterred. However, for firms with higher wealth, deterrence can be increased at no cost by increasing the fine. The fine for the higher wealth firms should not necessarily be extreme, because that might create overdeterrence given the fixed probability of detection.

If only imprisonment is used, the results may differ and cannot be stated in general terms. They depend on the relationship between wealth and the value of imprisonment (Polinsky and Shavell, 1984). Since a higher wealth indicates a higher income and hence a greater opportunity cost of imprisonment, imprisoning higher wealth individuals is more cost-effective. This may result in higher imprisonment sanctions for higher wealth groups, since it might be desirable to achieve a higher level of deterrence. However, it may also result in smaller imprisonment sanctions, because a smaller sanction will suffice to create the same level of deterrence.

---

76 See also Malik (1990). It may be optimal for enforcement authorities to invest in screening in order to learn about the magnitude of the compliance costs, because learning might reduce the costs of imposing sanctions or the costs that firms make avoiding enforcement (section 6.2).

Under the joint use of fines and imprisonment, it becomes optimal to impose imprisonment on firms with low wealth if the underdeterrence of these firms becomes severe enough. In such cases, it is optimal to punish wealthier firms with fines alone, and less wealthy firms with fines and imprisonment.

Polinsky (2006) discusses the situation in which wealth remains unobservable for the enforcement authority after detection. In this case, it might be optimal for the enforcement authority to offer a choice between two possible sanctions. Either a high fine or a low fine supplemented with imprisonment, with the latter as the overall most severe sanction.[77] This will induce high wealth firms to choose the high fine. Since low wealth firms cannot afford the high fine, they necessarily choose the combination of fine and imprisonment. The reason that the sanction for low wealth firms is higher is that deterring low wealth firms is more beneficial, because this saves the costs of imprisoning them. Note that, as discussed, such a policy is also optimal if wealth is observable and the enforcement authority uses both imprisonment and fines, so that it does not matter whether wealth is observable or not.

Of course, not only wealth, but also the maximum available imprisonment sanction can vary, due to such factors as differences in age, health, time discounting etc. The effect of these differences is similar to the effect of differences in wealth on fines.

Marginal deterrence – multiple offenses

So far, I have discussed a situation in which the same probability of monitoring applies to multiple firms each of which can choose whether to commit a single offense. A somewhat different situation arises when each firm must decide on a level of compliance or precautions. The monitoring system detects all levels of precautions and harm at the same rate. If harm is learned of after detection, offenses of differing severity may be prosecuted and punished at different rates. The penalty scheme is bounded by the maximum available sanctions. If the maximal sanction is imposed on low levels of harm, firms which still fail to comply have no incentive to choose minor levels of offenses, but will choose for the most harmful offense. The expected sanction for more serious offenses should be higher. Since the probability of detection is given, a higher expected sanction must be obtained through a more severe punishment. This case is extensively discussed in Mookherjee and Png (1994). Shavell (1992b) discusses a similar model in which an individual can choose between two acts that differ in the level of harm. Let me summarize their results.

1. First note that the wealth limit constrains how much deterrence can be provided by costless penalties. If this maximum is not binding for even the highest levels of harm, any desired pattern of deterrence can be achieved at minimal cost by trading off the monitoring probability and sufficiently

77 The high fine could also be supplemented with imprisonment if necessary. It is essential that the imprisonment sanction and the overall sanction on low wealth firms be higher.

large penalties. If the maximum fine is binding, the next points provide the optimal enforcement policy.

2. Under costless monetary sanctions, it is optimal to let the penalty rise with the harm in order to obtain marginal deterrence, with the highest available fine reserved for the most harmful offenses.
3. Under costless monetary sanctions, it is optimal to set marginal expected penalties everywhere at a point lower than the marginal harm (i.e. less than first-best deterrence). As before, since detection is costly, some underdeterrence is beneficial.
4. Under costless monetary sanctions, it is optimal not to punish minor levels of non-compliance, i.e. offenses that only cause relatively low levels of harm. The reason is that reducing expected penalties for minor offenses, reduces the costs of deterrence for larger offenses, because expected penalties should increase with the level of harm. The penalty for the highest level of harm equals wealth. For lower levels of harm, the sanction should be smaller or even, for the lowest levels of harm, zero. This allows the expected sanction to rise sufficiently steeply, so that the highest levels of harm need not be deterred by a large and costly probability of detection. It is optimal to permit some non-compliance, although the marginal costs of compliance are lower than the marginal harm, in order to reduce the costs of deterring greater harm.[78]
5. Under costly non-monetary sanctions, the results under 3 and 4 are strengthened for minor offenses. Since enforcement costs are higher, the marginal expected penalties will be even lower and more offenses will go unpunished. On the other hand, the marginal expected penalties for more serious offenses should be stiffer. This will induce firms to commit fewer serious offenses, which involve lower costs of imposing sanctions. The marginal expected penalties for serious offenses might possibly exceed the marginal expected harm (overdeterrence) if the costs of prosecution and punishment are high in relation to the costs of monitoring.
6. There is a remarkable difference between the case involving multiple firms committing a single offense (as discussed above, and referred to here as model A) and the case in which a firm is deciding about multiple offenses (this paragraph, referred to as model B). Consider only monetary sanctions. In both models, the optimal sanction for the most harmful offenses equals wealth and the sanction for offenses with lower levels of harm is lower (and rises with harm done). The highest degrees of harm are underdeterred because of the costs of enforcement. However, in

78 Heyes (2002) obtains a similar result in an otherwise completely different model for pollution emissions. Moreover, he considers the threshold as an either exogenous or endogenous policy variable. He shows that an increase in the threshold (the informal standard above which no penalties are imposed) induces non-serious violators to reduce their pollution, while it induces serious violators to increase their pollution if the penalty function is not steep enough. Therefore, more stringent enforcement might increase environmental damage and reduce social welfare.

model A, deterrence for lower degrees of harm is perfect, while in model B, all offenses are underdeterred, including the less important ones. In both models, one probability of detection applies to all offenses. This implies that in model A, firms that cause relatively small degrees of harm can be optimally deterred by a higher (costless) sanction. This does not affect other compliance decisions (by other firms). In model B, however, underdeterrence of less harmful offenses by choosing for a lower sanction is beneficial because this induces firms to choose less harmful offenses. This leads for instance to the following differences. In Polinsky and Shavell (1992, model A) the (marginal) expected penalties should always exceed the (marginal) harm because the expected penalty should equal the expected harm plus the fixed costs of imposing a sanction. Only when the costs of imposing sanctions increase rather steeply is some underdeterrence favorable. Moreover, the expected penalty should be independent of the monitoring costs, and if the prosecution/punishment costs are higher, the expected penalties should be raised for all acts. In model B, the marginal penalties are below the marginal harm. Moreover, there is a treshold below which violations are not prosecuted even under monetary sanctions. And higher prosecution/punishment costs increases penalties for some acts, but decreases for other acts.

3.2.3 *Investigation*

Suppose that spending resources on detecting offenders is necessary in order to catch them, and that information about the severity of the harm becomes available once an offense is committed (for example as a result of reports by victims), so that the probability of detection (the allocated resources) can be made dependent on this information. Then we can speak of specific enforcement, i.e. investigation.

If enforcement takes place through specific investigation, detection can be made dependent on the level of harm done. An expected sanction that is equal to the harm done will implement the first-best level of precautions. As noted by Becker (1968), it is possible to maintain deterrence but save on enforcement costs by lowering the probability of detection and increasing the sanction. Hence, the following conclusions hold (Polinsky and Shavell, 1984; Shavell, 1991 and 1992b; Mookherjee and Png, 1992):

1. If monetary sanctions are used, they should, according to this logic, be increased to the maximum. This holds irrespective of whether single or multiple offenses are involved and whether there are one or more firms. For all offenses, if the sanction is less than maximal, welfare can be increased by increasing the fine and decreasing the probability of detection, maintaining the same level of deterrence. Marginal deterrence (the expected sanction should rise with the level of harm) can be obtained by varying the probability of detection.

2. For all offenses, some level of underdeterrence is beneficial. The socially optimal probability of detection balances the marginal costs of detection and the marginal benefits of deterrence. If compliance approaches first-best behavior, the benefits of deterrence become very small and it is not efficient to eliminate all offenses. Moreover, for very small offenses, it is beneficial to refrain from spending any resources on detection. This reasoning is the same as used before for justifying underdeterrence.
3. If only non-monetary sanctions are used, they should be increased to the maximum if the marginal social costs are decreasing or constant to the imprisonment time (section 3.1.3) and if the private disutility of imprisonment is increasing or constant to the imprisonment time. Suppose that the probability of detection is lowered and the imprisonment time increased, so that the expected sanction, and hence deterrence, remains equal. The enforcement costs, however, do change: (1) the costs of detection decrease; (2) the costs of imposing sanctions on non-compliant firms increase; (3) since fewer firms are apprehended, fewer imprisonment sanctions have to be executed and the costs of imprisonment decrease. If both the marginal private and social costs are constant, the net effect of (2) and (3) is zero. If the marginal private costs are constant, the expected value of imprisonment, $p * s$, will not change, so the expected man-years in prison remain unchanged. If the marginal social costs are constant, the costs of imposing sanctions are constant. The effect of a decrease in p is only a decrease in the costs of detection. However, if the marginal social costs of imprisonment increase, the costs of prosecution and punishment increase, while the costs of detection decrease. In which case, no general statement can be made and extreme sanctions are generally not favorable since the costs of imposing sanctions also become extreme. Similarly, if the marginal disutility of imprisonment decreases for the individual, imposing higher sanctions becomes increasingly less costly for the individual. When detection is lowered, a more than proportional increase in the imprisonment time is needed in order to maintain deterrence. Therefore, the social costs of imprisonment rise (unless the marginal social costs decrease enough), and this increase in costs should be balanced with the decreased costs of detection.[79]
4. Furthermore, under non-monetary sanctions, the same remarks about

79 This analysis with respect to the marginal social costs of imprisonment was not relevant before, as a trade-off between the probability of detection and the magnitude of the sanction was not possible at the individual level. Under monitoring and certain enforcement, the question was whether increasing the imprisonment sanction was beneficial, a question answered by comparing the costs of imposing sanctions (and detection) with the benefits of increased deterrence (and therefore only $k(t)$ and $k'(t)$ were important). Under investigation, it becomes possible to consider how increasing the sanction and simultaneously lowering detection affects the costs. Here deterrence, hence the number of offenses, is held constant and it is considered whether the enforcement costs decrease if the sanction is higher but executed less often (and therefore $k''(t)$ is important).

underdeterrence apply due to the costs of detection (including no enforcement for small offenses). However, as discussed in the previous point, the costs of imposing sanctions can give rise to both under- and overdeterrence.

5. If both monetary and non-monetary sanctions are used, it is optimal to use the monetary sanction to the maximum as this can be implemented at no costs. Again, the probability of detection is such that this will result in some underdeterrence. For small offenses, below a given treshold, no enforcement is efficient. If the harm increases, some positive enforcement becomes beneficial and the sanction should equal wealth. It is not always necessary to increase deterrence with imprisonment. More generally, it is not optimal to impose the maximal non-monetary sanction. If an imprisonment sanction is supplemented with a fine, this reduces the possibility to trade-off the probability of detection and the imprisonment time. Suppose that the imprisonment time is doubled and the probability of detection is reduced. This also reduces the expected fine. Halving the probability of detection would leave the expected imprisonment time constant, but reduce the expected fine, hence reduce deterrence. Put differently, in order to maintain the same level of deterrence the probability of detection should be less than halved. This will, however, increase the expected man-years in prison and hence increase the costs of imposing imprisonment (unless the marginal social costs decrease sufficiently). The costs of detection are reduced, but the costs of imposing sanctions increase, so that the enforcement costs may rise. The trade-off between the probability of detection and the non-monetary sanction is not necessarily efficient. Moreover, it may be optimal not to supplement the maximal fine with imprisonment. This depends on the severity of underdeterrence.

Some remarks on investigation

Several remarks can be added to the results obtained:

1. If the costs of compliance are known before detection, the costs of detection will only have to be made where violation is inefficient. Firms for whom non-compliance is efficient will not be sanctioned, so detection is not required. Similarly, if the costs of compliance are not known before detection, but only afterwards, the costs of prosecution and punishment can be saved by not prosecuting firms for whom non-compliance is efficient. As under monitoring, this implies that a sanction that implements $pS \geq c$ for inefficient offenses suffices. Since for inefficient offenses $h \geq c$, this means that a lower level of pS suffices for deterrence, hence the trade-off between p and S can be further utilized.
2. If the maximum available fine (or imprisonment time) is not known at the time that the probability of detection is determined (which generally seems to be the case), and if it is binding, the authorities should base the probability on the expected maximum fine. In choosing the optimal probability of detection, a balance must be made between deterrence of firms

with small maximal sanctions and the costs of detection. This depends on the distribution of f^{max}.

3.2.4 *Joint use of monitoring and investigation*

In the analyses above, the information about the harm done before detection was either perfect (hence investigation is efficient) or zero (hence only monitoring is available). Generally, information about harm done is imperfect and may vary with the severity of the offense. For example, the reporting degree of offenses by victims is imperfect but increases with the severity of the offense. When the reporting degree is sufficiently high for all levels of harm and the costs of investigation are sufficiently low (relative to those of monitoring), it is always optimal to only enforce through investigation (Mookherjee and Png, 1992). If, however, these conditions are not satisfied, some levels of harm will be insufficiently deterred by investigation. If reporting is positive (although insufficient), so that some information about the harm done is available, it is optimal to combine investigation and monitoring. Given that some positive monitoring is desirable, efficient enforcement is as follows (Mookherjee and Png, 1992; Shavell, 1991, for the case of multiple firms):

1. Under monetary sanctions, there is no specific enforcement for firms causing relatively little harm (below $h/p \leq f^{max}$). Only general enforcement is used and the fine is chosen so that the expected sanction equals the harm ($f = h/p$). It is possible to optimally deter these firms. If the sanction becomes larger than the wealth ($h/p > f^{max}$), the firm cannot pay the fine and there is underdeterrence. When this underdeterrence becomes severe enough, it becomes optimal to supplement the measures with specific enforcement. The specific probability of detection rises with the harm done.
2. Under non-monetary sanctions, similar results apply. For low levels of harm, only general enforcement is used and the results of section 3.2.2 concerning the optimal magnitude of the sanction (which may be zero if the costs of imposing sanctions are large) apply. For higher levels of harm, this results in underdeterrence and it might be beneficial to supplement the measures with specific enforcement. Whether it is optimal to employ maximal imprisonment depends on the costs function of imprisonment as discussed in section 3.2.3.
3. If both monetary and non-monetary sanctions are used, monetary sanctions should be used first. This means that for low levels of harm ($h/p \leq f^{max}$), there is only general enforcement. The expected sanction equals the expected harm ($f = h/p$). This implies that above the treshold $h = pf^{max}$, the firm will not be able to pay the fine and it will be underdeterred. Hence, beyond this treshold, the monetary sanction is maximal and optimal imprisonment and optimal specific enforcement eventually become positive if

underdeterrence becomes severe enough. Which of these measures is used, first or jointly, depends on the specific (social) cost functions.

3.3 Interactions between the crime rate and enforcement

In the above subsections, I discuss the influence of enforcement on the crime rate. However, there might also be significant influences of the crime rate on (the effectiveness of) enforcement. It is therefore important to study the interaction between the compliance rate and enforcement and the existence of (multiple) equilibriums of compliance rates and enforcement policies. The offense rate might influence both the probability of detection and the magnitude of the sanction. The literature – see Bar-Gill and Harel (2001) as well as Glaeser et al. (1996), Persson and Siven (2006) and Conley and Wang (2006) – studies the relationship between the crime rate, in particular theft, and enforcement. Let me describe the most important interactions that can be translated to safety regulation:

- The probability of detection might depend on the number of offenses. If the number of inspectors or police officers is given (in the short run), and the number of offenses exceeds their possible work load, the probability of detection decreases as the number of offenses increases, making non-compliance more attractive. Any firm's individual decision to offend lowers the probability that any other firm's offense will be detected.
- The presence of offenders can influence others to become offenders as well (peer-effects). Alternatively, a lower compliance rate might decrease the stigma imposed on offenders, making non-compliance less costly. More generally, extralegal sanctions such as reputational effects might depend on the imposition of legal sanctions and the compliance rate.
- If the offense rate increases, both non-compliant firms as well as enforcement authorities might realize positive returns to scale. Firms might learn from each other or be better able to obtain the equipment required for violation. Enforcement authorities might be more capable of exploiting information.

In chapter 7, I discuss the influence of informal (extra-legal) sanctions and will demonstrate how their influence depends on the compliance rate. With respect to the effects of the interactions discussed here and the existence of multiple equilibriums, similar conclusions apply as those derived in chapter 7 about informal sanctions. The existence of multiple equilibriums might explain why compliance rates vary over time and place, even if social and economic conditions are comparable.

With respect to the optimal level of enforcement resources, Bar-Gill and Harel (2001) conclude that if a decrease in the offense rate raises [lowers] the probability of detection as well as the magnitude of the sanction, the deterrent effect of enforcement is stronger [weaker] than in the above analysis, hence the optimal level of enforcement resources will be higher [smaller]. If a lower offense rate has opposing effects on the probability of detection and the magnitude of the sanction, conclusions about the deterrent effects of enforcement depend on the magnitude of the interactions. Because of these interactions, it is possible that an increase in enforcement deters violation, but at the same time this decrease in the offense rate makes violation more attractive. This again provides an explanation for less than maximal sanctions.

3.4 THE VALUE OF SELF-REPORTING

Law enforcement often uses what is called self-reporting, i.e. the reporting by firms or individuals of their own compliance status to the enforcement authority. For instance, an individual may confess a murder or firms may report an excess of environmental emissions. Sometimes self-reporting is even required. Let me summarize the general benefits of self-reporting (Malik, 1993; Kaplow and Shavell, 1994a; Innes, 1999b).

1. Self-reporting allows enforcement authorities to save on detection costs. A firm can be induced to self-report a violation when the penalty for a self-reported violation is equal to or slightly less than the expected penalty for a violation that is detected by the authority itself. This will not affect deterrence, but will allow the enforcement authorities to save on the costs of inspecting these firms.

   The extent to which costs are saved depends on the nature of detecting violators. If detection is carried out by means of investigation, many resources can be saved. The problem is how to find the perpetrator of a single known violation (with known harm) such as an environmental spill. If the injurer self-reports, the whole detection problem is solved. On the other hand, under a pure monitoring system, for example when police officers patrol on streets or on highways, there will probably be no reduction in monitoring costs if offenders self-report. If monitoring is carried out by means of inspection of firms, without knowledge concerning their potential violations, monitoring costs can be saved because a firm that self-reports does not have to be inspected. However, firms that do not self-report still have to be inspected.
2. In general, the optimal penalty for a self-reported violation is smaller than the optimal penalty for dishonest reports. This is the case in the real world where dishonest reports are usually considered to be a criminal offense for which individual employees can be prosecuted. The logic behind this echoes that of Becker (1968). Enforcement costs for dishonest reports can be minimized by making the fine for dishonest reports as high as possible and reducing the probability of inspection (= detecting the truth of reports). However, this implies that the penalty for truthfully reporting a violation should be smaller than maximal in order to induce self-reporting.

   If the decision to commit an offense is a binary decision, the authority will only inspect those firms that report compliance, because a firm will have no incentive to wrongfully report a violation. If reporting is perfectly truthful, this implies that monitoring does not detect violations and is therefore seemingly ineffective. Inspections are required in order to induce truthful self-reporting. Inspections are not aimed at detecting violations of safety standards, but at detecting dishonest reporting and it therefore focuses on firms that report good behavior.

3. Not only monitoring costs, but also the costs of prosecuting and convicting a violation might decrease. If a firm confesses to a violation, enforcers will possibly have to spend less effort on trying to find evidence and less time in costly court proceedings that would guarantee sufficient accuracy. If a firm self-reports a violation, a sanction can be imposed immediately.
4. The costs of imposing sanctions can increase as well as decrease. If imposing sanctions is costless, self-reporting is always beneficial because it reduces monitoring costs. If there are fixed (administrative) costs involved in imposing a sanction, the costs of imposing sanctions will increase because every firm that reports a violation will have to be punished. Therefore, self-reporting is only beneficial if the reduction in monitoring costs outweighs the increase in costs for imposing sanctions.
   If the costs of imposing sanctions increase over the level of sanctions (like imprisonment), we have a different result. Because a lower sanction is required under self-reporting, the costs of imposing an imprisonment sanction decrease. On the other hand, the frequency with which imprisonment is imposed increases. If the marginal social costs of imprisonment are increasing, self-reporting decreases the expected costs of imprisonment and self-reporting is certainly socially beneficial.
5. Self-reporting indirectly reduces enforcement costs by reducing the reliance on imperfect monitoring technologies. If the actions of the enforcement authority do not detect all violations and/or the authority sometimes penalizes compliant firms, deterrence is weakened (section 6.1) and more resources are required to achieve the same compliance rate. Hence, the potential benefits of self-reporting are higher. Put differently, self-reporting is more likely to be desirable if the public enforcement authorities commit more errors. The more violations are self-reported, the fewer mistakes are made by public enforcement authorities.
6. Similarly, self-reporting is more likely to be efficient if the inspection costs are high, the maximum feasible sanction is low or the standard is high. All these factors increase the costs of public inspections, and therefore imply higher potential savings by self-reporting.
7. Self-reporting depends on the ability of individual firms to observe their level of compliance. This might require costly self-audits by firms. Friesen (2006) discusses that a firm will only self-audit if the costs of audits are sufficiently low relative to the gains of self-reporting in the form of a lower expected sanction. The higher the costs of audits, the lower the fine for reported violations must be. More importantly, if audit costs are

relatively large, it is possible for the enforcement authority to induce self-reporting, but the overall costs might be lower if the enforcement authority carries out the inspection itself. This is what happens if the benefits of self-reporting in the form of direct and indirect saving of enforcement costs are insufficient to outweigh the costs of auditing itself. Audits (by firms) and inspections (by enforcement agencies) are a kind of substitute. Therefore, before recommending self-reporting, it is important to investigate which party best observes violations relative to the costs: firms or public agencies.

A more complicated analysis is required if (non)compliance is continuous (Friesen, 2006). In that case, auditing does not only reveal whether or not a violation occurs, but also the level of violation and hence the level of penalty than can be escaped. This makes the benefits of auditing uncertain. A possible equilibrium might be achieved if firms were to audit, but only report small violations, while large violations would not be reported. In addition, firms may be reluctant to conduct audits because the violations the firm is aware of might be punished more severely.[80]

8. Friesen (2006) also discusses the problem of credibility (more extensively discussed in chapter 9). The problem with self-reporting is that if firms truthfully report their compliance status, the authority has no incentive to inspect firms. However, if the authority does not inspect, firms have no incentive to report violations truthfully. It is optimal for the enforcement authority to update its belief about the probability of a violation if it receives no report or a report that claims compliance. The enforcement authority decides on its inspection strategy *after* the firm submits a report. Generally, this implies that the authority inspects with a smaller probability than above because it believes that compliance-reporting firms are more likely to be compliant, but with a probability larger than zero, because otherwise firms would have no incentive to report.
9. Self-reporting is also beneficial when potential offenders are risk-averse. If they self-report, they face a certain sanction instead of an uncertain one. The risk-bearing of sanctions is eliminated by self-reporting.

3.5 Conclusion: Extreme sanctions?

This chapter starts with the logic of implementing the highest possible sanctions, as put forward by Becker (1968) and subsequent scholars. By lowering the probability of detection and increasing the sanction, enforcement costs

80 Pfaff and Sanchirico (2000) discuss a situation in which self-audits by a firm increase the probability of detection. If enforcement agencies cannot observe self-audits, this implies that the firm will not have a socially optimal incentive to self-audit and take precautions and/or remedy. Possibly, the authority can make fines dependent on observable characteristics that are positively related to self-audits, such as the mode of detection, self-reporting, or observed corrective actions after self-audits.

can be saved and deterrence remains constant. Within the restraints of the available maximum sanctions this inevitably reduces social costs and is therefore socially desirable. Moreover, increasing the sanction and/or increasing the probability of conviction should increase compliance.
Let me now summarize the arguments concerning the extent to which these conclusions are correct.

1. When prosecution and punishment are costless, this conclusion holds. It is optimal to impose the maximal sanction and the minimal probability of detection that is necessary to implement the desired level of deterrence.
2. When prosecution and punishment are costly, this conclusion depends on the more precise nature of the costs. If the marginal social costs of prosecution and punishment do not increase and the marginal private costs do not decrease, the conclusion still holds. Increasing the sanction to the maximum and decreasing the probability of detection keeps deterrence at least equal, it does not lead to an increase in costs of prosecution and punishment, but decreases the costs of detection, hence it is socially desirable. If the marginal social costs of prosecution and punishment increase or the marginal private costs decrease, implementing the maximal sanction is not necessarily optimal.
3. With respect to the probability of detection: because of the costs of detection some underdeterrence is desirable. It is not beneficial to eliminate all offenses. This implies for example that for low levels of harm, optimal enforcement is zero if detection can be made dependent on the harm done. However, under costly non-monetary sanctions, it may be optimal, in line with the previous point, to have overdeterrence.
4. If enforcement authorities have perfect information about harm and compliance costs, it is not necessary to carry out punishment if non-compliance is efficient.
5. Under monitoring (including certain enforcement), and if only harm is known, it is not optimal to impose maximum sanctions because the expected sanctions should rise with the harm (marginal deterrence). Hence, the magnitude of the sanction should rise, implying that for lower levels of harm, maximum sanctions cannot be implemented. The maximum sanction should be saved for the more harmful offenses. However, this implies that under monitoring, the more harmful offenses are underdeterred when multiple firms decide on a single offense. However, if firms decide on a level of offenses or precautions, all levels of harm are underdeterred, unless non-monetary sanctions are so costly that it is optimal to overdeter the most serious offenses. This induces firms to choose for less harmful offenses.
6. If the maximum (monetary) sanction is not known to enforcement authorities (before detection) the probability of detection should be large enough to deter firms with low wealth. Therefore, the sanction on firms with higher wealth should not be maximal.

However, there are also other factors which are relevant for the optimal enforcement policy and which have consequences for the trade-off between the probability of detection and the magnitude of the sanction. These include (1) time, (2) limited enforcement budgets, (3) counter-enforcement actions by firms and (4) informal sanctions. These are discussed in the following chapters.

4 Continuing compliance and enduring enforcement – time constraints

The analysis in the previous chapter assumes that an offense is committed only once and that the harm caused by an offense is immediately realized. This chapter discusses the time problems arising when offenses can be repeated (section 4.1) or when offenses or the resulting harm can last several periods (section 4.2).

4.1 Dealing with repeat offenses – increasing sanctions?

Firms can again commit the same, or at least a similar, offense in subsequent time periods. The optimal enforcement policy with respect to this problem is discussed in this section. It is common practice in most enforcement policies to punish repeat offenders more severely than first-time offenders. This section discusses whether or not it is optimal to increase the penalty for repeat offenses (given a constant level of detection). Attention is also given to the issue of incapacitation.

4.1.1 *The optimality of non-increasing penalties*

If the penalty is optimal for the first offense, there is no reason to increase the sanction for a second (identical) offense.

This is most rigorously argued for in Emons (2003).[81] He shows that if all offenses are undesirable – hence complete deterrence is desirable – and if wealth is large enough to deter the first offense, it is always optimal to set the monetary sanction for the first offense at the extreme, i.e. wealth. Because wealth is completely seized by punishment of the first offense, the sanction for the second offense is necessarily zero. However, because the expected sanction suffices to deter the first offense, there are no second or third offenses committed.

Generally, not all offenses are undesirable and due to the enforcement costs some underdeterrence is optimal. However, Emons (2003) argues that if the benefits and costs of potential offenses are identically distributed in time over the potential offenders, the same result holds. Firms that can commit offenses that society wants to deter, i.e. firms with relatively small costs of compliance and high levels of harm, are deterred from committing offenses

81 See also Polinsky and Rubinfeld (1991) and Shavell (2004).

by a maximal fine according to the same logic as above. Firms that can commit desirable offenses or some not very undesirable acts, i.e. offenses with high costs of compliance and low levels of harm, are offenses which society does not want to deter (given the costs of detection). These are committed and, if detected, sanctioned maximally in the first period[82] and still committed and not sanctioned (because of the absence of any wealth) in the next periods.

This argument is extended by Polinsky and Shavell (1998). Emons (2003) presupposes that wealth is a given maximum for all periods (e.g. a building or a car). Polinsky and Shavell (1998) presuppose that wealth is a maximum per period (e.g. income). In that case, the sanction for the first offense in the first period should be maximal, and can and should also be maximal for a repeat offense. They show, however, that it may be optimal to lower the sanction for first offenses in the second period. Such a lowering of the sanction has two effects. First, it reduces deterrence in the second period for firms which complied in the first period. Secondly, it strengthens deterrence for firms in the first period, because compliance in the first period implies a smaller fine and hence potential benefits of violation in the second period. Whether or not lowering the fine in this manner is beneficial depends on the magnitude of these two effects. Note that it can only be beneficial if deterrence is not first-best, i.e. there should be some underdeterrence. Otherwise there would be no benefits of increased deterrence in the first period.

Polinsky and Shavell (1998) provide the following interpretation of their results: it is optimal to sanction young ('first-period-of-life') offenders and old ('second-period-of-life') repeat offenders with the maximal sanction. Old first offenders are punished leniently. Similarly it can be argued that it is an efficient strategy to be tough on young companies that have just recently been established, as well as firms that have often shown to be in violation, while existing companies that have good compliance records should be punished only moderately.

4.1.2 *Increasing sanctions: offense as signal*

The reasons for increased penalties should be based on the fact that offenses reveal valuable information to the enforcement authorities.

Note first that in the above discussion, there is no reason for *specific* or: individual *deterrence*. Specific deterrence refers to the objective of punishing offenders in order to stop them from committing further offenses in the future. Specific deterrence implies that people who are sanctioned are more deterred than those who have not been sanctioned, or more deterred than they were before being sanctioned. On the other hand, general deterrence is

82 Of course, as explained in section 3.2, if the enforcement authority is informed about individual compliance costs and expected harm, it can decide not to sanction this type of offense.

the deterrence, through threat of penalties, of firms who have not (yet) been sanctioned, as in section 3.2. If rational and well-informed firms decided to commit a crime, there would be no reason for them to increase their belief in the expected sanction and to be deterred in the future. As long as the expected sanction is the same before and after apprehension, they will make the same decision about compliance (Shavell, 2004). Only in case of uncertainty about the probability of detection or the magnitude of the sanction will the firm rationally increase its belief in the expected sanction after punishment, leading to specific deterrence.[83]

Similarly if enforcement authorities were perfectly informed and determined the enforcement policy optimally, there would be no reason to increase their belief in the fact that a firm is dangerous and therefore to change the enforcement policy. The optimal policy would remain the same. An increased sanction would only create overdeterrence. It may be optimal to increase punishment, but only if enforcement authorities are imperfectly informed and committing offenses signals valuable information about the dangerousness of the offender. There should be some 'learning' by enforcement authorities. Let me give some examples.

In Polinsky and Rubinfeld (1991), committing an offense reveals information about the private benefits of crime. Some of these private benefits are assumed to be 'illicit', i.e. they are deemed to be socially worthless, and therefore are not counted in social welfare. But the higher the illicit benefits, the more likely the firm will be to commit the offense. First-best deterrence implies that offenses with low acceptable gains (which are inefficient) are deterred (in all periods), while offenses with high acceptable gains (which are efficient) are undeterred (in all periods). Suppose, however, that the benefits of an offense are unobservable for the enforcement authority. A uniform fine creates either under- or overdeterrence. If the fine is low, there is underdeterrence of firms with low acceptable gains but high illicit benefits. If the fine is high, all firms with low acceptable gains are deterred, but there may also be overdeterrence of firms with high acceptable gains and low illicit benefits. An increasing fine for repeat offenders might lower both the under- and overdeterrence. The overdeterrence is weakened because the fine in the first period is lower. Firms with low illicit benefits will commit the offense in the first period if the acceptable gains are high. The underdeterrence is reduced because the fine is higher in the second period. Firms with high illicit benefits and a detected first offense will comply in the next period if the acceptable gains are low. Since committing an offense shows that the illicit benefits are relatively high, it may be optimal to increase the fine in order to increase deterrence.

Two crucial assumptions are made in order to arrive at this result. First, some of the private benefits or costs are illicit, i.e. not counted in social welfare. Otherwise, a sanction equal to the harm done would induce the firm

83 A more precise and detailed discussion of specific deterrence can be found in Shavell (2004, pp. 515-518).

to perfectly balance the costs of compliance and the expected harm.[84] However, the construction of illicit benefits is a doubtful idea. It is an admission of weakness to a priori place some benefits or costs outside social welfare when explaining the undesirability of offenses is an important aspect of the enforcement theory. Functional economic explanatory theories are mixed with moral theories. Moreover, it is hard to think of good examples, especially in regulatory crime by firms. With respect to crimes by individuals, we might think of the pleasure of killing or raping a victim. An example for firms might be the higher turnover a horeca proprietor obtains by admitting more customers than allowed. But I do not understand why the festive joy of customers measured by this turnover should not count.

The second crucial assumption is that firms can have different acceptable gains over time (i.e. gains are stochastic), while the illicit gains are fixed for each firm. If both gains are stochastic, second-time offenders do not systematically differ from first offenders, so there is no rationale for a different punishment. The first offense does not predict high illicit benefits for the second offense. If both are fixed there is also no reason for changes in punishment. The (marginal) costs and benefits of punishment are equal in all periods. There is some unique optimal level of (under- or over-)deterrence for all periods. If the acceptable gains are fixed and the illicit gains stochastic, decreasing fines may be optimal. Committing an offense reveals that the acceptable gains of the firm are high, implying that the offense is less likely to be inefficient, allowing lower fines to decrease possible overdeterrence.

Rubinstein (1979) shows that it may be optimal to increase sanctions if it is uncertain whether or not an offense is committed deliberately or accidentally. Accidental offenses cannot be prevented and hence deterrence of these offenses has no social objective. If the enforcement authorities cannot observe the willfulness of an offense, they cannot make punishment dependent on this willfulness. Rubinstein (1979) shows that if the offense can be committed an infinite number of times, an equilibrium is created in which the enforcement authority does not punish offenders who have a "reasonable" offense record and in which the potential offenders refrain from committing deliberate offenses. Hence, there is some form of increased punishment for repeat offenders.

The distinction between accidental and deliberate offenses is related to the issue of erroneously sanctioning compliant firms. Chu et al. (2000) work out the argument that was already informally sketched by Stigler (1970, pp. 528-529) and others: punishing repeat offenders more severely is justified because the probability that an innocent firm is punished twice is relatively

84 If the firm also has some private illicit gains, it will compare its private illicit gains and compliance costs with the expected sanction equal to the expected harm, leading to a less than optimal outcome because more, or more severe, offenses than optimal are committed. The expected sanction should be higher than the harm but it is unclear exactly how much higher because the illicit gains are unknown.

low. Increased punishment therefore reduces the social costs of convicting innocent firms (type II errors).[85] Under any uniform sanctioning scheme, welfare can be improved by a slight reduction in the first penalty and an increase in the second penalty such that overall deterrence is held constant, but the social costs of convicting innocent firms are reduced. Note that – as discussed in section 6.1 – if the probability of erroneously convicting firms is positive, it may be inefficient to impose the maximal penalty (on first offenders). Only in cases where the sanction for first offenses is not maximal is there scope for an increase in the sanction in the next period(s), if enforcement authorities have more certainty concerning the firm's guilt.

4.1.3 *Non-monetary sanctions and the benefits of incapacitation*

All of the above references discuss the use of costless monetary sanctions. The question is whether imposing costly non-monetary sanctions would affect the results. If imposing imprisonment leads to underdeterrence (see section 3.2 for details), it can be optimal to increase deterrence in the first period by reducing imprisonment in the second period for first-time offenders (Polinsky and Shavell, 1998, p.314-15). In addition, the results derived in 4.1.2 hold analogously. However, in all analyses, we have to take into account the increase in the costs of imprisonment when the time that has to be served increases. These costs have to be balanced with the potential benefits of increased deterrence.

The analysis in section 3.2 assumes that the benefits of imprisonment only consist of the deterrent effect on potential offenders. However, once we consider possible repeat offenses we must acknowledge an additional benefit of imprisonment: the fact that imprisoned individuals are unable to commit offenses, the so-called *incapacitation effect*. As a result, even if someone cannot be deterred, imprisoning him will reduce offenses, because he will be unable to practice fraud, spill toxic substances or cause too many fire accidents, etc. If these offenses are inefficient, this is potentially socially beneficial. The higher are the number of offenses someone might commit, the higher the benefits of incapacitation.[86]

Suppose that someone cannot be deterred, and that the only function of imprisonment is incapacitation. In this case, the person in question should be imprisoned as long as the net harm he can inflict in a period exceeds the costs of incapacitating him that period. This implies that if the number of potential offenses is constant, a person should either be imprisoned for ever (because the harm from his potential offenses is higher than the costs of imprisonment) or not at all (because of the reverse situation). However, if the number of offenses decreases over time (as is the case for individuals who become older) a finite imprisonment time might be beneficial.

---

85 The costs of convicting innocent firms are discussed in section 6.1.
86 See for example Shavell (2004, pp. 531-535).

If imprisonment also has deterrent effects, we must recognize, in addition to the results derived in section 3.2, that the benefits of imprisonment are higher due to the benefits of incapacitation. This implies that it might be efficient to impose more or longer imprisonment sanctions. Incapacitation also influences the optimal probability of apprehension. To acquire a desirable level of incapacitation, the probability of detection should be high enough and generally be higher than a minimal probability (say 1%). However, given this probability of detection, the sanction for deterrence should be lower than the maximal imprisonment term, in order to prevent overdeterrence. Becker (1968)'s idea that sanctions should be maximal and probabilities of detection as low as possible no longer necessarily applies.

Incapacitation is not limited to imprisonment. Undesirable acts can also be banned by otherwise preventing an individual or firm from engaging in a particular activity, for instance the withdrawal of a license and/or forced closing of a firm (Shavell, 2004, p.531). A truck driver can be prevented from causing traffic accidents by taking away his driver's license. A restaurant can be prevented from violating fire safety prescriptions by threatening it with closure for a finite or infinite period of time. Of course, such measures are only successful if engaging in the activity without the necessary license is sufficiently enforced. The most extreme incapacitation sanction is the death penalty, which forever stops all offenses.

4.1.4 Increasing apprehension

So far I have discussed whether it is efficient to punish repeat offenders more severely. It was assumed that the probability of detection is constant. The fact that a firm or individual has been prosecuted before might influence by itself the future probability of detection, both positively and negatively (Polinsky and Rubinfeld, 1991). The probability of detection can decrease because repeat offenders are more experienced, and therefore less likely to be detected. Therefore, for deterrence to remain unchanged, the sanction should rise with the number of offenses. On the other hand, enforcement authorities might have learned about relevant characteristics or acquired information about the repeat offender, leading to a higher probability of detection and implying lower sanctions.

A different question is whether it is efficient to allocate more resources to detect repeat offenders than first offenders, assuming that the effectiveness of detection is equal. Is it efficient to increase deterrence by increasing the probability of detection? As discussed above, if there is no new information about the offense that might be valuable in deterring future offenses, there is also no reason to increase the expected sanction. However, if there is serious underdeterrence and perfect deterrence is not feasible, increasing the expected sanction as part of the so-called state-dependent enforcement policies (Harrington, 1988) might be valuable. This is discussed elsewhere, in a separate section (section 5.3).

4.2 Repair sanctions and the time spent in violation

The analysis so far assumes that the interest of the enforcement authority lies in reducing the extent of violations, i.e. the extent to which precautions fall short of the standard. Until now, these violations might have been committed over several periods of time, but every offense occurs at a single given moment. An offense is a discrete event. This applies to most crimes by individuals, such as murder and theft. For offenses of firms, examples of this type of offense are fraud, illegal trading or illegal transport of dangerous substances. However, in many situations, especially with respect to offenses by firms, (non)compliance is not a discrete event, but rather a continuing state of affairs (Veljanovski, 1984) which can last for a lengthy period of time, until it ends deliberately or naturally. Such an offense is spread over multiple periods. An enforcement authority cannot only reduce the harm from violation by reducing the extent of the violation, but also by reducing the *time spent in violation* (Nadeau, 1997). If the enforcement authority detects a violation, it can take action to induce compliance as quickly as possible.

The duration of a violation is examined in this section. First, I discuss the importance of the duration of an offense and the enforcement actions that can be taken to induce compliance. The second subsection discusses the importance of self-reporting in limiting the duration of violations.

4.2.1 *Duration of violation and 'repair sanctions'*

Nadeau (1997) provides two major reasons why enforcement agencies might wish to ensure shorter periods of non-compliance. First of all, longer periods of non-compliance are associated with higher levels of (expected) harm. Especially if the harm is cumulative over time, enforcement authorities should be concerned with reducing the time of non-compliance. Secondly, the existence of persistent violators is harmful for the enforcement authority's reputation and signals that its actions are ineffective. This can reduce the incentive to comply for other firms, especially if they believe that the enforcement authority will not have enough time and resources at its disposal to deal with violations of standards that the authority cares less about. Moreover, it might reduce political support and therefore the funding of the authority's program.

Remediation

Enforcement authorities might be interested in decreasing the period of non-compliance and take actions to return the violator to compliance. This return to compliance, labeled *remediation*, can happen in two different ways.

First of all, some offenses lead to damage that can (partially) be undone, such as (illegal) discharges of environmentally hazardous substances, illegal dumping of waste, etc. The enforcement authority can force the firm to clean the area and return to the state of compliance. Time is an important factor in this case because if an offense has been committed, it might be socially beneficial to have the damage repaired before it is too late. Possibly, the costs of clean-up or the harm done will increase if the offense persists for a longer time. Hence, the socially optimal moment of clean-up is probably immediately following the offense. I will call this type of remediation *clean-up*.

Secondly, some offenses lead to a continuing state of violation until something is done to stop the violation, for instance the breakdown of an abatement technology such as escape-route marks or emission filters, violating building prescriptions or using unsafe working equipment. The issue here is not repairing the damage, but 'repairing' the precautions that were taken and – willingly or unwillingly – have fallen below standard. If the enforcement authority finds out about such a violation, it might be socially beneficial to stop the offense from lasting any longer and the authority can force the firm, under threat of penalties, to return to compliance. I will call this type of remediation *repair*.

In both clean-up and repair cases, harm might result 'accidentally', beyond the control of the firm, and the offense is not so much that damage occurs, but that a firm exerts insufficient effort to limit this damage. An offense might not be referred to as efficient or inefficient, but there might be an efficient period of violation (possibly zero).

Repair sanctions and administrative law enforcement

The actions that an enforcement authority can take to force the firm to return to compliance are especially related to the administrative law enforcement process. By means of warnings, (legal) letters, re-inspections etc. the authority can induce the firm to comply. The sanctions that it can use to force compliance (at least as a threat) are called *repair sanctions* (contrary to punitive sanctions). The two most important sanctions are the *duty by penal sum* and the *duty by administrative coercion*. A duty by penal sum states that a specific firm must for example remove the waste it has dumped or stop dumping its waste before a certain given date, otherwise the firm will have to pay a penal sum (a fine) of, for instance, €500 a day. A duty by administrative coercion requires that the firm should carry out similar actions, but threatens that the enforcement authority will otherwise itself restore compliance and recoup the costs on the violating firm. Administrative coercion also includes the possibility to close the firm.

A penal sum has the benefit of delegating the decision to the firm. It leaves room for the firm to evaluate the best way to end the violation. If this is carried out by the enforcement authority, it will probably be more costly for two reasons. First of all, the firm is more likely to be experienced. Since these sanctions are usually incidental, the authority may not realize economies of scale. Secondly, the authority must satisfy further requirements, such as those of entering private property, opening closed gates etc. which leads to higher (administrative) costs. On the other hand, under administrative coercion, the enforcement authority is certain that the violation will end. This may be beneficial if there is, for example, a judgment proof problem with a penal sum. Even if the authority can not recoup the costs, administrative coercion can be efficient, because recoup is only a transfer of wealth. It is important which type of sanctions lead to the lowest costs of violation and ending the violation.

Remediation and enforcement

The presence of remediation benefits does influence the optimal enforcement policy (Innes, 1999b). Suppose that the enforcement authority can use repair sanctions to enforce a return to compliance while also using fines for deterrence. The optimal enforcement policy balances the (marginal) gain in deterrence, the increase in enforcement costs and the benefits of remediation. Trading off the probability of detection and the magnitude of the sanction (as proposed by Becker, 1968) is no longer always beneficial. A trade-off of the fine and detection which would keep deterrence equal decreases the frequency of socially beneficial remediation (because the probability of detection is decreased) and is therefore costly. If the probability of detection is larger than the minimal probability, following Becker's logic, the magnitude of the sanction for deterrence should be adjusted in order to prevent overdeterrence.

However, this analysis only holds if the enforcement authority cannot condition sanctions on clean-up or repair, because enforcement takes place before any possible clean-up or repair could be carried out or because the authority is unable to observe the timing of clean-up and repair (Innes, 1999b). If – as generally seems to be the case – enforcement can be made dependent on the level of remediation, the problem is reduced to one of marginal deterrence as discussed in section 3.2.2. The problem is to let the firm choose a level of compliance – consisting of both ex-ante precautions and ex-post remediation – in order to minimize the costs of compliance and the expected harm. The optimal enforcement policy would prompt firms to engage in pre-inspection clean-up or repair by offering a lower sanction if clean-up or repair is undertaken.

If the only objective of enforcement is remediation – for example because offenses are accidental – the optimal sanction is such that it will implement the efficient level and moment of remediation. If the sanctions are such that remediation only takes place if an offense is detected (as under the repair sanctions above, which begin with a warning), the probability of inspection should be high enough. However, a better solution would be to impose sufficiently high punitive sanctions on firms that fail to remedy. This is particularly relevant if the socially optimal moment of remediation immediately follows the offense.

4.2.2 *Remediation and self-reporting*

Section 3.4 summarized the general benefits of self-reporting. Self-reporting may be particularly beneficial in cases where offenses are accidental and/or there are benefits associated with remediation. Self-reporting allows the benefits of remediation to be realized with probability one, whereas non-reporters only clean up or repair if they are caught.

Self-reporting and accidental offenses

Let us first assume that the only benefits of enforcement are remediation, as in Livernois and McKenna (1999).[87] This is what happens if a firm has no control over its violations because these result from an accidental equipment failure, variations in input quality or process upsets. If there is a technical failure in an abatement device, which results in non-compliance, repairing the failure will restore compliance. If the enforcement authority finds out about a violation, it will impose a fine and order the firm to repair the device and return to compliance in the next period. It is assumed that the authority can do the latter by means of warnings and legally enforceable administrative orders as discussed above, but that returning to compliance takes some time and allows the firm to save on the costs of compliance for one period. If the abatement device fails, the firm has three options, where the optimal strategy depends on their costs of compliance (per period)[88]: (1) Repairing the device immediately and fully comply (optimal if the costs of compliance are small). (2) Reporting the violation to the enforcement authority, which will then impose a fine and order a return to compliance (optimal if the compliance costs are intermediate) (3) Remaining silent and await inspection, and only repair the device if the authority detects the violation, in which case a fine will have to be paid and compliance will have to be restored (optimal if compliance costs are large). Of course the optimal choice also depends on the fine structure. The influence of a fine on reporting firms (type 2) is ambiguous. Increasing this fine induces some firms to repair a failure immediately (type 1), instead of reporting and slow repairing (type 2). However, it also induces firms to no longer report, but to await inspection (type 3 instead of type 2). Depending on the distribution of compliance costs, which determines which effect dominates, decreasing the fine may be beneficial. In this case, the optimal policy for reported violations is a zero fine and only requiring a return to compliance, as this will maximize the incentive to report. No one will repair a failure immediately, but the number of firms that will report and then return to compliance is maximized.[89]

However, this is only true if the target compliance rate is sufficiently low (see section 5.1). The maximum feasible compliance rate under this strategy is the probability that the device does not fail. Firms that report truthfully comply only in the next period. The expected compliance rate if all firms report truthfully therefore equals the probability that there is no failure. If the target compliance rate is higher, it might be optimal to have some firms that repair immediately. In order to induce immediate repair, the fine for reported viola-

87 See also Malik (1993) and Friesen (2006).

88 Similar to Harrington (1988) and Raymond (1999), section 5.3.

89 Livernois and McKenna (1999) show that under a uniform distribution function this policy is efficient. Friesen (2006) assumes that the costs of remediation are lower for firms if they voluntarily repair or clean up than if they are forced to repair or clean up by the enforcement authority. Therefore, immediate repair will be beneficial for some firms, even if the fine for reported non-compliance is zero.

tions should be higher than zero. Since this will make reporting less attractive, the probability of inspection must be increased in order to deter non-reporting firms (type 3).

Accidental and deliberate offenses
If violations are not completely beyond the control of the firm but also depend on for instance the level and frequency of servicing, the enforcement policy should not only aim at remediation, but also at deterrence of technical failures. This might be problematic because deterrence and remediation are sometimes in conflict. The benefits of repair by self-reporting are only realized if the penalty for non-compliance is low enough (possibly zero) to induce self-reporting. However, a low penalty does not deter the firm and prevent technical failures. Livernois and McKenna (1999) show that if a firm can also deliberately disable the abatement device, the efficient policy depends on the self-reporting of deliberate offenses. If deliberate offenses are not reported (type 3), the optimal policy is the same as described above. The only difference is that some type 3 firms will now deliberately disable the device and only comply if inspected. If, however, truthful reporting of deliberate offenses dominates, the fine for reported offenses should be high enough to deter from deliberately disabling the device. If the fine for self-reported offenses is zero, every firm will disable the device and truthfully report they did so, hence the compliance rate will be zero.

The possible conflict between deterrence and remediation in self-reporting is best illustrated by looking at the consequences of self-reporting for large damages (Friesen, 2006). For large damages, the social benefits of remediation are relatively high, making self-reporting socially desirable. However, because self-reporting requires costly remediation by firms, the sanction should be lower for self-reported violations. This implies that self-reporting harms deterrence, which is worst for severe offenses with high damage. So the social benefits of self-reporting (remediation) as well as the social costs of self-reporting (reduce deterrence) are relatively high for large damages.

Innes (1999b) shows that self-reporting brings an additional benefit. When benefits of remediation are present, self-reporting allows for a lower probability of detection for deterrence. Recall from earlier in this section that if the sanction cannot be conditioned on the level of remediation, the optimal sanction (without self-reporting) may be less than maximal. Self-reporting undoes this effect. When self-reporting is introduced, it will be possible to increase the sanction for dishonest reporting to the maximum, following Becker's logic. If the fine is lower than the maximum, the same level of deterrence of dishonest reporting can be achieved by increasing the fine to the maximum and accordingly lowering the probability of detection. This does not affect the benefits of remediation, since these are obtained by inducing self-reporting. This implies that a given level of deterrence can be achieved against lower enforcement costs than without self-reporting. So, not only allows self-reporting to inspect only non-reporting firms, but also if there are remediation benefits to inspect non-reporting firms less intensively.

4.3 Conclusion

This chapter draws attention to the problem of time in enforcement policies. If there is sufficiently large underdeterrence, it may be optimal to punish repeat offenders more severely than first-time offenders. This limits the possibilities to trade-off the probability of detection and the magnitude of the sanction. Moreover, this trade-off is constrained by the fact that enforcement might not only be aimed at deterrence, but also at remediation.

It was assumed that the probability of detection remains constant over time. An alternative approach would be to increase deterrence for repeat offenders relative to first offenders by increasing the probability of detection and conviction. The next chapter investigates this approach.

5 Partial compliance and targeting enforcement – resource constraints

It might be efficient to base enforcement on past performance if there is sufficiently large underdeterrence. In section 5.1 I explain what might create such underdeterrence. Then I discuss the enforcement problem, in particular the probability of detection, in cases of underdeterrence. In section 5.2, enforcement concerns only one period. In section 5.3, I extend the analysis to multiple periods, allowing for so called state-dependent enforcement policies and warnings.

5.1 Partial compliance

The firm complies with a standard if the costs of compliance are lower than the expected sanction. Therefore the enforcement authority must have a certain level of resources in order to enforce compliance. The efficient level of resources balances the expected harm, the costs of compliance and the costs of enforcement. There are two reasons why full compliance should not be the enforcement authority's objective.

First of all, non-compliance can be efficient when the marginal costs of compliance exceed the marginal expected harm for some level of non-compliance. If this occurs – even in the absence of any enforcement costs – partial compliance is desirable. Full compliance is *not the aim* of the enforcement authority. This includes the point made in section 3.2 about partial compliance within the population when there are multiple firms with different costs of compliance. For example, for some cafés, compliance with the prescribed closing times is efficient (those located in residential areas), while for others compliance is possibly inefficient (those located outside the city). However, partial compliance can also apply to a single firm if their costs of compliance or the expected harm vary over time. For example, compliance with closing times is efficient during week days (when café attendance late at night is limited and people's night rest is important) but not during weekends (when many people want to go out and noise is less damaging). If this is the case, the optimal allocation of resources is such that it leads to partial compliance, and enforcement should target inefficient offenses. Offenses with limited harm might go unpunished.

A second reason why full compliance might not be an enforcement authority's objective follows from the fact that enforcement is costly. Even if all offenses are inefficient, full compliance might require more enforcement resources than is efficient, i.e. the marginal decline in expected harm does not outweigh the marginal increase in the costs of compliance and enforcement. I already argued in section 3.2 that this is a reason for underdeterrence. If the maximum (monetary) penalty is infinitely large, this problem will never occur. The enforcement costs can be minimized by maximizing the sanction and minimizing the probability of detection and therefore the enforcement costs. However, if the maximum penalty is binding, and especially if it is relatively low, full compliance will require significant resources since the probability of detection should be sufficiently large. Therefore, even if all offenses are inefficient, it might not be efficient to allocate sufficient resources to induce full compliance (Lando and Shavell, 2004). If the marginal costs of enforcement are constant or decreasing, it is always efficient to impose either full compliance or zero compliance. If offenses are identical and inefficient and if the marginal costs of enforcement are constant, an enforcement policy that is efficient for one group of firms or for one period of time must also be optimal for the other group or period. Hence, enforcement costs are either low enough to enforce full compliance, or too high to take action against any offense. However, if the marginal enforcement costs are increasing, the results might be different. If marginal enforcement costs are increasing, enforcing a higher level of compliance becomes increasingly costly, so that it is less costly to increase the level of compliance from 10% to 20%, than from 90% to 100%. It is efficient to increase enforcement as long as the marginal costs of enforcement are lower than the marginal benefits of the decline in expected harm minus the compliance costs. When the marginal costs of enforcement increase, a point might be reached where this condition is no longer satisfied, implying that it is efficient to allocate a level of resources that will induce partial compliance (both in time and over the population). Increasing marginal costs are common because of capacity constraints. For example, in the short run, the amount of technical equipment and the number of enforcement agents is given. Increasing the capacity would require a larger budget or the training of agents, both of which take significant time (Stigler, 1970). Suppose that some circumstances change, affecting the expected harm, the compliance costs or the enforcement costs. For example, an (unexpected) increase in the number of cafés will imply that more cafés have to be inspected on closing times, leading to a decrease in the expected probability of inspection per café if the resources are not adjusted. Alternatively, enforcement agents might become ill, requiring replacement. Changing circumstances such as these might possibly require an increase in enforcement resources. However, this might prove to be impossible in the short run because of capacity constraints, implying that full compliance can no longer be achieved. As a result, because marginal enforcement costs are increasing (at least in the short run), the enforcement authority will frequently find itself constrained in its resources with respect to the enforcement problem, which in turn will prevent it from enforcing full compliance.[90] Therefore, the second reason for partial compliance as an objective is that full compliance is *unachievable* due to limited enforcement resources and given the maximum penalty. This is significantly different from the first reason for partial compliance: if a higher level of deterrence could be obtained at no cost (against the same level of enforcement costs), it would be socially beneficial.

In the remainder of this section, I focus on situations in which full compliance is unachievable, ignoring the first reason for partial compliance (to which I return in section 6.1). In the remainder of this section, I assume that all offenses are identical and inefficient, and I examine how the highest level of compliance can be obtained given the level of resources and the maximum sanction, first when an offense is committed only once, secondly when an offense can be committed several times.

5.2 Targeting enforcement

Suppose that all potential offenses are identical and inefficient, and that constraints on resources and penalties are such that not all offenses can be deterred. In that case, the best thing the authority can do is to try and achieve the highest possible level of compliance. If the authority adopts a uniform enforcement policy in which it approaches all offenses in the same way, the consequence of incomplete deterrence will be that no one will ever comply.

Consider the example of enforcing closing times in cafés. Suppose that a café should comply with the rule concerning closing at 1:00 AM in order to keep the night's rest in the neighborhood. Every Saturday, each café must decide about compliance. If a given café complies, it foregoes a benefit of €200 of being opened until 2:00 AM. The maximum feasible fine equals €1000. If the municipality is able or willing to carry out a random inspection rate of only 10%, a café will violate (assuming risk neutrality) and remain open until 2:00 AM, since $10\% * €1000 = 100 < €200$. Hence the compliance rate will be zero.

In this case, it would be helpful if all available resources were allocated to some subgroup of the population (Lando and Shavell, 2004). This would induce this particular population group to comply, achieving some level of partial compliance. Of course, deterrence is lost for the part of the population that is not inspected, but this part would also violate under the uniform policy.

Consider again the example of enforcing closing times at cafés. If the available resources allow an average inspection rate of 10% and the maximum fine

90 A related reason for partial compliance as an objective is that the political process does not allocate efficient enforcement budgets, constraining the authority's objective to a maximizing of compliance given the available resources. Note also that although on paper, the objective of enforcement agents is often to achieve full compliance, in practice this means that they try to maximize compliance with a limited enforcement budget.

is €1000, the best policy would be to only inspect the cafés that are located in one half of the city (region A), and not inspecting all the other cafés (region B). This would allow the inspection rate in region A to be 20%, sufficient for deterrence, while the inspection rate in region B would be 0%. Hence, half the population will comply, leading to a compliance rate of 50%, which is an improvement on the 0% compliance rate of the uniform inspection rate of 10%.

Other examples, given by Lando and Shavell (2004), are tax audits which focus only on citizens in some regions of the country or citizens have names starting with A-M, or highway patrols for speeding that discriminate between highways.

Harrington (1988) provides a similar solution for enforcement in dealing with one firm during multiple periods. In this case, enforcement can target on certain periods. For example, cafés could be inspected during one half of the year (from July – December), and be left without control for the rest of the year (January – June). This would result in cafés complying half the time, leading to a compliance rate of 50% with an average inspection rate of 10%.

Another variant is offered in Heyes and Rickman (1999). They consider a situation in which an enforcement authority has to enforce regulations on several domains of a firm, for example several locations or several types of regulation. If the authority has insufficient resources and penalties, a firm might possibly incur violations in all domains. It can then be efficient to tolerate non-compliance in one domain in order to enforce compliance in the other.

For example, if a proprietor has two cafés (A and B) in which he can violate closing times as in the example above, he will, under a uniform policy, incur violations in both. If however, the enforcement authority follows a policy of not enforcing at location A, but enforcing at location B with probability 20%, the proprietor will at least comply at location B. In which case, the compliance rate will again be 50%, with an average inspection rate of 10%.

The recommended policy of targeting is only efficient under certain conditions (Lando and Shavell, 2004). First of all, the population must be aware of the targeting otherwise they will not correctly estimate the probability of inspection. Knowledge of the average inspection rate is insufficient. A proprietor should know whether or not his specific region will be inspected. An automobilist should know whether or not the highway he takes will be inspected. Otherwise, the compliance rate remains zero. Secondly, movement from one group to the other should not be possible for potential offenders, or it should at least be sufficiently costly. Otherwise, the potential offenders can still escape enforcement. Individuals cannot easily change their name to avoid enforcement when targeting is based on the first letter of a name (A-M), and in the example of inspecting closing times, cafés do not easily move to region B. Evasion can be discouraged by varying the inspected group in time. However, if a particular highway is targeted, some truck drivers might possibly take an alternative route, while in case of environmental offenses with high compliance costs there might be an incentive to locate the industry in a region with relaxed enforcement regimes.

5.3 State-dependent enforcement

In case of serious underdeterrence, due to full compliance being unachievable, it is possible to increase detection for repeat offenders. If the overall level of resources is given, deterrence is enhanced by differentiating inspection. This problem is dealt with in the literature studying the use of so-called *state-dependent enforcement policies*. These policies were first studied mathematically by Landsberger and Meilijson (1982) in a model of income tax evasion. Harrington (1988) applied the analysis to the enforcement of pollution standards and provided a more general framework.

5.3.1 *The benefits of a state-dependent enforcement policy*

Harrington (1988) shows that an enforcement regime in which the probability of inspection depends on the outcome of the most recent inspection is more cost-effective than a system in which the inspection rate is independent of past performance. An enforcement authority can enhance deterrence by dividing regulated firms into two groups according to their past compliance record. If a firm is in the first group and it is caught violating, the firm will be transferred to the second group. In this second group – also called target group – inspection is more frequent and the penalties are higher. If the firm in question again violates while in the second group, it will remain in this second group. If the firm is found to be in compliance, there is a probability that it is sent back to the first group.

Such an enforcement policy can only be beneficial in case the maximum feasible level of compliance is partial. If the resources for enforcement are insufficient – given the maximum sanction – for full compliance, a firm with costs of compliance that are too high will not comply under a uniform enforcement policy, i.e. a firm will fail to comply if $c > pf^{max}$. This does not necessarily imply that it is better to save costs and allocate no resources at all. I already demonstrated above that a targeting policy might result in partial compliance for these firms. Harrington (1988) shows that a state-dependent policy can obtain even higher levels of partial compliance.[91] Using a given level of resources, the enforcement authority can choose to impose a different probability of inspection for group 1 than for group 2. If the probability of inspection in group 2 becomes large enough, the firm will comply if it is in the second group. The probability of inspection in group 1 necessarily decreases, but no deterrence is lost because compliance in group 1 was already zero. Hence, partial compliance is gained, namely the time the firm spends in the

91 More specifically, Harrington (1988) shows that a given compliance objective can be obtained with a lower level of resources, which is in essence the same reasoning. I discuss the objective of maximizing compliance given a constrained level of resources.

second group. A state-dependent enforcement policy has two major benefits over the uniform targeting policy.

1. Under a state-dependent enforcement policy, a firm may be induced to comply in group 2 even if the compliance costs exceed the expected penalty. The reason is that there are some additional benefits from compliance in group 2. If the firm complies in group 2, it will possibly be returned to group 1 where it will receive the benefits from violation (avoiding compliance costs). Because of this incentive – called *leverage* – the expected formal sanction can be lower than the costs of compliance ($c > p^*_2 f_2$) allowing for a savings in resources which can in turn be used to increase the number of firms in the target group or the time the firm spends in the target group.
2. A state-dependent enforcement policy can result in at least a certain level of compliance even if the costs of compliance exceed the maximum feasible penalty ($c > f^{max}$) because of the same leverage effect. The maximum level of compliance that can be obtained is the ratio between the costs of compliance and the maximal penalty. To obtain this maximum level, all firms in group 2 must be inspected.[92]

In order to make the incentive to comply in the second group (the leverage) as high as possible, the difference in expected punishment between the two groups should be made as large as possible. This implies that the penalty in the second group should be maximal, while being zero in the first group. This maximizes compliance in the second group and has no effect on compliance in the first group, as these firms violate anyway. The optimal probability of inspection is also higher in the second than in the first group. The probability of inspection determines the time spent in a group. For example, the probability of inspection in the first group should be high enough to ensure that enough firms transit to the second group where they are forced to comply, and it should be small enough to let the firms in the first group benefit from violation, so that the incentive to comply in the second group is high enough.

Note that in this state-dependent enforcement policy, no penalties are ever collected, the expected sanctions are low (or at least lower than compliance costs), while the compliance rate is positive and can be quite high. As Harrington (1988) notes, this might provide an explanation for the enforcement policies which are actually observed, later called the 'Harrington Paradox' (for example Heyes, 2000; Harrington and Heyes, 2001). This paradox is the combination of the following observations. In environmental regulation (1) the frequency of inspection is low, (2) (even if a violation is detected) penalties are hardly ever imposed and are rather low, (3) yet the level of compliance seems to be pretty high.

Harrington (1988) also shows that by adding to this model a third group, in

---

92 I.e., if $c > f^{max}$, the maximum feasible compliance rate Z equals f^{max} / c. To achieve $Z = f^{max} / c$ requires $p_2 = 1$.

which all firms are inspected and no escape is possible, even more can be saved in enforcement costs. The main reason is that adding such a group increases the costs of violation in group 2, hence allowing the inspection frequency in group 2 to decline. We might also think of the threat of closing firms which violate in group 3. The problem with such a third group is that when enforcement agencies make mistakes, all firms end up in the third group.

5.3.2 *An example of state-dependent enforcement*

An example might be useful at this point. Consider again the café that should comply with the closing at 1:00 AM rule in order to keep night's rest in the neighborhood. Every Saturday the café must decide about compliance. If it complies, it foregoes a benefit of €200. The maximum feasible fine equals €1000. The municipality is willing or able to carry out inspections at an average rate of only 10%. As discussed above, under (random) uniform inspections, the café will violate and remain open until 2:00 AM (because 10% * € 1000 < 200). Hence, the compliance rate will be zero. Under a targeting scheme of enforcement during half of the periods, a compliance rate of 50% can be obtained.

A state-dependent enforcement policy can increase the compliance rate even further. Suppose that cafés are classified into two groups. First of all, observe the policy in which the municipality inspects cafés with a standard probability of 5% (group 1), which means that cafés will fail to comply. However, if a café is caught, it will be transferred to group 2 where it is inspected with a probability of 20% and, if found in violation, sanctioned with a fine of €1000. Once caught, the café will comply (because 20% * € 1000 = € 200). If, after a certain period of compliance, the firm is returned to the 5%-category, it will return to non-compliance, until it is caught again, etc. Now note that this possible return to group 1, where the café non-complies, presents some future benefits. If the café is returned to the first group, it will violate and receive benefits of €200. Because of these benefits, compliance in the second group remains attractive if the expected formal sanction is somewhat decreased. Suppose the probability of inspection in group 2 is reduced to 15%. If the café violates in group 2, it will have to pay a fine of €1000 with a probability of 15%, so the expected costs will be €150. If the café complies, it will forego a benefit of €200. Thus, non-compliance in group 2 saves €50. However, if the café violates, it will remain in the second group, yielding expected costs of €150 every week. If the café complies, it might also remain in group 2 (yielding opportunity costs of €200 per week), but there is also a chance that the café will be returned to group 1 where it benefits from violation by €200. If the expected future benefits of a possible return to group 1 are valued higher than €50, the café will still comply in the second group, even under an inspection rate of 15%.

Table 5.1 An example of optimal enforcement policies of closing times[a]

	Uniform		State-dependent			
	Random	Target	Basic	Variant 1 (r = 0,05)	Variant 2 (Fmax = 500)[e]	Variant 3 (Fmax = 100)
***Inputs*:**						
C	200	200	200	200	200	200
Fmax	1000	1000	1000	1000	500	100
β[b]	0.9982	0.9982	0.9982	0.9991	0.9982	0.9982
Iav[c]	0.10	0.10	0.10	0.10	0.10	0.10
***Results*:**						
Z	0	0.50	0.6784	0.6867	0.4663	0.2010
p1	0.10	0.10[d]	0.0118	0.0088	0.0118	0.0118
p2	-	-	0.1418	0.1416	0.2010	0.4505
u	-	-	0.0393	0.0282	0.0671	0.1042
f1	1000	1000	0	0	0	0
f2	-	-	1000	1000	500	100

a. Solutions are calculated with help of Excell 2000.
The example concerns a café that every Saturday can decide whether or not it complies as explained in the main text. Variables are defined as follows:
C = costs of compliance
Fmax = maximum feasible fine
β = discount rate per week = 1/((1+r)^(1/52))
Iav = average inspection rate = P(G1)*p1 + P(G2)*p2
Z = compliance rate = P(G2)
p1, p2 = probability of inspection in respectively group 1 and group 2
u = probability of transition from group 2 to group 1 if found to be in compliance
f1, f2 = fines in respectively group 1 and group 2
P(G2) = probability of being in group 2 = p1/(p1+p2u) = 1 – P(G1) = 1 – p2u/(p1+p2u)
The problem is to maximize Z given the following constraints: (a) Iav = 10%; (b) the firm complies in the second group, implying that the costs of compliance should be smaller than the expected sanction and the benefits of leverage:

$$c \leq p_2 f_2 + L = p_2 f_2 + \frac{p_2\, \beta u\, (p_2 f_2 - p_1 f_1)}{1 - (1 - p1)\, \beta}$$

(and the firm violates in the first group implying $c \geq p_1 f_1$); (c) variables are feasible so probabilities should be between 0 and 1 and penalties between 0 and Fmax .

b. An interest rate r of 10% per year corresponds to a discount factor β of 0.9982 per week. An interest rate of 5% per year corresponds to a β of 0.9991 per week.

c. It is assumed that the enforcement resources can and should result in an average inspection rate of 10%. Hence, Iav is held constant for all variants. Examination shows that under Iav of 5%, the uniform policy becomes of course increasingly impossible. Also, the possibilities to target compliance in the uniform policy become smaller and hence the state dependent policy becomes more favorable. Although of course the feasible Z becomes smaller under a state dependent policy too. Reverse results hold for an Iav of 18%.

d. This means the agency inspects the firm half of the periods (for example from July – December) with probability 20%. Hence average inspection probability is 0,5*0% + 0,5*20% = 10% and average compliance rate is 0,5*0 + 0,5*1 = 0,5.
e. The variant of a decrease of fmax from 1000 to 500 yields exactly the same results as an increase in c from 200 to 400. The effects of decreases in fmax are similar to those of increasing c.

A café will comply in group 2 if its costs of compliance are lower than the expected sanction plus leverage benefits (formally $c \leq p_2F_2 + L$, where L denotes the leverage[93]). This implies that the probability of inspection in the second group can be reduced to below 20%. Moreover, if this probability decreases, the firm can be kept in group 2 longer while keeping the average inspection rate at 10%. And if the firm spends more time in group 2, the compliance rate will increase. Calculation reveals that under a discount rate of 0.9982 (which corresponds to an interest rate of 10% per year) the optimal policy is to inspect firms in the first group with a probability of 1.2% per week, in the second group with a probability of 14.2% per week, and to return firms in the second group that are found to be in compliance to the first group with a probability of 3.9%. This policy will induce an expected sanction in the second group of €142 (which is less than €200), but it will achieve a compliance level of 68%. See table 5.1. Given the expected sanction of €142, the leverage benefits equal €58. These benefits are the additional benefits of compliance in group 2. They follow from the expected time the café spends in group 1 multiplied by the discounted expected penalty the café avoids when in group 1. The expected time spent in group 1 depends on the probability of being inspected in group 2 and returned to group 1 relative to the probability that – once the café is returned to group 1 – it will be inspected and transferred back to group 2, etc. If the café is in group 1, it will not face the expected penalty of €142. These leverage benefits are maximized by the enforcement policy (i.e. the fine in group 2 is maximal, the fine in group 1 is zero) so that the café will just decide to comply in group 2: €200 ≤ €142 + €58. Thanks to these leverage benefits, the compliance rate is 68%, which is higher than the feasible compliance level under the targeting scheme described above (50%, third column).

Table 5.1 also contains some variations which reveal further characteristics of the state-dependent enforcement policy. First, if the firm cares more about the future (i.e. the discount rate β increases), the probability of inspection in group 1 (p_1) and the probability of return to compliance (u) can be made arbitrarily low. Since this firm cares more about the future, even a low probability of return to compliance is promising. As a result, the firm can be induced to comply in group 2 through a low probability of return. If few cafés are returned to group 1, few cafés need to be targeted from group 1 to group 2,

93 Harrington (1988) shows that L equals $L = \frac{p_2 \beta u (p_2 F_2 - p_1 F_1)}{1 - (1 - p1)\beta}$ (see table 5.1 for an explanation of variables.

allowing the probability of inspection in group 1 to also remain low. Secondly, if the fine becomes lower, or the costs of compliance go up, the maximum feasible level of compliance becomes lower (given resources). However, even if the costs of compliance exceed the maximum fine (for example €100), leverage effects will occur and the café can be induced to comply partially. If the maximum fine becomes lower, the probability of inspection in group 2 should be higher in order to ensure compliance. Note that the solutions for the probability of inspection in group 1 do not depend on the level of the maximum fine. This follows from the fact that the fine in the first group is always zero and the probability of transition (u) determines the time spent in each group in order to meet the inspection rate of 10%. Finally, a different interpretation of u would be that the café is not returned to the first group with a probability u (after observing compliance in group 2), but instead the firm is returned to the first group after some t periods of being found in compliance. This yields the same results with $t = 1/u$, but is less flexible as these periods must have integer values. For example, in the basic variant the café could be returned to compliance after 8 inspections in which compliance is observed ($1/0.1418 = 7.052$ which is rounded upwards to 8).

Furthermore, we can show that the benefits of a state-dependent enforcement policy become smaller as the available resources increase. Greater resources allow for higher compliance levels. In order to have a higher level of compliance under a state-dependent enforcement policy, the expected time in the second group must be higher, implying that the possible savings in resources will be lower. On the other hand, if the available resources increase, the achievable level of compliance under a uniform targeting policy comes increasingly close to unity. If the available resources allow for an average inspection rate of 20%, the optimal policy reduces to a uniform policy with a fine of €1000 and an inspection rate of 20%. If a uniform enforcement policy already obtains a high level of compliance, the possible benefits of a state-dependent policy are lower.[94]

5.3.3 *Improvement and discussion of the state-dependent enforcement policy*

In later contributions, the results of Harrington (1988) have been extended and criticized. The most important criticism is directed at the specific objective function of the enforcement authority. In Harrington's model the objective of the enforcement authority is to maximize compliance given the available resources and the maximum feasible penalty, and assuming that the authority knows the firm's compliance costs. Given this structure of the model, Harrington shows that a state-dependent enforcement policy (1) can achieve the same compliance level at lower enforcement costs, (2) can achieve at least some compliance even

94 Formally, Harrington (1988) demonstrates that if the compliance rate under a uniform targeting policy is α, the maximum state-dependent compliance rate Z will have the property $Z \geq \alpha \geq Z^2/(1+u)$.

if the costs of compliance exceed the maximal available penalty. Later contributions show that these results crucially depend on the structure of the model.

1. Harford and Harrington (1991) criticize the fact that the standard in Harrington's model is fixed (which is probably correct for tax evasion models), while in environmental and safety regulation, the standard can often vary. The higher the standard imposed, the higher the costs of compliance, hence the lower the social welfare. From a social point of view, it is not only important to minimize enforcement costs given the level of compliance, but also to minimize the costs of compliance with the standard, given the level of compliance. This is obtained when all firms realize the same marginal costs of compliance for all periods.
   In Harrington (1988)'s model, firms with identical cost functions pollute at different levels, implying that this condition is not satisfied. If – as is usually the case – it becomes increasingly costly to satisfy a higher standard, the costs of compliance increase more than proportionately with the level of compliance (i.e. the marginal costs are rising). If these costs rise sufficiently[95] and if a uniform enforcement policy cannot enforce the (legal) standard at the optimal level, the best response is not to impose a state-dependent enforcement policy, but to lower the standard somewhat and to uniformly enforce this lowered standard. Because this lowering of the standard implies that compliance costs will decrease, the expected sanction for non-compliance can be smaller, so that enforcement resources can be saved or can be used to enforce a relatively higher level of compliance.
   Consider the example discussed above. Suppose that customers leave at different times and cafés become increasingly empty, so that reducing closing time from 2:00 AM to 1:40 AM costs €30 and reducing from 1:40 AM to 1:20 AM costs €60, while the costs of reducing the closing time from 1:20 AM to 1:00 AM costs €110 (total costs €200). In this case, a café can be forced to comply under a uniform policy if it is inspected only after 1:20 AM (10% * € 1000 = € 100 > 60 + 30 = € 90). It will then comply for forty of the sixty minutes, in other words at 67%. It is better to always close a café at 1:20 AM (costs €90) than to close it at 1:00 AM 67% of the time and at 2:00 AM 33% of the time (expected costs €134 as in the example above under a state-dependent enforcement policy[96]). For the same level of compliance and hence of expected noise ("67%") and for the same average inspection rate (10%), the costs of compliance are lower under the

95 Or, analogously, if the marginal harm decreases sufficiently in the level of compliance.

96 For the sake of simplicity, I am rounding off the compliance rate of 0.6784 (table 5.1) to 0.67 or 2/3.

uniform policy of inspecting only after 1:20 AM (€90) than under the former state-dependent policy (€134).[97]

Of course, the numbers in the example are such that after 1:20 AM, full compliance can be achieved and compliance after 1:20 AM will in strict terms be more beneficial. However, the general point is that the decline in compliance costs under a uniform policy and lowering of the standard can be higher than the decline in monitoring costs under a state-dependent policy. Therefore, uniform inspections and certain compliance after, for example, 1:30 AM (a compliance rate of 50%) might be socially more desirable than a state-dependent enforcement policy in which the café complies at 1:00 AM with a probability of 67% and fails to comply at all with a probability of 33% (a compliance rate of 67%).

Thus Harford and Harrington (1991) show that the uniform enforcement policy is superior if the enforcement authority sets both the standard and the monitoring strategy simultaneously. However, if setting the standard is a separate decision, then Harrington's (1988) results apply. Once a standard has been selected, a state-dependent enforcement strategy is the most cost-effective way to achieve a given level of compliance with that standard. This shows that there might be efficiency gains from simultaneous decisions.[98]

2. Harford (1991) discusses a model in which the levels of compliance are measured with some random error and in which the enforcement authority can differentiate standards, fines and inspection rates. If these errors are large enough, it will be beneficial for firms to choose a level of compliance that is different from the standard, but higher than zero.[99] Harford (1991) shows that in such cases it is optimal to have a lower standard and a lower inspection rate in the first group. This makes compliance in group 2 more attractive, allowing savings of monitoring costs. However, under reasonable conditions, it is optimal to have maximum fines in both groups. At the very least, the fine for group 1 should be larger than zero in order to offer sufficient incentive for some firms to choose a positive level of precautions.
3. Raymond (1999) analyzes the situation when enforcement authorities fail to observe the costs of compliance. As Harrington (1988) shows, firms can be categorized by their level of compliance costs. If the costs of compliance are low, a firm will comply under a uniform enforcement policy

---

97 Following note 95, it could also be beneficial to lower the standard if the marginal compliance costs are constant, but the marginal harm is increasing. If noise from cafés increases sufficiently, society will be better off always having the café closed at 1:30 AM (achievable under uniform enforcing at this time) than sometimes (probability of 67%) at 1:00 AM and sometimes at 2:00 AM (probability 33%) (under a state-dependent enforcement policy).

98 This is examined further in section 9.3.1 where I discuss the optimal discretion and organization of enforcement authorities.

99 If there are no errors, it will be optimal, as in Harrington (1988), to choose for either zero compliance or the standard.

($c \leq pf$ with $f \leq f^{max}$). If the costs of compliance are higher, a firm will violate under a uniform policy. For very high compliance costs, a firm will always violate under a state-dependent policy too. However, for intermediate values of compliance costs, the firm can be induced to partially comply under a state-dependent policy. Harrington (1988) assumes that the enforcement authority knows the level of compliance costs and he then only discusses firms with such intermediate compliance costs, since the other two cases are not interesting. If the enforcement authority cannot distinguish between firms, it will have to apply the same enforcement strategy to all firms, regardless of their compliance costs. If the enforcement authority switches from a uniform policy to a state-dependent enforcement policy with zero fines for the first group, this will have two opposing effects: (1) It will be beneficial because compliance will be gained for firms with intermediate compliance costs which would violate under the uniform policy and at least partially comply under the state-dependent policy; (2) It will be costly because compliance is lost for firms with low compliance costs which would always comply under the uniform policy and now only partially comply because the fine in the first group has been reduced. Whether or not the benefits are higher than the costs depends on the distribution of compliance costs over firms. If the proportion of firms with low compliance costs is high, setting the fine for the first group at zero is not a good idea. Depending on the distribution, it might be optimal to impose a maximal fine for the first group and choose for a uniform enforcement policy.
In the example of cafés, cafés with compliance costs below €100 will comply under a uniform enforcement policy. If the enforcement authority switches to a state-dependent enforcement policy, these cafés will only partially comply. Therefore, the enforcement authority will have to balance the decrease in compliance (from 100% to some partial compliance rate) for cafés with costs below €100 with the increase in compliance (from 0% to some partial compliance rate) for cafés with compliance costs above €100.

4. Friesen (2003) shows that in the state-dependent enforcement policy, further savings in monitoring costs can be realised. Since firms incur violations in the first group and only comply in the second, inspection in the first group will not serve deterrence but is only required to select the firms that will be moved to the second group. By randomizing the movement of firms from group 1 to group 2, the enforcement authority can avoid the costs of inspection in group 1 without affecting the firm's incentive to comply. Therefore the optimal choice will be to set the inspection rate in group 1 at zero and randomly select the firms that will be moved to group 2. These results follow from the fact that Friesen (2003) extends Harrington's model by making all possible transitions between the two groups endogenous. In Harrington's model, only the transition probability of moving from group 2 to group 1 when in compliance is endo-

geneous. Besides the savings in inspection costs in group 1, other costs are also saved. The incentive to comply in group 2 is maximized by setting $u=1$ (the firm will always be transferred to group 1 if found in compliance in group 2). Because the leverage benefits are higher, the expected formal sanction can be lower. Therefore $u=1$ will allow the probability of inspection in group 2 to be smaller, leading to lower enforcement costs, which can be used to increase compliance. The random transition from group 1 to group 2 is adjusted to meet the constraints on the inspection rate and/or compliance level. In contrast, in Harrington (1988), the transition from group 1 to group 2 is given (and equal to one if a firm is inspected and found in violation), so u has to be adjusted in order to meet the constraints. Hence, the optimal u is generally smaller than 1, implying that leverage is not maximized.[100]

In the example used above, the same compliance level of 0.6784 can be obtained by setting $p_1=0$, $u=1$, $p_2=0.1360$ and randomly moving cafés from group 1 to group 2 with a probability of 0.7132 (table 5.2). In that case, the average inspection rate will only be 0.0922, as compared to 0.1000 in the example above. This might seem a small difference but it implies a decrease in the inspection rate (and hence enforcement costs) of nearly 8%.

Therefore, in Harrington (1988), in a "past compliance targeting" scheme, transition from group 1 to group 2 as well as from group 2 to group 1 is based on past performance. In contrast, Friesen's (2003) "optimal targeting" scheme uses past compliance records only to enable a transfer from group 2 to group 1. As such, it has a similar drawback to the one involved in the case of repeat offenders. It is not optimal to base enforcement on past violation records and to increase enforcement on offenders. However, it is optimal to reward compliant behavior by relaxing enforcement, which in this case means no inspections at all.[101]

---

100 From this, Friesen (2003) concludes that the past performance policy of Harrington (1988) is feasible for higher levels of target compliance rate and/or higher levels of compliance costs. The random optimal targeting policy is not feasible if the probability of inspection in group 2 exceeds 1 or if the probability of transition from group 1 to group 2 is smaller than zero. However, she seems to continue to assume that $u=1$. If she allows $u<1$, the random targeting policy remains feasible. The general point of saving inspection costs in group 1 seems to always apply. It is always possible to randomize the inspection probability of Harrington (1988) since this probability has no other function than to determine the transition to group 2.

101 The differences in the targeting schemes are further discussed and evaluated in experiments in Clark et al. (2004). Cason and Gangadharan (2006) also provide experimental evidence. Empirical research based on field data is given in Helland (1998) and Eckert (2004). Both provide support for Harrington's policies, but also present counter-effects. Helland (1998) finds that firms in the pulp and paper industry which violate environmental regulations experience a one- or two-quarter period of more frequent inspections. However, he also finds that the return to a weak enforcement regime is especially based on self-reported violations rather than demonstrated compliance. Empirical research is hampered by the absence of (reliable) data on individual reporting behavior and uninspected firms.

Table 5.2 State-dependent enforcement policy of closing times improved by Friesen (2003)[a]

	Uniform		**State-dependent**	
	Random	**Target**	**Basic**	**Target (Friesen)**
Inputs:				
C	200	200	200	200
f^{max}	1000	1000	1000	1000
β	0.9982	0.9982	0.9982	0.9982
I_{av}	0.10	0.10	0.10	0.0922
Results:				
Z	0	0.50	0.6784	0.6784
p_1	0.10	0.10	0.0118	0
p_2	-	-	0.1418	0.1360
u	-	-	0.0393	1.00
a	-	-	-	0.7132
f_1	1000	1000	0	0
f_2	-	-	1000	1000

a. See the notes to table 5.1. In addition:
a = probability of transition from group 1 to group 2.

5.3.4 *Warnings*

In the state-dependent enforcement policy, a firm caught incurring a violation in group 1 will not be penalized, but transferred to a target group in which a stricter enforcement policy applies. Only if it again incurs a violation will it be punished. This is similar to the use of warnings, a system in which the enforcement authority detects a violation, issues a warning and only punishes the firm if the violation is repeated or continues.

The use of warnings is very common. Most repair sanctions discussed in section 4.2 are in effect warnings. If a firm is found to be in violation, a warning is issued that it will be penalized if it does not return to compliance and/or repair the damage caused. At first sight, warnings seem to be inefficient (Nyborg and Telle, 2004). If potential violators know that they will be given a second chance, will they ever comply before receiving a warning? And why should regulators not sanction violators immediately, thereby providing stronger incentives to comply (recall the discussion about repeat offenders in section 4.1)?

There are several answers to these questions. First of all, warnings are a form of state-dependent enforcement with the state depending on whether a warning has been issued. As explained above, if full compliance is unachievable, overall compliance may be higher if the enforcement authority uses a state-dependent enforcement policy, in which the penalty in the non-target group is zero.[102] The problem is that the regulator faces a limited budget and *monitoring* is costly. Hence, targeting inspections may be more beneficial than random inspections.

A different (although related) answer is that warnings can be beneficial if the enforcement authority faces a limited budget and *prosecution* is costly (Nyborg and Telle, 2004). This implies that the authority can prosecute only a limited number of violators. Therefore, the ability to impose sanctions on an individual violator is decreasing with an increasing number of violators.[103] When the number of violators exceeds a critical threshold, the expected sanction becomes insufficient to deter violation. The outcome therefore depends on a firm's expectation about (non)compliance by other firms. Firms might possibly form their expectation on past compliance rates. Nyborg and Telle (2004) show that if violation can be both accidental and deliberate, the enforcement authority is likely to lose control at some point. This is what happens if, due to accidental violations, the compliance rate exceeds the threshold in one period, following which every firm will begin to violate deliberately, leading to an equilibrium in which the enforcement authority spends all its resources on enforcement, yet most firms violate. Nyborg and Telle (2004) also show that warnings may substantially improve the effectiveness of enforcement. Since warnings are less costly for the regulator to impose, they offer more possibilities for stopping the breakdown of compliance. For firms, delayed compliance is more costly than immediate compliance (due to the costs of extra inspections for example). This explains why firms might comply under warnings and not always adopt a "wait-and-see approach". However, it often pays to delay compliance (due to discounting effects or because monitoring and detection are uncertain), in which case warnings do not necessarily help the regulator to retain control, since many firms will start delaying compliance. Still, I suggest that even if firms delay compliance in reaction to warnings, warnings can nevertheless be beneficial. Since warnings are cheaper than formal sanctions, the enforcement authority will have more resources available to monitor firms under warnings. If the enforcement authority's budget competes for both monitoring and prosecution, using warnings might save pros-

102 Fenn and Veljanovski (1988) for example argue that if the limits on the enforcement authority's budget and penalties are such that firms will fail to comply ($p^*S^{max} < c$), it might be more effective, after an inspection, to offer the offenders the option to comply and avoid prosecution, where the fine will be certain if non-compliance is chosen for. This might induce some firms to comply after inspection ($S^{max} \geq c$). Therefore, as in the state-dependent enforcement policy, the outcome will depend on the distribution of compliance costs. If this distribution is such that many firms will choose to delay compliance ("wait-and-see") instead of complying immediately, warnings would clearly harm social welfare. If this distribution is such that many firms will choose to at least comply after inspection, instead of not complying at all, it might improve social welfare.

103 This is an example of the interactions discussed in section 3.3.

ecution costs and hence the monitoring rate (before and after a warning) can be higher, which will increase the compliance rate. Warnings are beneficial if the limits on the authority's budget and sanctions are such that immediate punishment will enforce a lower compliance rate than the situation in which firms only comply after inspection ("wait-and-see").

Two final points can be made concerning the desirability of warnings, both of them relating to the fact that warnings may well be a necessary element in an effective enforcement process. This is the case, for instance, when inspections are used to cheaply inform the regulated firms about the rules, following which, a procedure of enforcement and prosecution is started. Or, alternatively, warnings might be needed to sustain the firm's internal motivations to comply. These issues are discussed in the following two chapters.

5.4 Conclusion

This chapter explains that immediate and equal punishment of all (identical) offenses – as is the case in Becker's model – is not a good idea in cases where the level of resources and the penalty are sufficiently restricted. The threat is that the enforcement authority will spend many resources without effectively enforcing compliance because the expected sanction is too small. It would then be optimal to target enforcement resources. The optimal targeting strategy is one in which enforcement depends on past compliance, for instance by issuing warnings. By tolerating a given level of non-compliance, or by sometimes tolerating non-compliance, the enforcement authority can 'buy' or 'negotiate' partial compliance. It can also reward compliance in the target group by transition to a non-target group. The prospect of the possibility of violating and thereby avoiding the costs of compliance will offer the firm a higher incentive to comply.

6 Flexible compliance and cooperative enforcement – information constraints

Until now it was assumed that firms are perfectly informed concerning their own compliance costs as well as the standard. It was assumed that the legal standards are known to all parties and that a firm is fully aware of the enforcement authority's request as well as the consequences of non-compliance. This may, however, not necessarily be the case. I now discuss the consequences of uncertainty over the legal standards. There are two issues to be discussed: firms might claim ignorance of the relevant standards or the way to comply with them (section 6.1) and firms might be able to exploit the uncertainty concerning the standards by arguing that they are innocent (section 6.2).

6.1 Uncertainty about standards – flexible compliance

Especially for technically complicated regulations, firms might not be costlessly and perfectly informed about the standards they should obey or about their actual level of compliance. Moreover, they might be unaware of the costs of all alternatives. According to Veljanovski (1984), there are two kinds of uncertainty for firms:

1. Uncertainty concerning the law: Given the number of (detailed) standards, firms will often have an incomplete perception of the law, their obligations and the possible sanctions.
2. Uncertainty concerning the least-cost method of compliance: Firms do not always have perfect information on the best compliance techniques, particularly not small firms that do not possess the necessary expertise.

Shavell (2004, pp. 562 – 564) discusses two related problems. First, a firm may claim that it was unaware of acting unlawfully (*ignorance*). Secondly, a firm may choose a level of precautions that it believes to be sufficient although it actually violates the standard (*mistake*).

In this section, I discuss the problems of ignorance and mistake and the optimal reaction to claims of this type. I then investigate in further detail the sources of these types of uncertainty. This will lead to a discussion of the advantages of cooperative enforcement strategies.

6.1.1 *Ignorance and mistake, and an information exchanging enforcement policy*

Firms may claim that they are ignorant of the relevant regulations. Generally, such claims should be ignored in order to provide firms with an incentive to learn about the law and adhere to it. If, however, a reasonable effort is insufficient to learn about a legal rule, it is best to permit firms to escape (costly) penalties, since a firm can be deterred by possible sanctions only if it knows which levels of precaution lead to the imposition of sanctions (Shavell, 2004).

Following Friesen (2006, see section 3.4 about self-reporting), it can be argued that firms only have an incentive to obtain information if the costs of finding information are not too high in relation to the benefits. These benefits are that if a firm is informed and finds out that it is in violation, it can take appropriate action to escape penalties for non-compliance. Therefore, more stringent enforcement creates a higher incentive to obtain information. Generally, imposing sanctions equal to harm done gives firms the optimal incentive to invest in self-audits in order to investigate (non-)compliance.[104] Therefore, ignorance should not be considered a valid excuse for sanctions. However, (still following Friesen, 2006) if auditing costs are high relative to public inspection costs, it would be socially beneficial for firms not to self-audit but instead for public enforcement agencies to inspect the firms' compliance level. A firm's incentive to obtain information might be inadequate. This may occur if standards are highly technical and change rapidly, or if there are many standards and they possibly run together. In these cases, information requires effort or the development of special expertise (rather than this expertise being a natural by-product of operating business) and/or the benefits of information development are non-exclusive. If this is the case, firms – especially small firms – might be ignorant, even under reasonable costs, of the standards that apply to them and/or of the way to comply with the standard.[105] Under such circumstances, public agencies might produce and share the relevant information at less cost than the firms themselves. For example, the fire brigade might be better informed about taking precautions to prevent fires in cafés than the proprietors.

A different problem is that a firm may choose a level of precautions that it believes to be sufficient although this actually violates the standard (mistake). Generally, there should be no sanctions in such instances, (Shavell, 2004) especially if imposing sanctions is costly, because someone cannot be deterred from non-compliance that he is unaware of. For example, a café proprietor who incorrectly believes that he bought the right inflammable decorations cannot be deterred by sanctions. At the same time, claiming unawareness of

104 Recall the discussion in section 2.3.5 concerning the incentive to obtain information under liability. Shavell (1992a) and Pfaff and Sanchirico (2000) demonstrate that if sanctions equal harm and if self-audits do not influence the probability of apprehension, firms will have the socially optimal incentive to self-audit, take precautions and remedy.

105 Brehm and Hamilton (1996) do find evidence for the existence of ignorance. They argue that non-compliance may be due to ignorance rather than evasion. They provide empirical evidence that violation is better understood by those variables associated with the likelihood that a firm is ignorant of reporting requirements, than by those associated with evasion.

the requirement to hang only inflammable decorations – the problem of ignorance – is insufficient to escape sanctions. A related problem is that a firm may knowingly be non-compliant, but believe this to be either more or less harmful than it actually is. The conclusions concerning whether sanctions should be based on actual or on perceived harm are not uniform (Shavell, 2004). On the one hand, to which extent a firm can be deterred depends on what the firm thinks it is doing, not on what it is in fact doing. On the other hand, sanctions can be imposed both on actual harm and on average (expected) harm. The latter may be most closely related to ex-ante expected harm. However, if firms are able to falsely convince courts that they thought they were doing little harm, sanctions will be too low and deterrence will be weakened.

If problems of ignorance or mistake persist and/or if information about the law and its applications is better produced by public agencies, a strict enforcement approach might be inappropriate. Veljanovski (1984) argues that enforcement activities may be used by the inspector to freely provide information to the firm and buy greater compliance. If firms are ignorant of the law, the inspector's visit can be used to provide the firm with information on its legal obligations, possibly at zero cost. Instead of the firm incurring the costs of fully acquainting itself with the law, the inspector will provide the firm with information as part of his routine activities. If firms are unaware of the least-cost method of compliance, while the inspector possesses the necessary expertise, a similar information transfer may take place. It may be welfare-improving for the inspector to adopt the role of advisor rather than policeman.[106]

Of course, this implies that the probability of detection must be sufficiently large in order to realize the benefits of advice. Penalties can be decreased accordingly, if they are necessary for deterrence at all. Enforcement then consists of two stages. First of all, with sufficiently high probabilities of detection, the firm is informed about the relevant regulation. After that, an enforcement policy as proposed by Becker (1968) can be used. Note that such a policy can only be efficient if either the firm fails to remember the information or if the information quickly becomes outdated. If, following an inspection, a firm knows for quite some time what they have to do, the optimal enforcement policy would be to inform it only once and after that to enforce severely according to the guidelines discussed in chapter 3.

The discussion here pertains to the case in which firms do not know what the standards are and/or how they can comply with it. This uncertainty can be solved by somehow supplying this information, possibly through inspections. A somewhat different problem occurs when firms know what the standards are, but are uncertain of how these standards apply to them in a particular situation, because the law does not precisely describe what should happen in all possible situations. Such uncertainty might be more persistent. This is discussed next.

106 See also Law (2006).

6.1.2 *Incomplete law, sources of uncertainty and cooperative enforcement*

The literature so far assumes that law is complete, i.e. that all potentially harmful actions are unambiguously stipulated by law. Generally, however, law is incomplete for two major reasons (Xu and Pistor, 2002)[107]. First, law uses ambiguous open-ended norms, implying that the boundaries of the law are not clearly circumscribed (this is called type I incompleteness). As a consequence, it is not unambiguously clear which actions will actually be punished, leading to uncertainty in choosing precautions. Secondly, the law can try to be highly specific, but because the law-maker will never be able to capture all actions that may result in harm, some relevant applications will be missing (this is called type II incompleteness), especially in cases of a high rate of technological change, socioeconomic development or institutional transformation. Gaps will always remain and there will always be actions which the law fails to address, and which prove, in hindsight, to be equally harmful to the ones stipulated in the law.[108]

As a consequence of these two types of uncertainty, the law is both under- and overinclusive (Veljanovski, 1984). The law is written for all firms and for all possible situations. It therefore may fail to adequately address a particular situation in a particular firm. The law is overinclusive because it forbids activities that are efficient or prescribes levels of precaution that are inefficiently high. The law is underinclusive because it fails to forbid activities that are inefficient or prescribes levels of precaution that are too low.

If the law were complete, a deterrent approach as proposed by Becker (1968) would efficient. Both firms and the enforcement authority would agree on a single interpretation of the law. In such a world, the law could be enforced to the letter provided evidence was established. Hence, full deterrence could be implemented if the expected penalty was large enough. Provided that rules are efficient, achieving compliance would be desirable and punishing all detected offenses would be efficient (refrained from the influence of enforcement costs). However, if the law is incomplete, punishing every violation is not efficient because of overinclusion. Moreover, due to underinclusion, society will sometimes require more compliance than the minimal level of compliance that satisfies the standard. The problem is therefore how to implement socially optimal levels of precaution, provided that the law is incomplete. If the law is incomplete, the rules do not necessarily prescribe efficient solutions for all particular situations and a suboptimal outcome may result if the enforcement authority punishes all detected viola-

107 See also Pistor and Xu (2003).

108 An implication of this distinction is that if the law attempts to be specific, it becomes very detailed and numerous and firms may claim that they are ignorant of the law. If the law contains open-ended norms, firms may claim that they thought they were in compliance and satisfied the standards. Therefore, the incompleteness of the law may be a source of ignorance and/or mistake.

tions on the basis of observed harm. Even if ex-post enforcement, for instance by courts, can determine perfectly whether or not an offense is efficient, the problem remains that firms are uncertain ex-ante about the evaluation of their behavior. Therefore, the law may both under- and overdeter (Xu and Pistor, 2002). It overdeters [underdeters] if firms expect the enforcement of the law to punish more [fewer] actions than is actually intended by the law and/or if they choose high levels of precautions in order to be certain of complying with the law.

Xu and Pistor (2002) argue that there are two possible solutions. In one solution, the law would remain incomplete and courts would restrain the level of punishment in order to prevent excessive punishment of relatively harmless actions. By implication, punishment would become insufficient to deter non-compliance. A different solution would be to delegate law-making powers to the enforcement authority (a regulator), who is then able to specify the law for individual firms and particular situations ex-ante. This may be done for example by means of licenses. Alternatively, especially if the rate of technological and institutional change is high, a different possibility would be to carry out a regular inspection in which the firm is told which specific precautions it must take. Subsequent inspections would then be held to investigate whether the firm complies with these specifications. This is an advantage of enforcement by regulators instead of courts.[109] However, this solution depends on an enforcement authority that has sufficient information about the relevant costs and benefits to implement socially optimal levels of precaution.

But suppose the enforcement authority is not informed about the compliance cost function of the firm. In general, there is two-sided uncertainty (Ricketts and Peacock, 1996). Due to the incompleteness of the law, regulators have better knowledge of the applicable standards than firms. However, firms have better knowledge of their production process, which governs the detailed application of regulations and which determines the costs and possibilities of achieving compliance. In such a situation, firms and the authority are able to overcome problems of over- and/or underinclusion if they exchange private information. This requires the kind of cooperative enforcement strategy that is proposed by Scholz (1991) and Ricketts and Peacock (1996).

In essence their reasoning is as follows: since the enforcement authority has no specific information, it can choose for a strict enforcement policy, punishing every violation based on the observed harm. The authority will then impose penalties on all detected violations and require all violations to

109 This argument is further explored in section 9.3.2. Arguedas and Hamoudi (2004) discuss the importance of the timing of decisions. Will the firm decide on investment in compliance before or after the enforcement policy (consisting of standards and inspections) is announced? If the policy is announced afterwards, the firm has an incentive to over-invest in compliance technologies. Over-investment is a good signal for the enforcement authority, which might in turn decide not to inspect the firm at all.

be abated regardless of cost or the firm's particular safety problems. Since all firms face the same probability of being caught and punished in every period, the firm will choose minimal compliance, i.e. it will choose a level of compliance that minimizes the firm's expected costs of compliance and penalties. Both the enforcement authority and the firm benefit if they behave more cooperatively. Ricketts and Peacock (1996) argue that both benefit if they reveal their private information. If firms reveal their information, it will allow the enforcement authority to enforce the socially optimal level of precautions. The firm will benefit if it implements the required precautions, because compliance with this (adjusted) standard will relieve it from penalties. This will allow the enforcement authority to bargain for higher precautions, which is beneficial if the law is underinclusive. In essence, this type of information-sharing allows the enforcement authority to switch from strict liability to a negligence approach that favors both the firm and the enforcement authority.

Scholz (1991)[110] argues that rather than having firms and authorities reveal their information, firms and authorities could be induced to forego a strict, inflexible approach if they interact with each other repeatedly. A voluntary compliance equilibrium is preferred by both the firm and the authority. In this equilibrium, the authority can reduce its monitoring and prosecution costs with flexible enforcement, overlooking minor technical violations (due to overinclusion) in recognition of the firm's extralegal safety efforts to reduce greater hazards not directly addressed in the regulations (due to underinclusion). If, in response, the firm chooses flexible compliance, this will allow it to tackle its worst health and safety hazards with the most efficient methods available rather than spending money on complying with safety standards that are less important and efficient in the firm's particular situation. Hence, flexible compliance produces greater safety at lower costs.

Therefore, in both Ricketts and Peacock (1996) and Scholz (1991) the enforcement authority is able to handle the over- and underinclusion problem by being more flexible. The enforcement authority foregoes a legalistic, to-the-letter enforcement of regulations that are inappropriate for a particular firm in return for extralegal efforts by that firm to work towards the policy objective of damage prevention. By tolerating offenses of overinclusive regulations, the authority can induce the firm to comply with standards not included. While both the firm and the enforcement authority eventually benefit in such a system, the problem is that in the short run both have an incentive to deviate (Scholz, 1991). If the authority chooses flexible compliance, the firm will be better off choosing minimal compliance instead of flexible compliance. Since enforcement is weak, it might in this way avoid most or all compliance costs by choosing very low levels of precaution and ignor-

110 A more extensive formal treatment is given in Scholz (1984). See also Fenn and Veljanovski (1988) and Arguedas (2005). Informally Scholz's (1991) argument is also presented by Veljanovski (1984).

ing even severe safety problems. If the firm chooses flexible compliance, the authority will benefits from choosing more stringent enforcement, because this will achieve even greater safety gains. Therefore both have an incentive to 'cheat' on the voluntary compliance equilibrium. This is an example of the Prisoner's dilemma. Scholz (1991) shows that a voluntary compliance equilibrium may result if both parties are sufficiently concerned about the future to resist the short-term incentive to obtain higher gains. The best strategy for the enforcement authority might possibly be to choose a tit-for-tat strategy as proposed by Axelrod (1984), where the authority chooses for maximal enforcement against firms that have established a record of minimal compliance in the past, and flexible enforcement if the firm has a record of flexible compliance. Observe that this cooperative strategy is similar to other strategies already discussed. It is also based on past behavior, it is related to the discussion about reasonable compliance records and it is only possible if the authority and the firm interact with each other repeatedly. Under this cooperative strategy, enforcement is characterized by negotiations (Veljanovski, 1984). The enforcement authority will attempt to persuade the firm that compliance, sometimes extralegal compliance, is the cheapest way to achieve the highest safety gains. The firm will try to persuade the authority that compliance is infeasible or extremely costly in order to minimize compliance costs.

Since the law is incomplete, there is uncertainty concerning the application of the law to individual cases. Hence, the incompleteness of the law opens the possibility for a discussion concerning enforcement actions, and firms might try to contest an enforcement authority's decisions. This is the subject of the following section.

6.2 Accuracy of enforcement technologies

In section 3.2, it was assumed that enforcement authorities detect violations perfectly and that the enforcement authorities are perfectly able to convict only firms that really violated the standards. The problem of the conviction of innocent firms has already received some attention. It is further investigated in this section. First of all, the general problem of the accuracy of enforcement is discussed. I then examine the firms' incentives to invest in, respectively, adjudication to escape penalties (non-conviction), and hiding violations from detection (non-detection). Finally, I discuss the optimal response to the problems of non-conviction and non-detection.

6.2.1 *Court's mistakes and the problem of the conviction of innocent firms*

Detection, prosecution or conviction of offenders is generally not perfect. The resources that are required to obtain perfection are not worth the benefits. However, this implies that sometimes, innocent firms will be convicted (type

II errors) or that guilty firms will escape sanctions even when inspected (type I errors). The conviction of innocent people leads to personal and social costs. Both personal and social resources are wasted in prosecution and execution of punishment, and the result is a feeling of injustice and distrust in the legal system. Moreover, both types of error harm the deterrent effect of enforcement because they reduce the difference between the costs of compliance and the costs of the expected sanction.[111] Errors of type I reduce deterrence because they lower the expected fine for violation. Errors of type II reduce deterrence because they lower the benefits of compliance.[112] Note that mistakes can also lead to a certain degree of overdeterrence. Since compliant firms that just comply with the standard may be judged to be in non-compliance, firms may have an incentive to take excessive precautions in order to prevent erroneously being found in non-compliance.

Most likely, the costs of convicting innocent people will increase with the magnitude of the sanction. If they increase more than proportionately with the sanction, it might not necessarily be desirable to increase the sanction to its maximum and to trade-off the probability of detection and the sanction as far as possible. For example, in many European, countries capital punishment was abolished, (partly) because of the impossibility to restore mistakes that are made. The probability of detection can be both larger and smaller than in the absence of mistakes. Since deterrence is weakened, a higher probability of inspection is required to achieve a given level of deterrence. On the other hand, inspection is more costly because mistakes make inspection less effective.

The presence of mistakes also explains why the standard of proof is smaller for lower or less severe penalties. Imprisonment is a severe sanction because it removes personal freedoms and imposes large costs (including stigma) on the convicted individuals. Therefore, also from a social welfare point of view, the benefits of preventing the conviction of innocent people are large and the standard of proof in criminal law is high. On the other hand, imposing small administrative penalties does not lead to many other costs than the costs of the fine and some feelings of injustice if there are mistakes. Therefore the standard of proof is much smaller and the level of inspection and prosecution needed for conviction can be lower both quantitatively and qualitatively. If a higher or more severe sanction is imposed, this will increase the standard of proof and hence increase the costs of imposing sanctions. A regulatory system with a low burden of proof (hence a high probability of conviction) and low penalties might be more efficient than a system with a

111 See for example Polinsky and Shavell (2000a, section 8). The precise consequences of errors, especially wrongful conviction, for deterrence are analyzed in more detail in Lando (2006).

112 To see why this is the case, suppose that (following inspection) the probability that a compliant firm is convicted equals α, and the probability that a non-compliant firm escapes conviction is β. Then compliance leads to costs $c + p\alpha f$ (suppose: fines) and non-compliance $p(1 - \beta) f$. Hence, the firm will comply if $c \leq p(1 - \alpha - \beta) f$. As a result, the higher α or β, the smaller the deterrent effect of enforcement.

high burden of proof (hence low probability of conviction) and high penalties (Garoupa, 2004).

Of course, the probability that mistakes are made (by inspectors, public prosecutors or courts) is not given, but depends on the resources (in time or money) spent on a case (see Kaplow and Shavell, 1994b). I will not discuss the full implications of such models. Generally speaking, it is optimal to spend resources in such a way that the marginal costs equal the marginal benefits, which implies a probability of mistake that is larger than zero, but decreases with the magnitude of the sanction. I now focus on the fact that the level of errors not only depends on public resources, but also on a firm's private resources for adjudication.

6.2.2 *Firms' investment in escaping penalties*

The fact that enforcement authorities make errors in enforcement creates a possibility for discussing the decisions they make. Firms can object to and appeal against the decision to impose a penalty. The discussion can deal with the question of whether and if so to which extent the firm violates a standard, which standard actually applies and which sanction or which magnitude of the sanction is appropriate. If the violator contests a decision successfully, the contemplated sanction will not be imposed. Therefore, the higher the sanction, the more resources the firm will use to try and escape sanctions. Khambu (1989) demonstrates that since enforcement is contestable, an increase in the expected sanction or an increase in the standard can have an adverse effect on the level of compliance if, as a result, firms make more use of procedures of objection and appeal. This effect is even stronger when such procedures are costly for enforcement authorities, because of the time spent on collecting evidence, concluding objections, conducting prosecution, etc. On balance, raising the regulatory standard and/or raising the sanction can decrease the realized expected sanction and can cause the level of compliance to decline. In cases of full compliance, a firms' level of precautions almost always increases when the standard is raised. However, in cases of non-compliance, raising penalties or standards will sometimes cause compliance to drop because non-compliant firms will have a stronger incentive to contest regulatory outcomes.[113] Khambu (1989) therefore argues that contestable enforcement explains why enforcement agencies often use informal, lowered standards.[114] Non-compliance with the statutory standard that is not penalized does not necessarily imply that the enforcement authority is lax

113 A similar point is made in Nowell and Shogren (1994) and Raymond (2004).

114 A related argument was used in section 5.3 about relaxing the standard (Harford and Harrington, 1991). In both papers, in cases of non-compliance, relaxing the regulatory standard will always lead to a higher level of compliance.

or captured. On the contrary, by enforcing only up to the lowered informal standard, the regulator is able to obtain a higher level of compliance than under 'strict' enforcement. In many cases, a firm will have two possibilities for reducing sanctions: (a) by taking more precautions, or (b) by spending money on 'good' lawyers to 'talk down' the offense in court, thereby reducing the magnitude of the ultimate sanction.[115] Increasing the expected sanction or the standard will increase both incentives. It cannot be argued that (a) will always dominate (b).

In a subsequent paper, Khambu (1990) distinguishes between contesting the fine imposed and the regulatory standard against which compliance is measured. This distinction refers to two different regulatory systems. When enforcement takes place through direct controls of the regulator by means of directives and commands, the firm may challenge the standard applied. When enforcement is incentive-based, for example through fines, the firm can challenge the magnitude of the sanction imposed. Khambu (1990) shows that enforcement by direct controls is superior if the enforcement powers of the authority are weak, i.e. if the probability of successful contest by firms is high. Enforcement through direct controls has the benefit that, if the enforcement power becomes weak, the enforcement authority can mitigate the standard, which will lower the incentive to contest and result in at least a 'reasonable' level of compliance. Under an incentive-based regime, the standard is immutable and weak enforcement powers therefore cause enforcement to be increasingly less effective in achieving compliance with the standard. A decrease in enforcement power has stronger negative effects under enforcement through direct controls than through fines. However, Nowell and Shogren (1994) show that an incentive-based regime is still favorable if the government subsidizes compliance. For a situation involving dumping of waste, they show that a decrease in the legal costs of waste disposal unambiguously reduces illegal dumping because it does not affect a firm's incentive to evade enforcement. Although it might increase total waste, a greater percentage of the waste will be disposed of in a legal manner.

The presence of possible contest actions explains why enforcement policies with severe sanctions are, as predicted by Becker (1968), ineffective. They are subverted ex-post (Glaeser and Shleifer, 2003). The market structure is relevant here (Glaeser and Shleifer, 2003; Nowell and Shogren, 1994): if a firm's assets become larger, the incentive to avoid paying damages increases.[116] In a perfectly competitive market, a firm operating at the margin will often not have sufficient assets to rigorously contest enforcement. On the other hand, monopolists

115 One paragraph in Khambu (1989) is labeled "Raise your standards and you'll hear from my lawyer". See also Heyes (2000, section 3.1).

116 Note that reducing the wealth constraint may have two consequences. As discussed above, this may imply an increase in precautions. However, it may also imply an increase in enforcement procedures.

or oligopolists generally have numerous assets at their disposal, and therefore the possibility to use them to reduce the effectiveness of penalties.

6.2.3 *Firms' investment in non-detection*

I argued above that the effective penalty is an endogenous variable depending on a firm's incentive to invest in ways to escape penalties. These investments can be made after inspection during the adjudication process (investment in non-conviction), but firms can also invest ex-ante in hiding an offense in order to reduce the probability that a violation is detected during inspection, which Heyes (1994) calls an investment in "uninspectability" (see also Malik, 1990). For example, illegal dumping of waste can be performed at night or at an inconspicuous place; automobilists can install radar detectors; café proprietors can try to quickly remove beer crates from the emergency exits; or firms can simply increase the number of objects or the area that has to be inspected. Similarly to the results above, a stricter enforcement or regulation might on balance decrease the level of compliance, because it becomes more attractive to invest in hiding violations, and it is no longer necessarily optimal to set the fine at its maximum.

6.2.4 *Public investment in detection*

Heyes (1994) demonstrates that if regulatory outcomes are contestable (ex-post or ex-ante) it becomes relatively more beneficially to invest in the thoroughness, rather than the number of inspections. Investing in the thoroughness of an inspection increases the likelihood that a detected violation will also be convicted and punished (such as collecting more samples or using more sophisticated monitoring equipment), and thereby makes investments in appeal procedures by firms less profitable. Alternatively, investments in thoroughness might/will increase the probability of detection itself (by for instance increasing the time spent on inspection), thereby making investments in hiding violations less attractive. Increasing the thoroughness of inspections unambiguously raises the level of compliance, because it increases the probability of conviction and reduces the marginal productivity of investments in uninspectability. When the frequency of inspections or the level of fines is increased, it becomes more valuable for firms to invest in uninspectability to avoid sanctions. A decrease in compliance may result if the direct positive effect on compliance from a higher expected sanction is smaller than the indirect negative effect on compliance through reduced inspectability. Khambu (1989) made a similar point when he argued that investments in monitoring are effective if they not only raise the expected penalty (which will increase the evasion effort), but also reduce the marginal productivity of a firm's evasion effort.

If firms can evade enforcement, it is optimal to invest in the thoroughness of inspections. A two-stage inspection can then be beneficial because in such a procedure, not every firm has to be inspected thoroughly. Hence, on average, inspection costs can decrease. For example, Heyes (2002) discusses an inspection process that consists of two stages. First a firm is inspected. If the observed precautions of a firm fall short of a certain "trigger" the firm is audited, which implies a more thorough inspection of the firm. If the audit reveals that the firm is in violation, it will be penalized. Jost (1997a) discusses a similar model, where investigations are executed when a firm lodges an appeal against the imposition of a fine. Note that this type of two-stage enforcement process is a sequential combination of monitoring (general enforcement) and investigation (specific enforcement) as a variant on the joint simultaneous use discussed in section 3.2.4.

6.3 Conclusion

If the law were complete and perfectly predictable, a strict enforcement policy as proposed by Becker (1968) would be efficient – provided that the expected sanction is large enough. If, however, the standards are uncertain, a more flexible form of enforcement might be needed. Public inspections might be required to inform firms about the standards and the compliance techniques, so that enforcement is aimed at advice, in addition to or instead of deterrence. Cooperation might be required to buy or negotiate precautions not included in the regulation in exchange for tolerating violations of other standards. Moreover, the trade-off between the probability of detection and the magnitude of the sanction is hampered by the incentive of firms to try to escape sanctions. Informal enforcement might on balance achieve higher levels of compliance than strict punishment of every offense.

7 Voluntary compliance and persuading enforcement – informal constraints

Until now I have considered the compliance decision to be one in which the firm balances its costs of compliance with the expected formal sanction. However, compliance does not only depend on the formal sanction that is imposed by the enforcement authorities. Apprehension, prosecution and/or conviction may have other consequences for the individual or firm, which are identified in this section. Special attention is given to the interaction with public enforcement.

7.1 Informal sanctions: social and moral control

A sanction does not only include the formal sanction imposed by the enforcement authorities, but also the informal sanctions imposed by private institutions. Apprehension, prosecution and/or conviction may harm an individual's or firm's reputation in their social network. For example, crime may lead to disapproval by family members, friends or neighbors or to a decrease in future career possibilities. If a firm incurs a violation, consumers might decide to turn to another producer, or governmental agencies or sector organizations may change their attitude towards the firm.[117] Besides, violation may be accompanied by feelings of regret, guilt, remorse, moral dilemma, etc. Therefore, these informal sanctions will increase the personal, moral or social costs of non-compliance (section 2.1.1).

The informal sanctions can be of two types: in economics, a distinction is made between extrinsic and intrinsic motivations. Extrinsic motivations refer

117 For example, regulators are in a position to make use of a firm's compliance history when determining whether a permit should be granted. Decker (2003) shows that firms with a bad reputation with governments have to wait longer for the necessary permits. He investigates how long it takes for environmental agencies to process and issue new source construction permits pursuant to the Clean Air Act regulations and new industrial discharge permits pursuant to the Clean Water Act regulations. He finds that plants (or firms) with fewer instances of non-compliance receive permits for major projects more quickly. For New Source Review (NSR) permits, according to one model specification, "one additional violation increases the time it takes to receive a permit by roughly 23 percent. For the average (median) major NSR permit, this translates into an increase of about 52 days" (p.120). For Industrial Discharge Permits (IDP), similarly, "one additional violation at any one of the firm's plants will increase subsequent permitting times by 15 percent. For the average (median) major IDP, this translates into an increase of about 81 days" (p. 124).
See also Karpoff and Lott (1993) and Karpoff et al. (2005) who demonstrate that reputational penalties are important for criminal fraud but not for environmental offenses.

to incentives from outside the individual and are imposed on him by others. These include the informal sanctions, imposed by relatives, consumers, neighbors, employers or employees, etc. which lead to loss of (future) earning capacity and other financial costs or shame, stigma and other psychological costs. These types of informal sanctions might be labeled *social control*. They follow from the fact that others find out about a violation. Intrinsic motivations refer to incentives inside the individual that do not depend on rewards or penalties by others. These costs (or benefits) are realized irrespective of whether others are informed about the offense. These include informal sanctions such as guilt and regret. These types of sanctions might be labeled *moral control*.[118 119]

Informal sanctions, just as formal ones, are seen as costs for (potential) offenders and are taken into account when comparing the expected costs and benefits of (non-)compliance. The costs (or benefits) of informal sanctions should be counted from the moment that they are realized, which might vary from the moment of violation to the time of conviction or execution.

There are three important points to be made regarding the interaction of informal sanctions with public enforcement. First of all, public detection might enable private informal sanctions to be carried out. Secondly, formal and informal sanctions might reinforce each other. This happens when informal sanctions depend on the level of compliance by others, hence on public enforcement. Thirdly, formal sanctions might crowd out the intrinsic motivation to comply, leading to a conflict between formal and informal sanctions. Let me discuss these three points in more detail. The first interaction presumes that the presence of informal sanctions has no effect on the *effectiveness* of formal sanctions and public enforcement, while the other two forms of interaction refer to cases where informal sanctions increase, respectively decrease, the effectiveness of public enforcement.

7.2 The deterrent effect of informal sanctions

First of all, the costs of informal sanctions increase the sanctioning risk and/or make compliance more attractive. This makes it easier to induce compliance. Even if the costs of compliance are higher than the expected formal sanction, the

118 A review of norms and possible sanctions can be found in Posner and Rasmusen (1999). Elffers (2005*) discusses the importance of social and moral control from a rational choice perspective.

119 Sugden (1989) explains how a system of norms and values can evolve and, once sufficiently supported in the population, can reproduce itself. Certain conventions arise because people need to coordinate their potentially conflicting behavior. When a sufficiently large number of people obey the convention, these conventions can become norms – written (law) or unwritten – because people expect and demand that other people should obey them too. Violation of the norms will lead to disapproval and possible sanctions by other people, not only by those who are directly harmed by the violation. See also Ellickson (1991).

firm might comply because of the social or moral costs. Acceptation of norms does not imply that the rules are always obeyed. However, violation of the rules might produce guilt, regret or a loss in reputation, all of which make violation less attractive. A firm will comply if the compliance costs are lower than the expected formal sanction and the disutility that results from social and moral control.[120] Since informal sanctions make violation less attractive, a lower expected formal sanction suffices to enforce compliance. It is possible to strengthen the effect of informal sanctions – especially the extrinsic social control that depends on others finding out about the violation – through, for instance, publicity, public criminal records, etc.[121] However, the presence of informal sanctions also implies that it is less attractive to trade-off the probability of detection and the magnitude of the formal sanction, in line with Becker (1968). A lower probability of detection decreases the effectiveness of the informal sanctions, at least the extrinsic informal sanctions which are only imposed if violation is detected.

7.3 Formal and informal sanctions as reinforcing instruments

A second important point concerning the interaction of formal and informal sanctions is that the level of informal sanctions might depend on the level of compliance by other firms. A report of the WRR (2003*) provides a clear analysis of the implication of this dependency for enforcement policies, based on the work of Akerlof (1980).[122]

Both moral and social control lead to psychological costs which depend on the proportion of individuals complying. Individuals may agree on the rules but to a different degree. Some individuals may not accept the rules at all. If the rules are accepted, violation of the rules may produce regret, guilt and other moral costs. If fewer individuals comply, the acceptation of the rules might decrease. Especially those who regret violation least might lose their belief in the rules. Hence, a decrease in compliance leads to a decrease in the costs of regret for at least a proportion of the individuals. The same thing happens with informal sanctions such as social control. Individuals who do not accept the rules and/or fail to comply are unlikely to address others on their non-compliant behavior or switch their business to a more compliant firm. This implies that the level of costs in loss of reputation depends on the proportion of firms who comply.[123]

---

120 Put differently, a social or moral norm (acceptation of the rules) is treated as a preference. Individuals try to maximize utility, which implies that they have to balance several preferences given the limited resources available. See WRR (2003*, chapter 4 and its appendix).

121 See further Posner and Rasmussen (1999).

122 See also Lai et al. (2003) for a more specific application to firms. See also Conley and Wang (2006). In a different context a similar point is developed by Wittberg (2006).

123 See note 119 for an explanation of how such informal sanctions can arise and at a given moment in time can be at an equilibrium in which they reproduce themselves at a certain level of public and private enforcement.

If informal sanctions depend on the proportion of individuals who comply, there is an important implication for public enforcement. If the enforcement policy is weakened, this not only directly decreases the level of compliance, but it also breaks down the acceptation of the norm and the social control of the norm, which further reduces compliance. A "snowballing" effect is engendered so that more and more individuals are induced to violate. This process only ends if the internalization of the rules in part of the population is so strong that it can stop the decline of the acceptation. Therefore, weakening public enforcement also weakens the effect of informal sanctions.

Another implication is that deterioration and recovery of compliant behavior may be asymmetric processes.[124] A weakening of enforcement will reduce compliance. Returning to the old enforcement policy will not suffice to restore the old compliance level. The decrease in compliance will have led to a decay of moral standards and social control that can only be made undone through an increase in the sanctions which more than off-sets the weakening of the policy. This is the only way to induce compliance by those individuals who no longer accept the rules. If this procedure is successful, so that the proportion of individuals who comply increases, the belief in the rules may possibly be restored. This will, in the long run, increase social control and the costs of regret. However, the restoration of norms may possibly take a much longer time than the decline, especially if it has to be restored by means of formal control. Therefore, a small decrease in public control – for example the removal of supervisors in public transport, the change to self-service in shops – can lead in the long run to a large reduction in compliance. Another inference is that informal and formal control are not only possible substitutes for each other, but also have complicated interactions. As long as rules are widely accepted and violation brings large losses in reputation, a low level of public enforcement may have a large impact on (non-)compliance. If, however, the belief in rules and informal control breaks down, enforcement may have a much smaller impact on (non-)compliance.

7.4 Crowding out of intrinsic motivation

In the economic literature, it is well-known that extrinsic rewards might crowd out intrinsic rewards. For instance, voluntary blood donation decreases both in quantity and quality if a monetary reward is given to those who donate, because this might decrease or crowd out the altruistic motives for blood donation.[125] Some authors have applied a similar analysis to penalties.

Gneezy and Rustichini (2000) analyze the introduction of a fine on late-coming parents in a day-care centre. They found that after the introduction of a monetary fine the number of late-coming parents increased significantly.

---

124 See Mulder et al. (2005) for a similar point.
125 The classic reference is Titmuss (1970). See Frey (1997) for an overview.

However, once the fine was removed, no reduction in late-coming occurred. Lin and Yang (2006) explain these findings on the basis of the social norm that individuals should be on time and not late. Violation of this norm leads to a psychological cost resulting from guilt (intrinsic) and shame (extrinsic). The introduction of a fine for late-coming parents is assumed to reduce this psychological cost directly (again following Akerlof 1980 and psychological theories such as "equity theory"). This erodes the effectiveness of the social norm in preventing late-coming. An increase in the fine has a deterrent effect, because it imposes a financial cost on violation, and a norm-psychological effect, because it reduces the costs of guilt and shame. If the fine is below a certain threshold an increase in the fine will increase late-coming because the norm-psychological cost dominates the deterrent effect from the fine. Suppose there is no fine. Then two equilibriums are possible: one with few parents coming late, because of the high psychological cost, and one with many parents coming late, because of low psychological costs ("everyone else is doing it"). In the low late-coming equilibrium, the introduction of a small fine can start a "snowballing" effect. The fine reduces the psychological costs, which (if the fine is below the threshold) will induce parents to come late. This will increase the proportion of parents coming late, further reducing the psychological costs, increasing late-coming, increasing the proportion late-coming parents, etc. In the end, a new equilibrium will be established in which many parents come late.[126] Once the small fine is removed, it will increase late-coming somewhat because of the deterrent effect. Since in the meantime, late-coming has become widespread, the social norm will be too weak to generate substantial psychological costs on late-coming parents and the new equilibrium will still be one with many late-coming parents. Note that there is once again an asymmetry, but a somewhat different one from the one above. Here, an *increase* in enforcement is not comparable to revoking this increase, because in the meantime this increase has reduced norm support. Above, a *decrease* in enforcement is not comparable to revoking this decrease, because in the meantime this decrease has reduced norm support.

The important lesson is that imposing formal sanctions can reduce or even remove the effectiveness of informal sanctions. Once formal sanctions crowd out the intrinsic motivation to comply, public and private enforcement will have opposite effects. Put differently, the effectiveness of public enforcement might be reduced because of the presence of informal sanctions.

126 Note that the main difference between Lin and Yang (2006) and WRR (2003*) is that WRR (2003*) examines what happens when for some reason there is a *weakening* of enforcement. This weakening leads to a reinforced (downward) snowballing effect because compliance depends on the proportion of individuals complying. In Lin and Yang (2006) a downward snowballing effect results from the *strengthening* of enforcement due to a direct effect of the fine on informal sanctions, and is further increased by the dependency of informal sanctions on the proportion of individuals complying. Here the snowballing effect is counterproductive.

7.5 Conclusions on informal sanctions

Taking informal sanctions into account allows us to acknowledge that firms do not make decisions in isolation, playing against nature or the authorities, but in a social context, playing against their neighbors, customers, etc. (Lin and Yang, 2006). Informal sanctions make compliance more attractive. However, informal control and intrinsic motivation to comply might not only increase the deterrent effect of (formal) enforcement, but also impose limits on it. First of all, informal control may only work when there is some public enforcement, limiting the possibilities of reducing detection. Secondly, informal sanctions may depend on the proportion of individuals or firms complying, hence strengthening the effectiveness of public enforcement. This may be problematic if enforcement is weakened. Thirdly, formal control can diminish the intrinsic motivation to comply, which may be so strong that an increase in formal control leads to a decrease in compliant behavior.

If the crowding out of intrinsic motivation is very severe, imposing even a small sanction breaks down the moral or social intention to comply. A policy in which the authority refrains from formal sanctions and only tries to stimulate the intrinsic motivation (through relaxed inspections, informing etc.) may possibly result in a higher level of compliance than a policy in which a violation is immediately sanctioned with a (limited) formal sanction. At least, the efficient enforcement policy should recognize the impact on intrinsic incentives. A related issue is that in the presence of important extrinsic informal sanctions, sufficient to achieve compliance but depending on detection, public enforcement can be limited to inspection and detection and can possibly avoid almost all (costly) formal sanctions.

This may give rise to enforcement strategies that are not based on penalizing offenders, but on strengthening the informal control and intrinsic motivation. However, Lin and Yang (2006) show that enforcement based on deterrence has not lost its power. The optimal policy is to "fine enough or don't fine at all". If enforcement is severe enough, it is always possible to enforce the desired level of compliance. The optimal response in case of the day-care centers is not to remove the fine, but to increase it. This already appeared in the model of WRR (2003*), where compliance could be enforced by means of a sufficiently low (i.e. negative) extrinsic reward, and this is recognized in Gneezy and Rustichini (2000) when discussing the use of fines in day-care centers in the United States. As long as the formal sanction can be made large enough, it is efficient to induce compliance by means of sufficiently high sanctions irrespective of the effects of informal sanctions.

8 Economic explanations of compliance strategies

The previous chapters dealt with the economic analysis of optimal public enforcement policies. I discussed the possibility to save resources by trading off the probability of detection and the magnitude of the sanction, and investigated the conditions for and limits of this trade-off. This chapter attempts to summarize the analysis and to bring together the various conclusions. I do this by starting from a somewhat different perspective than before. In practice, enforcement authorities, especially those concerned with enforcing safety standards in firms, often do not use strict enforcement policies of immediately punishing violations with sanctions, and even less often do they apply extreme sanctions. Instead, they often use a more relaxed approach by bargaining with the firm, issuing warnings and only penalizing following a very long enforcement process. This approach is sometimes referred to as a *compliance* strategy. This chapter discusses, making use of the insights from the previous chapters, how these policies can be interpreted from an economic point of view and why they might be efficient.[127]

First, I discuss in further detail the issues concerning deterrence and compliance strategies. I then provide the economic rationale for compliance strategies and examine in further detail how compliance strategies might be designed from an economic point of view. I also attempt to explain why these strategies might be particularly helpful in cases involving regulatory crime by firms. Finally, I relate the economic analysis to the approach used in other social sciences.

8.1 Compliance strategies: what is it all about?

It has been observed that in practice, many enforcement authorities, especially when dealing with regulatory crime by firms, do not adopt a policy involving severe penalties. Becker's (1968) recommendation of maximizing the sanction in order to minimize the probability of detection is not often followed. Most of the time enforcement authorities issue firms found in violation with a warning. With the first discovery, this is often not a legal warning (such as the repair sanctions discussed before), but only an informal warning that an enforcement process will be started if the firm does not return to compli-

127 In Suurmond (2007), I analyze the discussion on compliance and deterrence strategies both theoretically and empirically (strategies of enforcing fire safety regulation in horeca establishments, chapter 13).

ance. In addition, there is room for a bargaining process between the firm and the enforcement authority. The parties negotiate about the time the firm may spend in violation, the standard the firm has to obey and/or the solutions to ensure that the firm will comply with the standard. Firms are not sanctioned, but informed about the danger of non-compliance. Such enforcement actions seem to harm the deterrent effect of enforcement. Firms can use the space that is given to (partially) fail to comply or at least to delay compliance.

Yet, firms are thought to be in compliance most of the time. This is referred to (for instance in Heyes and Rickman, 1999) as the *Harrington paradox*. This is a label for the following combination of facts, observed by Harrington (1988) with respect to the enforcement of safety standards at firms:

Fact 1: The frequency of surveillance is usually low.
Fact 2: Fines or penalties are hardly imposed, even if a violation is detected.
Fact 3: Yet, compliance with the standards seems to be pretty high.

Similar observations have inspired scholars, particularly non-economists, to distinguish two different enforcement styles, a *deterrence* or *penalty* strategy and a *(negotiated) compliance* strategy (table 8.1).[128] The central idea of a deterrence strategy is that tracing offenders and punishing their violations with a punitive sanction deters them from committing a violation. A deterrence strategy assumes that the potential offender makes a conscious decision by comparing the costs and benefits of (non-)compliance. Compliance is obtained by instilling a fear of the consequences of violation: a penalty. It is a punitive, repressive method that is aimed at achieving general deterrence or prevention. The government bears the main responsibility for compliance. The deterrence strategy strengthens the extrinsic incentives to comply with the rules. Becker (1968) and more generally the economic models and theories are considered to be the main defenders of this approach.

This strategy is often criticized in literature, beginning with Bardach and Kagan (1982) and Hawkins and Thomas (1984). According to these authors, enforcement should not be based on the assumption that potential criminals make conscious comparisons of costs and benefits, but on the moral and social value they attach to compliance. Individuals or firms do not comply because of the costs associated with non-compliance, but because of a moral intention to do so. People morally adhere to rules, and violations are considered immoral. Compliance is mainly spontaneous and enforcement should consist of moral persuasion to comply. Non-compliance is either the result of a mistake or misinformation, or of a lack of norms and values. A deterrence strategy has some important disadvantages. It stimulates people to think in

128 See for example Hawkins and Thomas (1984), especially the chapter by Kagan and Scholz (1984), Hawkins (1983 and 1984), Huisman (2001*), Van Stokkum (2004*), Ponsaers and Hoogenboom (2004*) and Van Rooij (2006). This literature and debate is inspired by a number of volumes by Bardach and Kagan (1982), Braithwaite (1985), Tyler (1990), Ayres and Braithwaite (1992), Hutter (1997) and many others. A few economists who participated in this debate are Veljanovski (1984), Fenn and Veljanovski (1988) and Garvie and Keeler (1994). See also Ogus and Abbot (2002).

terms of costs and benefits, reducing moral intentions and hence compliance. A deterrence strategy is also reactive, hence coming into action too late, when the harm has already occurred. A pro-active approach would be more beneficial. If people are confronted with non-compliance at an earlier stage, the enforcement authority can inform them about the fact that they are in non-compliance, as well as about the dangers of non-compliance and about the available methods for restoring compliance. Therefore enforcement primarily takes the form of negotiation, convincing and advice rather than sanctioning. Imposing penalties is only seen as a final possibility for restoring compliance if negotiations break down. It has also been argued that in our modern society, rules are not only dictated by the government ('command-and-control'), but that many forms of alternative regulation hold with joint responsibility for the government, firms, interest groups, etc. A deterrence strategy is characterized as 'going by the book', a detached approach in which every detected violation is sanctioned, without any reference to circumstances or motives for non-compliance and/or the objective of the rules. However, enforcement should strengthen the intrinsic incentive to comply by moral persuasion.

It should be stressed that these strategies are stereotypes and that no scholar argues that a single one of them can be the sole means of solving the enforcement problem. For instance, in a compliance strategy imposing a sanction is necessary if all other means of enforcement have failed. However, there is a clear distinction between the two different strategies and between those who favor them.

Table 8.1 Deterrence versus compliance strategies

Deterrence:	**Compliance:**
Fear of consequences non-compliance	Moral approval and rejection
Cost-benefits, rational	Spontaneous compliance
Command-and-control	Alternative regulation
Punitive, sanctioning	Bargaining, persuading, advising, informing
Repressive, reactive	Pro-active
Criminal law, government action	Confidentiality, social cooperation
Generic prevention (deterrence)	Fight causes (repair)
Going by the book (to the letter of law, bureaucratic)	Motives and circumstances
Police-officer	Advisor, relief worker
Extrinsic incentives	Intrinsic incentives
Economists, Becker	Hawkins and Thomas, Braithwaite, Scholz, etc.

8.2 Compliance strategies from an economic perspective

In this section I show that economists do not necessarily favor deterrence strategies. I also explain why a deterrence strategy may fail and why a compliance strategy might be desirable.

8.2.1 *Why and when deterrence may fail*

The deterrence strategy described above is based on the assumption that enforcement is aimed at deterring rational firms from choosing for non-compliance. Following Becker (1968), extreme sanctions can lead to enforcement of compliance and minimization of enforcement costs. One of the implications of Becker's (1968) approach is that, given the sanction and budget constraint, an enforcement authority should devote its resources to detecting and subsequently prosecuting as many breaches of the law as possible, weighted by some measure of the severity of the offense (Fenn and Veljanovski, 1988). The deterrent effect of enforcement relies critically on the expectation by potential offenders that a detected offense will be prosecuted, i.e. on the authority's reputation for consistently prosecuting wrongdoers. The enforcement authority must punish every violation, immediately. In the previous chapters, I discuss two major categories of reasons that might make such a strategy inefficient. First of all, a failure of deterrence results from the fact that the rules are incomplete and/or inappropriate. Secondly, there are important limits on the possibility of trading off the probability of detection and the magnitude of the sanction, because in many cases increasing the sanction is not costless. These reasons are especially relevant if the maximum achievable compliance rate is partial (section 5.1) and/or if there is serious underdeterrence.

Incomplete regulations

The first type of failure of a deterrence strategy is related to problems with the prescribed standards. Regulation is generally insufficiently appropriate to deal with a particular problem. As explained in section 6.1.2, the law is inherently incomplete. This has a number of consequences:

1. If firms are ignorant about the law or compliance status and methods, strict enforcement will not lead to compliance (section 6.1.1). If public authorities have a natural informational advantage (especially in the case of technically complex regulations), it might be favorable for them to visit firms in order to offer advice. Enforcement officials then behave as consultants or advisors rather than as policemen. However, I would like to repeat that generally speaking, firms themselves should acquire information about the relevant law. Ignorance can also be the result of weak enforcement.
2. More importantly, since the law is incomplete, regulations are both under- and overinclusive. Therefore, the enforcement authority should

forego a legalistic enforcement policy to the letter of the law ("going by the book") but instead consciously determine whether or not an offense is efficient (section 6.1). Furthermore, in order to reduce underinclusion, the enforcement authority should be granted law-making powers so that it can impose the most efficient standard on a particular firm and enforce it properly.
If it is impossible for the enforcement authority to determine the efficiency of an offense, the authority should bargain with the firm in order to improve the application of the law for a particular firm, in a particular situation. Both parties might benefit from such a cooperative approach in which they adopt a flexible attitude towards the standards.
3. Since rules do not fully address every particular situation, they are potentially conflicting or contradictory. In such cases, firms are unable to comply, hence high sanctions will not be effective. Firms will not comply under a strict enforcement policy, especially if they are uncertain about how the enforcement authority will deal with this conflict (with respect to its law-making powers). In such cases, the firm might possibly wait until it receives an official statement from the enforcement authority which states which standards must be satisfied. A similar point can be made when standards are handled inconsistently in time, for example if inspectors vary from year to year.
4. More generally, if the rules are wrong, i.e. if they prescribe inefficient levels of precaution, imposing high sanctions, thus forcing compliance, is inefficient. Moreover, the efficient level of precautions takes the enforcement costs into account. Since enforcement is costly, the enforcement authority should also forego minor inefficient violations not worth enforcing. As became apparent throughout the analysis, a certain degree of underdeterrence is favorable because of the costs of inspection.
5. In some cases, rules are not aimed at deterrence, but at remediation. If offenses are beyond the control of the firm (due to technical failures), the major task of the enforcement authority is not deterrence but remediation. For remediation, extreme sanctions might be inefficient if enforcement cannot be conditioned on the level of remediation performed (section 4.2).

Imposing sanctions is costly
Even without the problems resulting from incomplete or inappropriate law, i.e. even where an offense is most certainly inefficient, a deterrence strategy may be in- or counter-effective because increasing the sanction is costly. If imposing a sanction is sufficiently costly, extreme sanctions are inefficient. Moreover, if sanctioning leads to very high costs, refraining from sanctions and choosing for different enforcement actions can be beneficial. This situation was already described in the case of the direct costs of sanctions, as under imprisonment. However, I have also identified many other reasons why increasing a sanction is not without costs. Increasing the sanction is

costly if it somehow reduces overall deterrence. The enforcement authority then has to balance the effects of a penalty on direct deterrence of the offense at hand with the importance of overall deterrence. This might explain why the enforcement authority should not choose a deterrence strategy in which every violation is punished.

1. Increasing the sanction for minor offenses is costly if it reduces deterrence for serious offenses (the problem of marginal deterrence, section 3.2.2). If the harm from serious offenses is sufficiently high, it will be efficient to tolerate small offenses in order to deter the more serious ones. The enforcement authority should then be lenient towards minor violations.
2. Increasing the sanction is costly if it makes compliance in another period less attractive. If there is serious underdeterrence, the optimal penalty scheme depends on past compliance records. First, as explained in section 4.1, the optimal sanction for first-time offenders should be lower in the second than in the first period. Increasing the sanction in the second period decreases deterrence in the first period. For example, the authority should treat existing firms with good compliance records more leniently than young companies or persistent violators. Secondly, as explained in section 5.3, if the target compliance rate is partial, the efficient enforcement policy is based on past compliance, implying that enforcement is lax for one group and severe for the other. Tougher measures might reduce overall deterrence. In order to make the incentive to comply in the target group as large as possible, firms in the non-target group should not be penalized. Increasing the sanction in the non-target group will reduce deterrence in the target group. In essence, a warning functions as a transition to the target group.
3. When self-reporting exists, increasing the sanction is costly because a higher sanction for self-reported violations will reduce self-reporting and thereby increase monitoring costs (as well as reduce the other advantages of self-reporting, see section 3.4). Self-reporting as such may resemble cooperative enforcement. If self-reporting concerns technical failures beyond the control of the firm, a zero punishment policy may be efficient, even if the firm does not immediately return to compliance (section 4.2.2).
4. Increasing the sanction is costly if the enforcement process also convicts some innocent firms (section 6.2). This might be especially important in the case of high and severe sanctions. Moreover, higher sanctions require a higher standard of proof, which will increase prosecution costs.
5. Increasing the sanction is costly if firms are able to contest the enforcement authority's decision and/or to invest in the non-detection of offenses (section 6.2). This problem might be so severe that decreasing the sanction will increase compliance. Therefore a zero sanction can be optimal, especially if investments in the thoroughness of inspections are costly.
6. Increasing the sanction is costly if it crowds out the effectiveness of informal sanctions, especially the intrinsic moral intention to comply (section

7.4). If this crowding out is very severe, imposing even a small sanction will break down the moral or social intention to comply. No sanction might then possibly be the most efficient policy. A related point is that, due to the presence of extrinsic informal sanctions, public enforcement might possibly limit itself to detecting violations and refraining from (costly) formal sanctions.

7. Similarly, increasing the sanction may be costly due to some negative relationship between the decision to comply and the compliance rate, leading to a negative interaction between public enforcement and the compliance rate, as described in section 3.3. Therefore, increasing the sanction may have a negative counter-effect on deterrence. One example of such an interaction is that the enforcement authority faces a limited budget. If prosecution is costly, this implies that it can only prosecute a limited number of violators (section 5.3.4). Therefore warnings, which are cheaper to impose, might be a better enforcement technology, especially if delayed compliance is more costly for firms than immediate compliance (due to the costs of extra inspections, for example).

8.2.2 *The design of compliance strategies*

I identified above a number of reasons why a deterrence strategy may be ineffective or inefficient and why the enforcement authority should adopt a compliance strategy. If increasing the sanction is sufficiently costly, a non-extreme sanction is optimal. Even a zero sanction can be efficient. In cases of incomplete or inadequate rules, a legalistic deterrence approach is also inefficient.

There are a number of possible reactions to the failures of a deterrence strategy. Instead of immediately sanctioning a violation of the standard, the authority might choose to apply one of several forms of compliance strategy. These have already been discussed in the previous chapters; they are not mutually exclusive, and can be used simultaneously.

1. *Adjusting standards (tolerance).*[129] If the legal (statutory) standard cannot be enforced, it is generally optimal for enforcement authorities to adjust the standard downwards and to enforce some informal standard. This implies a combined use of a deterrence and compliance strategy. It qualifies as compliance because the enforcement authority does not legalistically enforce every violation of the standard but instead foregoes minor violations. It is a deterrent, however, because the authority can follow Becker (1968) in enforcing the adjusted standard.

   If the standard cannot be adjusted, it may be optimal to target compliance

---

129 See, for example, Khambu (1989), Harford and Harrington (1991), Heyes and Rickman (1999) and Lando and Shavell (2004), as well as sections 6.2 and 5.2. See also section 3.2 in which I discuss the fact that if the enforcement authority is perfectly informed, inefficient offenses should not be punished.

for some subgroup of potential offenders, and forego violations of the remainder of the population or rules. This garantuees compliance by at least some part of the population or of at least some part of the rules. This system however more strongly resembles the deterrent approach, except that it does not lead to any enforcement in the non-target group at all, neither by compliance nor by deterrence.

2. *Warnings and enforcement on past compliance.*[130] If there is some underdeterrence, it is often favorable to apply enforcement based on past compliance. The most explicit case involves a state-dependent enforcement policy in which a firm caught in violation is not penalized but transferred to a target group in which a more strict enforcement policy applies, i.e. if a first violation is detected the firm receives a warning. If offenses can also occur accidentally, it will be efficient to make punishment dependent on the compliance record. If this record is reasonable, the enforcement authority should forego punishment. This is closely related to well-known enforcement pyramids implying that the most severe enforcement actions should be kept for persistent non-compliers.
3 *Self-reporting*. Self-reporting might play an important role in a cooperative enforcement strategy (section 4.2.2) in which the firm reveals its offenses to the enforcement authority in exchange for a more flexible enforcement policy. Furthermore, self-reporting is particularly effective if offenses occur accidentally, beyond the control of the firm, and/or if remediation is also an objective of enforcement. Note, however, that self-reporting is not only beneficial if there are deterrence failures, but can also enhance deterrence as such because it allows savings in monitoring costs (section 3.4). Therefore, self-reporting can be an element of both compliance and deterrence strategies.
4. *Advice.* If firms cannot reasonably obtain information about the relevant standards or if obtaining and sharing information about compliance is cheaper through public agencies, inspectors should behave as consultants rather than policemen, at least in the first instance (section 6.1.1).
5. *Negotiation.* If the enforcement authority has insufficient information about the socially optimal level of precautions, it might be optimal to choose for a cooperative, bargaining strategy in which information is exchanged between the authority and the firm. Both may benefit from a flexible, cooperative approach (6.1.2). The firm will try to persuade the enforcement authority that compliance is too costly. The enforcement authority will try to convince the firm that non-compliance will be very costly.

130 See for example Harrington (1988), Rubinstein (1979), Livernois and McKenna (1999) and Nyborg and Telle (2004), as well as sections 5.3 and 5.3.4, 4.1 and 4.2.

8.2.3 An important remark on the desirability of compliance strategies

I discussed several reasons why a deterrence strategy might be ineffective or inefficient and why the enforcement authority should adopt a compliance strategy. However, it should be stressed that these reasons only hold if full compliance is *not the aim* of enforcement or is *unachievable* (section 5.1).

Full compliance is not the aim of enforcement if not all offenses are inefficient. If the law is incomplete, this implies that some offenses will be efficient, or that the precautions satisfying the standard may still be insufficient. Therefore, the enforcement authority should consider whether an offense is efficient and only prosecute if it is not. Non-compliance is efficient if the marginal costs of compliance exceed the marginal expected harm for some level of non-compliance. In the absence of other failures and costs, the enforcement authority can use a deterrence strategy against inefficient offenses. However, if it lacks the information required to determine the costs and benefits, the authority might try to cooperate with the firm. A compliance strategy can only be beneficial if full compliance is *not the aim* of enforcement (section 5.1) and if the authority does not have sufficient information or sufficient law-making powers to enforce socially optimal levels of precaution. If the law is complete and/or offenses are always inefficient, it is efficient to try to deter as many offenses as possible.

More importantly, full compliance can be *unachievable* due to the enforcement costs and the limits on the level of sanction. A compliance strategy is only favorable if the enforcement authority's budget is limited. If there are no limits on the level of sanction or on the enforcement resources, the expected sanction can be made large enough to enforce full compliance. A deterrence strategy in which every detected inefficient violation is punished, is efficient (see for example Fenn and Veljanovski, 1988). If everyone complies, there are no costs of imposing sanctions, including no adverse effects on deterrence. Most of the reasons discussed above for assuming that increasing the sanction is costly only hold given a limited budget and limited sanctions. If these are sufficiently limited, underdeterrence, possibly severe underdeterrence, will result, which can be reduced by choosing a different enforcement strategy. However, the optimal response is generally to increase the budget or (if possible) to try and increase the maximum sanction. This point has by now repeatedly been made. As discussed in section 7.5 (Lin and Yang, 2006), if the sanction is extremely high, the deterrent effect always dominates over a possible negative effect associated with informal sanctions. Similarly, a state-dependent policy (section 5.3, Harrington, 1988) is only favorable if partial compliance is the target and the budget is limited. Moreover, punishing repeat offenders more severely is only efficient if first-time offenders cannot be perfectly deterred (section 4.1). Warnings are only a beneficial response to costly prosecution if the enforcement authority's budget is not increased (section 5.3.4, Nyborg and Telle, 2004).

There are two important reasons for assuming that the budget is not increased (section 5.1). First of all, increasing the budget may take a long time. Therefore, in the short run, the enforcement authority's budget is given and a compliance strategy can be favorable. Secondly, increasing the budget may be inefficient because enforcement becomes increasingly costly. In that case, a partial compliance rate is the maximum feasible compliance rate and compliance strategies might be more effective in enforcing this level than a deterrence strategy. If a certain level of underdeterrence is desirable, a compliance strategy might be a better response than a deterrence strategy, due to the reasons discussed above. In cases involving a given level of non-compliance or underdeterrence, we can distinguish two types of firms: compliant firms and violating firms. The enforcement authority can try to exploit this fact by using compliance strategies. The enforcement authority and the firm can – implicitly or explicitly – bargain for a compliance strategy. The enforcement authority might profit if the rules are better, even though not perfectly, adhered to. Firms might benefit if they face smaller expected sanctions or are allowed an occasional violation. If there is full compliance, such a trade-off is impossible.

The focus in much literature lies on the effectiveness of enforcement: which policy will achieve the highest level of compliance, given the enforcement authority's budget? However, the efficiency question is still relevant. We must keep asking whether, if a higher compliance rate is desirable, it might not be more efficient, instead of adopting a compliance strategy, to somewhat increase the budget (or, if possible, the sanctions) and pursue a deterrence strategy.

The results of Garvie and Keeler (1994) most clearly summarize when and why a compliance strategy is favorable.[131] They show that compliance strategies are related to weak enforcement (or regulatory) powers and relatively inexpensive monitoring. Deterrence regimes are consistent with expensive monitoring and the existence of strong societal sanctions for violations. If monitoring costs are higher, this makes a trade-off in the way proposed by Becker (1968) more favorable. High monitoring costs require that the probability of detection should be kept low, making aggressive sanctioning necessary. On the other hand, if monitoring is relatively inexpensive, a high probability of detection can be realized if accompanied by moderate sanctions. The enforcement authority's powers depend on the political, legal and social attitudes towards the regulation being enforced. Severe offenses are largely objected to and can be enforced by low detection rates and vigorous enforcement. Offenses regarded more ambiguously by society will probably not be punished harshly and will therefore be characterized by frequent contact with regulators and relatively mild sanctioning.

131 See further Law (2006).

8.3 Explaining the scope for compliance strategies

As a result, compliance strategies might be desirable if an enforcement authority's powers are relatively weak. This may explain for which offenses and in which field compliance strategies are best applicable. Garvie and Keeler (1994) argue that a deterrence strategy is more likely to be efficient if the consensus about the desirability of higher compliance is broader and non-compliance is regarded more as a criminal act. Where such a consensus fails to hold, because there is broad disagreement about the desirability of regulation or the definition of compliance, compliance strategies are more efficient.

It is claimed that these conditions are satisfied in the case of regulatory offenses by firms. Several authors have identified important characteristics of regulatory offenses as opposed to criminal offenses (see for example Veljanovski, 1984, or Kagan, 1984). However, most of these studies do not discuss why these characteristics should make the use of compliance strategies more effective (as opposed to a deterrence strategy). Any explanation of this kind should link these characteristics to weak enforcement powers and the broadness of societal disapproval for the offenses in question. I will attempt to provide such an explanation with respect to the characteristics of regulatory offenses identified by Veljanovski (1984).

1. "Regulatory offenses do not involve a positive act, but rather a failure to act." It might be that failures to act are more difficult to observe and prove, and are less often reported by civilians or victims. Hence, enforcement is more costly and the boundaries on the enforcement authority become more pronounced. In addition, positive acts may be deemed morally more unacceptable. Moreover, ignorance and incomplete law is more problematic in cases of failures to act.
2. "A regulatory offense is not a discrete event, but a continuing state of affairs." Therefore, remediation is important, and the offense is something that can still be bargained about. It also implies that enforcement is a continuing problem, implying that constraints on the enforcement authority may become more problematic.
3. "Regulatory offenses normally occur within an organization where the responsibilities for compliance are often diffused, and non-compliance is the result of either ignorance or a conscious profit-maximizing decision." As discussed, if potential offenders are ignorant of the law, a cooperative enforcement strategy is possibly more effective. In addition, if no one feels responsible for achieving compliance, warnings or bargaining might help. If there is no clearly responsible offender, society might disapprove of severe sanctions because such measures would be considered 'unfair' or because there is a high probability of convicting innocent persons. Moreover, if no one is clearly responsible, informal sanctions such as regret and shame might be less effective, requiring higher formal sanctions. That an offense results from profit maximization does not seem to have any particular relationship with a deterrence or compliance strategy.

4. "Common crimes destroy or redistribute wealth, but regulatory offenses are frequently the by-product of some otherwise socially beneficial activity" (*mala in se* versus *mala prohibita*). This implies that in the case of regulatory offenses, the danger and social costs of over-compliance and overdeterrence are higher. In common crime, the efficient activity level is (almost) zero, while the problem in regulatory crime is not to deter efficient activities but to enforce efficient precautions. If these precautions are taken, every activity is efficient. Therefore, for common crime, full compliance is often the aim (although possibly unachievable). Moreover, this distinction implies different moral attitudes towards an offense, both by the offender and by society.
5. "In regulatory enforcement personality and offense-specific factors are (more) important in the enforcement process and the decision to prosecute. Detection is automatically linked to the identity of the offender and creates immediately a bilateral relationship between the authority and the offender, with the possibilities of bargaining." The availability of more specific information allows the authority to determine whether prosecution is desirable and to forego punishment. Possibly, too, if further circumstances are known, society might better 'understand' the offense and oppose severe punishment.

   I add to these factors three of my own:
6. Since firms as such cannot be imprisoned but only fined, the maximum sanction is more likely to be binding. Of course, responsible managers or other employees can be imprisoned, but this might lead to different problems, as discussed in section 2.3.6.
7. Regulators and firms interact with each other repeatedly. While common crimes may be committed more than once, they usually do not imply frequent interaction with the same inspector or other enforcement officials. Repeated interaction makes compliance strategies possible and/or more favorable.
8. Since firms, especially the larger ones, are wealthier than individuals they are better able to contest enforcement actions (section 6.2). More generally, Garoupa (2004) argues that corporations can more easily 'corrupt' enforcers, regulators and judges, as well as politicians and the media. Corruption harms deterrence and might influence societal attitudes towards the offense and the standard.

8.4 Concluding public law enforcement: Economics and other social sciences

This section began with a description of the debate surrounding *deterrence* and *compliance* strategies. Non-economists have criticized the deterrent approach of economics. As a final reflection, I wish to make a number of comments on this debate. These comments summarize some important lessons to be learned from the chapters on public law enforcement.

8.4.1 *Some comments on the debate concerning enforcement strategies*

Economics and compliance
First of all, one-sidedly linking economics to deterrence strategies is inappropriate. As should be clear from the above analysis, economists do not necessarily favor a strategy in which every violation is punished. The (modern) economic analysis of public enforcement began with Becker (1968), who recommended maximal sanctions. However, the literature has moved far beyond this analysis. It has identified important limits to the trade-off between the probability of detection and the magnitude of the sanctions. Moreover, there are many reasons why enforcement should make use of informal standards, warnings, compliance records, negotiations and/or advice. Therefore, economists cannot be reduced solely to defenders of deterrence strategies.

As Becker (1968) notes, from an economic perspective the social objective of law enforcement is not deterrence, but minimizing the social costs of crime and enforcement (i.e.: maximizing social welfare). These include the costs to (potential) victims (the harm done and the costs of enforcement) and the costs to potential offenders (the costs of compliance). In choosing enforcement, we should balance the benefits of deterrence (less harm but higher costs of compliance) against the costs of enforcement. It should be stressed that (maximal) deterrence is not the objective of optimal economic policy.

Rationality and morality
Secondly, many compliance theorists assume that if a firm behaves rationally, a deterrence strategy is favorable, while if a firm is irrational, a compliance strategy is most effective. In addition, a direct relationship is made between rationality and morality. Rational firms are assumed to be amoral. For example, Kagan and Scholz (1984) and, following them, Huisman (2001*) make a distinction between three different types of firms and argue that for each type of firm a different enforcement strategy is effective. First of all, firms can be amoral calculators who are motivated only by profit-seeking. For such firms, a deterrent, aggressive approach is favorable and the regulator should behave as a policeman. Secondly, firms might behave like a civilian who is inclined to obey the law as long as it is reasonable. Law-breaking follows from disagreement with the regulations. In these cases, the regulator should behave as a politician, bargaining with and persuading the firm about the relevant and optimal law. Finally, the firm may be inclined to obey the law but is organizationally incompetent to do so. Law-breaking in such cases follows from organizational failure. The regulator should serve as a consultant, educating the firm about compliance.

From an economic perspective, this scheme does not make sense. It is useful as far as it provides a number of reasons for and forms of the compliance strategies discussed above. However, rationality does not exclude morality and is certainly not restricted to profit-maximization. As explained above, even for rational, profit-seeking firms, compliance strategies can be favorable, while moral inclination to obey the law is insufficient reason for failure of deterrence. Economics does not require any assumptions concerning the morality of firms, but explains the desirability of compliance strategies as the result of the relevant costs and benefits, i.e. the costs functions of expected harm, costs of compliance and costs of enforcement. These costs as well as their interaction (the effectiveness of enforcement in reducing offenses) determine which level of compliance is desirable and feasible and whether this level can be obtained by means of a deterrence strategy, consistently assuming that firms or individuals behave in a rational manner. The costs functions include the possible crowding out of moral motivations and the incentive to learn the relevant law and its application. The desirability of compliance strategies follows from a difference between deterrence and compliance strategies in social costs and benefits and in the incentives resulting from these strategies. For example, whether enforcement should include advice to the firms about compliance depends on whether the firm itself or a public agency is better at acquiring information about compliance.

The importance of marginal analysis

Thirdly, there are sometimes differences in interests between the different fields. While non-economists tend to largely explore the motivations to comply among the whole population, economists are more interested in a marginal analysis. The main interest of most economic literature is not with identifying the average or aggregate motives to comply and enforcing average compliance. The main interest is in how the marginal individual or firm can be induced to comply. In addition, (increased) enforcement by public authorities is often precisely the extra incentive required to induce the marginal individual to comply. This might explain why economists focus more on incentives from enforcement than on social or moral intentions. Precisely the lack of such intentions in the marginal individual explains why such individuals fail to comply. As discussed in section 2.2, the social and moral concerns are usually insufficient to maintain compliance. If this were not the case, the enforcement problem would be absent because the efficient outcome would already be implemented. Moreover, the fact that most firms do comply because of their moral intentions to do so is not sufficient to justify a compliance strategy. A deterrence strategy might be required to impose extrinsic penalties on marginal offenders.

A lack of knowledge and experience

Generally speaking, it is impossible to make resolute statements on the desirability of both a compliance and deterrence strategy because of the limited empirical evidence concerning the explanation of regulatory behavior and concerning the effects of enforcement strategies.[132] Empirical research is hampered by the lack of data, especially on non-reported violations. Moreover, it is inappropriate to make firm statements if one of the strategies has never been tried. For example, Van Stokkum (2004*) quotes others on the fact that in the Netherlands, a truly punitive enforcement has never been seriously tried.

Enforcement failures

So far it was assumed that the enforcement authority tries to deal with the enforcement problem as well as possible, i.e. to maximize social welfare. However, the enforcement authority may also fail in achieving deterrence. This is more extensively discussed in chapter 9. The observation that the law is not vigorously enforced does not necessarily show the ineffectiveness of deterrence strategies, it might also indicate government insincerity and capture, if not corruption. Alternatively, deterrence strategies may fail to be implemented because bureaucrats are mostly interested in budgets and employment, while deterrence strategies as described by Becker (1968) require few resources.

8.4.2 *The appropriate distinction between compliance and deterrence strategies*

As can be seen from the discussion in section 8.1, the debate about compliance and deterrence involves many factors, such as morality and rationality, the timing of enforcement (ex-post or ex-ante; re-active or pro-active), information problems, alternative regulation etc. Therefore the debate about deterrence and compliance strategies is often confusing and does not adequately describe the difference.

The *labels* of a deterrence and a compliance strategy are confusing because they do not adequately address the difference. Both enforcement strategies are aimed at improving compliance (and deterring violations), or even better at the goal behind compliance, namely maximizing social welfare. What differs is the way in which they attempt to obtain this. In a deterrence strategy, this is carried out by means of directly punishing violations. In a compliance strategy, it is carried out by means of negotiations. When a violation is detected, no sanction is imposed, but instead some kind of warning is issued. Persuasion is by information rather than through sanctions.

The distinction between compliance and deterrence strategies is often analyzed confusing because it compares several factors simultaneously. For example, ex-post criminal law enforcement is compared with an ex-ante com-

132 See for example Huisman and Beukelman (2007*) and Ponsaers and Hoogenboom (2004*). The latter state (p.179) that, due to a lack of empirical knowledge, "the discussion breaks down in an ideological war of positions rather than an empirical founded, rational argumentation".

pliance strategy (Ponsaers en Hoogenboom, 2004*), and/or ex-post enforcement is assumed to be repressive. However, ex-post is nothing more than enforcement after some damage has actually occurred (harm-based) while ex-ante enforcement is enforcement of standards that attempt to prevent damage (act-based). There is no direct relationship between the distinctions ex-post/ex-ante and compliance/deterrence. For instance, remedying damage with a compliance strategy is ex-post, while strictly enforcing automobilist speeding is ex-ante. The distinction ex-post/ex-ante might explain some of the characteristics that are usually assigned to a compliance strategy but may as well be applied to a deterrence strategy and vice versa. When enforcement takes place ex-ante, there are no clearly identifiable victims since no damage has occurred. Ex-post enforcement is re-active (which is not the same as repressive) while ex-ante enforcement is pro-active.

It is important to limit the distinction between deterrence and compliance strategies to a distinction between the enforcement *styles*. The style describes what an enforcement authority will do once it detects a violation. A deterrence style is a style in which every violation is immediately punished and the enforcement authority bases its decisions solely on the information it has obtained itself. A compliance style is a style in which the enforcement authority might forego some violations (for some time) if this induces the firm to reveal important information necessary for enforcement. In addition to the enforcement style, the enforcement authority must also decide how the inspection process will take place (monitoring versus investigation) and whether it will be harm-based or act-based (ex-post or ex-ante). These three factors together determine the enforcement policy and they should not be confused or connected.[133]

8.4.3 Conclusion: Integrating different views

Note finally that economic and non-economic views on enforcement may not diverge as much as they seem to at first sight. What should be clear from this chapter is that economics has moved far beyond Becker (1968). Economists do not necessarily favor harsh, deterrent, legalistic enforcement, but have broadly investigated the conditions that determine the optimal policy, i.e. the most welfare-enhancing policy. Their conclusions concerning optimal enforcement policies do not always deviate from those of other disciplines. The economic description and explanation of deterrence and compliance strategies fits fairly well in the discussion at the beginning of this section. Both economists and non-economists generally favor a pyramid of enforcement in which the most severe enforcement action is saved for the most severe violations and most persistent violators. Moreover, the three important determinants of an enforcement policy as identified in Kagan and Scholz (1984) and Huisman

133 Chapter 10 further investigates the optimal structure of enforcement.

(2001*) can be interpreted economically. First of all, enforcement should be reasonable, which in economic terms implies that it should balance the social costs and benefits. Secondly, it should take into account whether the firm is willing to comply, which in economic terms refers to the costs and benefits of internal motivations and external incentives from the environment. Thirdly, it should take into consideration whether the firm is able to comply, which in economic terms refers to the formulation of standards and technological barriers. An important contribution that economists can make to this discussion is their investigation of the interaction of these factors with and the dependency of these factors on enforcement itself.

9 Discretion, credibility and objectives of the enforcement authority

The previous chapters describe the socially optimal enforcement policy which enforcement authorities should adopt in order to obtain socially optimal levels of precaution. Two important assumptions were made that have not, so far, been discussed. First of all, it was assumed that the enforcement authority can commit itself to the enforcement policy that it has announced. Secondly, it was assumed that the enforcement authority will seek to maximize social welfare. Both assumptions may be problematic and are therefore examined in this chapter. Essential to this analysis is the recognition that the decision to prosecute a violation is not fixed, exogenous, but depends on the firm's decision to comply (Tsebelis, 1989). The enforcement authority is a strategic player with its own objectives and incentives. Therefore, the firm's decision to comply is the result of a strategic interaction between the firm and the enforcement authority.

Section 9.1 investigates the credibility problem. Section 9.2 investigates the objectives of law enforcers. Section 9.3 discusses the role of courts in controlling enforcement authorities.[134]

9.1 Credibility and commitment

It cannot be taken for granted that the enforcement authority will be able to commit itself to the efficient enforcement policy. First, commitment is difficult if circumstances have changed so that expected harm and compliance costs have changed and so that punishing or remedying non-compliance is no longer efficient (section 9.1.1). Secondly, such commitment may be impossible because enforcement is costly (section 9.1.2).

9.1.1 *The problem of commitment when the costs and benefits of non-compliance change*

Kydland and Prescott (1977) show that it is sometimes better not to allow policy makers discretion to act in accordance with what is in society's best interest, but to provide them with long-term rules which are difficult to change.

134 The analysis in this chapter is presented in Suurmond and Van Velthoven (2006*) while discussing the desirability of a duty to enforce for municipalities [*beginselplicht tot handhaving*].

Otherwise, a time-inconsistency, or credibility problem may arise. A credibility problem arises when citizens realize that the government may, in future, use its discretion to make a new policy decision which differs (for good reasons) from the one in place when the policy is announced. This problem is well-known in macroeconomics. Kydland and Prescott (1977) argue that the best outcome can be achieved if policy makers are not given discretionary powers to determine the best policy for every period, but (fixed) rules, as for instance is the case with an independent central bank which is solely concerned with fighting inflation.

Credibility problems also occur in the case of enforcement policies. Consider the following example: suppose municipal by-laws prescribe that buildings must be built in a fire-proof manner which requires that an emergency exit can be reached within 50 meters. It will probably be efficient for café proprietors to comply with this rule when they decide to build or rebuild an establishment, because the extra costs of compliance are small relative to the reduction in expected harm. Violation would lead to a loss in welfare. However, once the café is built in such a way that an exit can only be reached within, say, 60 meters; a different balance of costs and benefits is achieved. In order for the proprietor to comply, he will first have to break down the existing walls and exits, and then to rebuild the café in such a way that it satisfies the requirements for emergency exits. In the meantime, the café will be closed, while the expected harm is only reduced by "10 meters". It may well be that these costs of compliance are higher than the benefits, so that requiring compliance would in this case not be efficient. As a result of this balance of social costs and benefits, the municipality may refrain from enforcement.

However wise it may be to refrain from enforcement ex-post, the deterrent effect of enforcement is seriously weakened. If proprietors know and expect that the municipality will not enforce the rule of exits within 50 meters, they will probably not comply with this rule. If citizens expect that the enforcement authority will, in the end, not enforce the rules, they will take this freedom to act against the rules. This is socially undesirable if the offenses in question are ex-ante inefficient and the administration ex-ante wants to enforce the rules, but cannot commit itself to the promised policy. It may be difficult for enforcement authorities such as municipalities to carry out enforcement based on ex-ante costs and benefits because citizens are interested in current, ex-post, results. The tendency to refrain from enforcement will be higher when the ex-post costs of compliance are high, the competition for the limited enforcement budget is larger and offenses occur infrequently. Then social pressure against enforcement ex-post will be large, due to for instance the fear of loss of employment or loss of consumption of pleasant activities, and/or due to the underestimation by citizens of the expected harm that is inflicted.

Situations in which such credibility problems occur are very common in environmental law, especially building law, where remediation is a relevant policy objective. With remediation, the costs of compliance are higher after an offense than before.

9.1.2 *Commitment when enforcement is costly*

A credibility problem is present when the decision to tolerate a violation is different ex-post and ex-ante because there has been a change in expected harm and/or compliance costs once an offense has been committed. However, the credibility problem is much broader and, in principle, concerns every offense. Once an offense has been committed, the enforcement authority will not want to realize costs of enforcement because the welfare loss of the offense can not be made undone (Boadway et al., 1995; Baker and Miceli, 2005). Once a criminal has decided whether to commit a crime, given the government's announced punishment plan, the social benefits from the enforcement policy will have been completely realized with respect to that individual or firm. The reason for realizing enforcement costs is that enforcement creates deterrence of potential offenses that might be committed in the future. Therefore, the announcement of enforcement is only credible if society and the enforcement authority are sufficiently interested in the future. Boadway et al. (1995) demonstrate that if the future is sufficiently important, the enforcement authority will have an incentive to build a reputation of strict enforcement. If the enforcement authority is not sufficiently interested in deterring future crime, it may refrain from enforcement activities ex-post.

Baker and Miceli (2005) show that credibility is especially problematic if detection and imposing sanctions are costly. Consider a static policy (i.e. without credibility problems) of some optimal probability of detection. This policy will result in enforcement costs in realizing this probability. When sanctioning happens by means of fines, the optimal dynamic policy (which takes into account the problem of credibility) causes the optimal probability of detection to fall; hence crime rates will rise, reflecting the greater weight society attaches to minimizing current enforcement costs compared to deterring future crimes.[135] If, however, jail is included as a possible punishment, credibility can result in either more or less aggressive detection and punishment policies, as compared to the optimal static policy. The optimal strategy depends on which policy is more likely to lower the expected enforcement costs.[136]

Tsebelis (1989) obtains similar results.[137] However, he does not consider

135 Similar results were derived in section 3.4 concerning self-reporting. Malik (1990, p. 243) recognizes the assumption that the enforcement authority is committed to its announced policy even though it realizes that the firm must have made an honest report. Friesen (2006) discusses the optimal enforcement policy when the enforcement authority updates its belief about the probability of a violation on the basis of the reports it receives. This leads, as in Baker and Miceli (2005), to a smaller probability of inspection because the enforcement authority believes that compliance-reporting firms are more likely to be compliant. See also Franckx (2002).

136 Along the lines discussed in section 3.2 that under imprisonment both under- and overdeterrence may occur.

137 See also Bianco, Ordershook and Tsebelis (1990) and Chang et al. (1998).

the situation in which the enforcement authority and the firm act sequentially (because detection takes place through inspection), but simultaneously (detection by monitoring). The enforcement authority and the firm have to simultaneously decide whether to inspect the firm respectively whether to comply with the law. The enforcement authority may prefer to enforce the law if it is violated, but due to the enforcement costs the enforcement authority will prefer not to enforce the law if not violated. Under these circumstances, the optimal strategies of the enforcement authority and the firm are interdependent. If the authority enforces the law, the firms will stop violating assuming that the expected sanction is high enough; if firms stop violating the law, the authority will stop enforcing it; if the authority stops enforcing the law, the firms will violate it; if the firms violate the law, the authority will enforce it; if the authority enforces the law, then the firms will stop violating it, and so on. Tsebelis (1989) shows that under these circumstances an equilibrium can only be found in mixed strategies, i.e. if the enforcement authority carries out inspections with some probability, and if the firm violates with some probability.

The results of Tsebelis force us to recognize that the probability with which an enforcement authority takes action is the result of a rational calculation, and not an exogenously given probability. Therefore, the decision of firms to comply given the probability that they will be inspected is not a simple decision under uncertainty. The probability that the firm will be inspected is not an exogenous value, which can be determined by the government at the desired level, but it results from the interaction between the firm and the enforcement authority. Moreover, as Tsebelis (1989) discusses, an increase in the sanctions on non-compliance might – in equilibrium – not lead to an increase in the compliance rate, but to a decrease in the probability of inspection. In the short run, higher sanctions may lead to some deterrence, resulting in a higher compliance rate. However, once the enforcement authority observes that the law is better adhered to, it will decrease its enforcement actions, so that deterrence is weakened and the initial compliance rate is restored. If society wants to obtain a higher compliance rate, the pay-offs to be modified are not those of firms (through sanctions), but those of enforcement authorities. It is therefore important to discuss the incentives and objectives of enforcement authorities, as well as the discretion given to these authorities.

9.2 Enforcement authorities' objectives

Up till now, it has been assumed that law enforcers try to maximize social welfare given the rules, the allocated budget and the available sanctions. In addition, the assumption was that budgets are allocated with the goal of maximizing social welfare. However, decisions concerning enforcement policies and actions are not made by "society", but by bureaucrats and politicians. Politicians allocate a budget, control the detection and prosecution policy, and – in their role of law-makers – determine both the rules and the maximum level and design of sanctions. Given the budget and the rules, a law-enforcement authority stipulates a detailed enforcement plan, including priorities and a production level, and carries responsibility for executing the enforcement policy.

As law enforcers also happen to be human beings, they are not necessarily seeking efficient enforcement, but may have different, even counterproductive, objectives. "A law is enforced, not by "society", but by an agency instructed with that task. That agency must be given more than a mandate (...) to enforce the statute with vigor and wisdom. It must have incentives to enforce the law efficiently" (Stigler, 1970, p.531). Therefore, this section discusses the objectives of law enforcers as well as the financing issues they face.

9.2.1 *The objectives of law enforcers*

Cohen (1999) and Firestone (2002) have identified six different theories concerning the behavior of law-enforcement officials:

1. *Social welfare maximization*. As assumed up till know, this objective implies that the enforcement authority compares the social costs and benefits of compliance and enforcement.
2. *Violation minimization*. Under this objective, the enforcement authority tries to reduce the number of offenses (or maximize the compliance rate), without reference to the costs and benefits of compliance. The enforcement authority is interested in achieving (general) deterrence. This objective corresponds to an authority with a limited enforcement budget which must compete with other functions of the authority in question. Given the budget, the authority tries to maximize compliance. The enforcement style best fitted to this objective is a deterrence strategy.
3. *Harm minimization*. The enforcement authority tries to maximize the benefits of compliance without reference to the compliance costs. It allocates its limited budget to action taken against offenses which cause relatively high degrees of harm. In contrast to the goal of violation minimization, which involves focusing on 'easy' enforcement targets even if these yield relatively little harm, an enforcement authority whose goal is harm minimization focuses on firms with the highest harm reduction per euro enforcement effort. However, this type of authority tends to ignore the costs of compliance.
4. *Case maximization*. The enforcement officials behave like (self-interested) bureaucrats who try to maximize budgets (following theories of bureaucratic behavior inspired by Niskanen). This implies excessive inspections instead of the minimizing of inspections proposed by Becker (1968), although larger budgets may also be used to create 'paperwork' instead of

carrying out actual enforcement actions. Such authorities will probably follow a compliance, rather than a deterrence strategy.

5. *Political support maximization*. The enforcement authority tries to maximize the difference between the number of supporters and the number of detractors of its enforcement policy (following for example Pelzman). For example, the authority puts the least regulatory burden on well-organized, concentrated parties. The possible consequences for employment may be an important argument for decisions on enforcement actions. Less stringent enforcement is demanded when the compliance costs are high and there is a danger of plant closing and loss of employment.
6. *Bias*. Collective decision-making may be decided by the median voter. Therefore, enforcement policies are designed so that they most closely match the needs of the median voter. For instance, the representative voter is not informed about enforcement effort and compliance costs and will therefore choose relatively stringent regulation and enforcement. On the other hand, voters are also less able to observe the extent to which the government is enforcing its standards. We might therefore expect more stringent regulations than are socially optimal, and less than stringent enforcement to compensate.

9.2.2 *The financing of enforcement authorities*

The goal of enforcement officials is also determined by the financing and evaluation of the enforcement authority. If the "success" of the enforcement authority is measured by the number of cases commenced, the authority is more likely to maximize the number of cases. Similarly, if the budget of the authority is based on the budget or costs of the previous year(s), the authority may have an incentive to increase the number of cases, hence increasing its budget. In addition, the extent to which enforcement is "politicized" depends on the institutional environment. Generally, public prosecutors and courts are quite independent. Regulatory agencies may also behave as independent bodies. On the other hand, administrative enforcement by the board of mayor and aldermen might be highly influenced by citizens' opinions.

Marette and Crespi (2005) discuss three different financing systems of enforcement authorities. First of all, the authority may be financed by general taxes. Secondly, the enforcement authority may be financed by user (or: monitoring) fees, paid by all potential violators. All regulated firms pay a fee regardless of their actions. Thirdly, an enforcement authority may be financed by the benefits of the fines paid by violating firms. Only firms that violate the rules pay. Marette and Crespi (2005) show that if it is expected that there will be some non-compliance, it is optimal to cover the authority's enforcement actions by penalties such as fines. This provides an optimal incentive to comply because non-complying firms also pay for the enforcement costs. As the monitoring costs increase and a higher budget is required, the financing has to be augmented with a lump-sum tax, rather than a fee. A fee would result in a negative incentive for complying firms. If, however, all firms are expected to comply with the standards, clearly no money can be collected through a penalty. In this case, a per-firm fee would be the optimal method of regulatory financing. As the monitoring costs increase, the financing has to be augmented with a lump-sum tax.

These results may also help to understand the kind of enforcement authority which may emerge. Suppose that a public authority cannot receive penalties and a private self-regulatory body cannot receive public taxes. It then follows that if only fees are required to finance the authority, there will be no difference between a private and a public organization. Both will be able to collect the necessary budget. It may be efficient to rely on private self-regulation if the per-firm monitoring cost is relatively low. However, if no penalties are raised, and hence no money can be collected in this way, a public agency would be better because its financing can be supplemented with taxes. Marette and Crespi (2005) also discuss the fact that the way a given authority is financed may help explain which enforcement style they choose. It may make perfect sense for an enforcement authority to allow some firms to remain in violation if the revenue generated from these firms in the form of fines can be used to finance inspections for the rest of the industry in a way that encourages more efficient firms to mitigate their pollution. In other words, a pure command-and-control scheme (deterrence strategy) in which all firms must comply with the same standards may be suboptimal as compared to a flexible-financing scheme (compliance strategy) in which a regulator will use the revenue generated through fees and penalties on a subset of firms, in order to pay for a program that will raise overall welfare.

An important warning must be made at this point. Marette and Crespi (2005) assume that the enforcement authority is acting in the public's best interest, given the budget constraint. However, as discussed above, enforcement authorities do not always maximize social welfare. Moreover, as shown in for instance Benson et al. (1995), the financing system may also create perverse interests and seduce the authority to seek other goals than social welfare. Benson et al. (1995) show that the US police resources allocated to drug enforcement haven risen dramatically since 1984, because local police bureaus which cooperated with federal drug enforcement authorities have a share in the money and/or property confiscated from drug criminals. As a result, police attention to other types of crimes, such as property crimes, was reduced, leading to an increase in these types of crimes.[138]

9.3 THE ROLE OF COURTS IN CONTROLLING LAW ENFORCERS

The previous sections identify two problems with respect to enforcement

138 See also Benson et al. (1998).

authorities. First of all, it may in practice be impossible for enforcement authorities to commit to the policies they announce. Secondly, enforcement authorities and their employees may not be trying to maximize social welfare. This section investigates how these problems can be solved. Special attention is given to the role of courts in controlling enforcement authorities.

9.3.1 *The optimal level of discretion*

Discretion for enforcement authorities is needed because law is necessarily incomplete (section 6.1) and an enforcement authority should balance the desirability of its enforcement actions. However, discretion also has its disadvantages.

Consider the time inconsistency problem discussed in section 9.1. An enforcement authority which adopts an enforcement policy involving strict enforcement may be unable to commit to this policy ex-post. This credibility problem can be mitigated by limiting the discretion of the enforcement authority. If the enforcement authority is not granted discretion to act in society's best interest, but is instead given a strict enforcement rule (such as violation or harm minimization), time-inconsistency may be less of a problem. Although such rules are less flexible and force the authority to also take action against efficient offenses, overall social welfare may in the end be higher.

Moreover, I argue in chapter 6 that under certain circumstances, the optimal enforcement policy is to target enforcement resources and to try to maximize compliance (although full compliance will never be achieved). Generally, however, inefficiencies result if enforcement officials do not maximize social welfare. Discretion allows enforcement officials to exercise their powers in seeking their own objectives. Ignoring the motivation of enforcement authorities might lead to the wrong outcome in making normative policy descriptions (Cohen, 1999).

Therefore, the optimal level of discretion will balance the flexibility required to evaluate the desirability of enforcement with the problem of time-inconsistency and the problem of abuse of powers. There are two ways in which the discretion of enforcement authorities can be controlled. Ex-ante, the actions of the enforcement authority can be constrained by the rules. Ex-post, the actions can be controlled by courts.[139]

---

139 In Suurmond (2006*) I discuss the optimal level of discretion for the Ministry of Justice in taking precautions against fire damage in prisons. This issue came to the fore after a fire in a detention centre at Schiphol airport killed 11 people. In the article in question, I also present another solution for controlling the discretion of politicians and bureaucrats: independent investigation boards which examine accidents and investigate whether rules and procedures are obeyed.

Ex-ante discretion by rules

First of all, the level of discretion of enforcement authorities is determined by law. If the law were complete, no credibility problems would arise because the law would prescribe exactly which offenses are inefficient and would force the authority to take action against all offenses. An enforcement authority which, given a sufficiently high budget, will try to minimize violations or expected harm, would be efficient.

However, the law is generally incomplete, and it therefore leaves room for interpretation of rules, facts and circumstances. It is important to take into account the costs and benefits of compliance and enforcement. This requires discretion for enforcement authorities. Discretionary powers allow the enforcement authority to consider whether an offense is efficient. Moreover, in cases of underdeterrence, welfare may be improved if the enforcement authority uses informal standards and warnings (section 8.2). On the other hand, incomplete rules may create credibility problems. A rule stating that a proprietor should use his establishment in a fire-proof way may be subjected to more time-inconsistency problems than a rule stating that the proprietor is not allowed to hang any decorations, although the latter is less flexible and applies to all firms in the same way (Viscusi and Zeckhauser, 1979). Moreover, the more rules are open to interpretation, the more enforcement authorities are able to use this room to seek their own objectives. There is therefore a trade-off between the informational advantages of discretion – that law enforcers can act on information not included in the rules – and the disadvantage of discretion – that law enforcers use this discretion to seek objectives that are different from society's (Shavell, 2005).

Discretion can also be controlled by, for instance, minimum (and maximum) penalties prescribed by law.

Ex-post control of discretion by courts

Given the level of discretion sustained by the rules, a second control of discretionary powers takes place through the courts. Credibility problems can be reduced by victims and courts. If a crime leads to serious social concern and disapproval, the social pressure to prosecute an offense will be large and the enforcement authorities will be forced by this pressure to take action. Therefore, credibility problems are more serious in (act-based) administrative law enforcement than in (harm-based) criminal law enforcement.

More importantly, the credibility problem can be reduced if an independent court uses a duty to enforce (following a request of the victims). For example, in the Netherlands, in cases of environmental law, if a municipality decides not to take action against a violation, (potential) victims can ask the court to convict the enforcement authority and force it to take action. Such a duty to enforce the law in fact reduces the discretion that enforcement agencies have in deciding whether or not to enforce. As a consequence, potential violators know that they cannot get away with violations of the rules, but that the enforcement authority will be forced to take action against violations (even if it does not want to do so), and potential violators will therefore comply with the law. A duty to enforce may thus be beneficial when the enforcement authority would use her discretion to be lax on violations as a consequence of the credibility problem. Again, such a duty to enforce also has an important disadvantage. Since the law is necessarily incomplete and not all violations are inefficient, a duty to enforce all offenses will also imply that inefficient compliance is enforced. A rigorous duty to enforce will force the enforcement authority to take action regardless of the social costs and benefits of enforcement.

Therefore, the optimal level of the court's (marginal or full) evaluation of the decisions of enforcement authorities and hence the optimal level of discretion granted to the enforcement authority balances the problem of time-inconsistency with the flexibility required to evaluate the desirability of enforcement. Note that it is important in this case for the court to be independent from politics (as in the case of the central bank in Kydland and Prescott, 1977), so that it is not subjected to the same credibility problem as the enforcement authority.[140] Bishop (1990) explains that this type of independency can arise because the judiciary faces different incentives than enforcement authorities. "The constraints that hedge a judge tend, by accident or design, to the same end: confining anyone appointed to the bench to one, and only one, role – to life as a judge. (…) to remove the possibility of using the office for any purpose other than acquiring judicial fame – that is, the approbation of judges and others whose esteem is valued" (p. 492).

Courts may also be helpful in controlling enforcement authorities' decisions because enforcement authorities abuse their discretion to seek their own objectives. If enforcement officials are reluctant to enforce, a duty to enforce the rules may improve social welfare. Moreover, this may improve rule-making since society is confronted with the obligation to really enforce the rules it imposes. If, however, enforcement officials are too much inclined to enforce the rules, a duty to enforce will be ineffective in controlling enforcement officials.

The next subsection elaborates further on the relationship between courts and regulators. First, the arguments of this subsection are further analyzed by considering whether social welfare is always best served by an independent enforcement authority or whether it might be optimal to have non-welfare-maximizing enforcement officials, even if their discretion is not controlled for. Secondly, what, if any, is the optimal level of control of law enforcers' actions by courts?[141]

---

140 Feld and Voigt (2003) have shown that the arguments of Kydland and Prescott (1977) can analogously be applied to the enforcement of private property rights. Ex-post, there is a danger that the government will expropriate these rights. Only if the citizens have sufficient confidence in an independent court will the promise not to expropriate their property rights be credible.

141 Bishop (1990) provides the most extensive theory on the role of courts in regulation.

9.3.2 *The relationship between courts and regulators*

In section 9.2, several objectives and incentives of law enforcers have been identified. One step further would be to examine which type of enforcement authority and which incentives and goals lead to the socially best results. Who should enforce a legal rule? In reality, we can observe different types of enforcement authorities. For common crime, enforcement is in the hands of a general public prosecutor and court who are instructed to balance the costs and benefits of taking action. For regulatory crime by firms, there are several national specialized agencies which enforce certain kinds of regulation, such as competition regulation or environmental regulation. At the local level, administrative enforcement is in the hands of administrative bodies such as the Board of Mayor and Aldermen.

These different authorities may differ in their objectives. Several arguments have already been offered to justify enforcement authorities having aims that diverge from social welfare maximization. If the credibility problem is severe, it may be optimal to create an enforcement authority which tries to minimize offenses or the harm from offenses. Enforcement can best be delegated to a specialized regulatory body with the specific task of detecting, prosecuting and convicting offenses, rather than to a general public prosecutor and court who are instructed to balance the costs and benefits of taking action. Such an agency can also be efficient when the maximum feasible compliance rate is a partial compliance rate. In such cases, targeting enforcement resources and trying to maximize compliance will lead to the best results.

In principle, society would wish to have a social welfare maximizing enforcement authority, for example a very professional and motivated judiciary which will have both a sufficient incentive to investigate a case and a strong interest in doing justice (Glaeser et al., 2001). However, such a perfect welfare maximizing authority may simply not exist.

Courts versus regulators

It may be impossible to have very professional and motivated law enforcers, because law enforcers also have their own objectives. Doing justice is personally expensive because it requires investigation into the facts, circumstances and laws of the case. Judges must be able, and more importantly willing, to read complicated contracts, to verify the events and clauses and interpret broad and ambiguous language. Society only has limited control over the incentives of law enforcers (Glaeser et al., 2001; see also Bishop, 1990). "Doing justice" is largely unverifiable. Many rewards for carrying out justice are intangible (such as self-esteem, or respect from one's peers). The government can only improve this reward in the long run. In reality, in many countries, the courts are underfinanced, unmotivated, unsure as to how the law applies, unfamiliar with economic or technical issues, or even corrupt. Such courts cannot be expected to engage sufficiently in costly verification of the facts and the laws. On the other hand, the government has the ability to politicize the enforcement of particular legal rules by rewarding the enforcers for certain outcomes such as finding violations (through future career concerns or budgets).

Generally, the government must choose between two alternatives for delegating the enforcement of a rule: judges and regulators (the specialized agencies described above). There are two important differences between them (Glaeser et al., 2001). First, regulators can be more easily provided with an incentive to punish violations of particular statutes. Judges, in contrast, are by design more independent and therefore harder to motivate. Regulators may be better motivated to invest in understanding the laws and circumstances of a case, but are also more likely – if over-motivated – to reach politically desirable decisions at the expense of doing justice.[142] A second distinction between courts and regulators is that regulators may have lower costs of investigation because they are more specialized.

Therefore, the best enforcement strategy – particularly when investigations are personally expensive – may be to have a regulator. Providing the regulator with a sufficiently strong incentive to punish violators may induce them to search for evidence and convict on the basis of this evidence, while courts may decide not to search for evidence and acquit regardless of violations. However, there is also a danger of abuse of power by regulators. If they are over-motivated, they might choose not to search for evidence and to punish regardless of violations.

Note that these results are consistent with the observed enforcement practices. Regulatory inspectorates which are responsible for the regulation and enforcement of a certain rule are mostly found in administrative law enforcement for firms. This is due to the fact that doing justice is more important in criminal law enforcement and therefore requires more independent judges. In addition, in regulation of firms, specialization is more important.

Courts and regulators

The analysis in Glaeser et al. (2001) focuses on whether enforcement should be in the hands of courts *or* regulators. Generally, enforcement is shared by regulators and courts together. Even if a regulatory agency is qualified to impose penalties, a firm can always appeal against this decision in a court. This is important because this control by courts deters enforcement authorities from abusing their discretion to seek their own objectives. Moreover, appeal procedures may help to correct enforcement errors as well as credibility problems.

There are three different legal procedures with respect to the relationship between regulators and courts (Jost, 1997b) when the regulatory agency is in charge of the detection of offenses. (1) Conviction takes place through a court trial. The regulatory agency is instructed to detect offenses and, if there is sufficient evidence, to prosecute the case in a court trial. (2) The regulatory

142 Note that regulatory bodies may also be 'captured' by the regulated industry so that they will not, in contrast to courts, be willing to either enforce or to search for information.

agency may have the power to penalize a violating firm. A firm which has been penalized by the agency has the right to lodge an appeal. The question whether the firm complied will be reconsidered. There are two different procedures at this point. (2a) The case is decided by the court, and the regulatory agency is required to provide sufficient evidence for its decision. (2b) The agency reconsiders its decision and decides whether to press charges in court.

Suppose the regulator can choose between two levels of investigation. A full investigation will certainly reveal (non)compliance. Alternatively, the regulator can choose a lower-scale type of investigation, called preliminary investigation, which will provide insufficient, sufficient or strong evidence for an offense. Under system (1), the regulatory agency will choose for a full investigation of firms. Otherwise, it will not be able to prosecute the case in a court trial. Given a full investigation, the detection of an offense will lead to prosecution and conviction. In system (2), the agency can save on enforcement costs by choosing only preliminary investigations, because the procedural and evidential requirements for penalties are lower for regulators than for courts. The regulator can punish a firm if there is sufficient or strong evidence of an offense. In (2a), the regulator is required to fully investigate the case if the firm appeals against its decision to penalize. A firm will lodge an appeal if it did in fact comply with the law, since further investigation will perfectly reveal its legal behavior. If the firm was non-compliant, it will not appeal because that would prove its offense. Therefore, for guilty firms, the regulator only has to carry out the relatively cheap preliminary investigations. On the other hand, for innocent firms, the regulator must perform additional investigations and prepare a court trial.[143] If the costs of additional investigations are low and the accuracy of preliminary investigations sufficiently high, the regulatory agency will sanction an individual if its preliminary investigations provide sufficient or strong evidence of non-compliance. The savings in enforcement costs can be used to achieve a higher compliance rate. This may be especially helpful if the compliance rate is low due to institutional conditions (such as limited budgets or low maximal sanctions), because the number of appeals will be lower if the compliance rate is lower.[144] In (2b), the situation may be somewhat different. If a preliminary investigation provides strong evidence, the regulator never drops the case, but carries out a full investigation once a firm has lodged an appeal. If a preliminary investigation provides sufficient evidence, the regulator will drop the case if the costs of additional investigation are high and the reliability of the investigation low. The system in (2b) saves costs as compared to (2a) and it is therefore always preferred over (2a). From an economic point of view, it is preferred that the regulatory agency is allowed to make cost considerations during the proceedings.

143 Note that it is assumed that the regulator can commit to its policy.

144 The regulator's policy under (2a) is more extensively discussed in Jost (1997a). See also Reinganum (1988) and related literature.

Courts plus regulators

The co-existence and relationship of regulators and courts can be characterized in two ways (Garoupa, 2004). Firstly, the enforcement of regulators and court may be interdependent: firms may want to dispute regulatory outcomes and consequently appeal to the courts. This is efficient for two reasons. First, it corrects for errors ex-post and a reduction in enforcement errors enhances efficiency. Secondly, it can produce an adequate incentive for regulators to perform efficiently ex-ante, for instance in order to mitigate credibility problems or abuses of power.

A second justification, not discussed so far, for the existence of courts in addition to regulators is that courts and regulators may exist as independent enforcement authorities at the same time. Both have their own procedures and may impose penalties. The imposition of a criminal sanction on top of a regulatory penalty (and violation of *ne bis in idem*) may be efficient if (a) a regulatory penalty is relatively cheap to impose, and (b) there are high agency costs due to delegating law enforcement, many legal errors are made by regulators or there is collusion between the regulator and the regulated industry, i.e. if regulatory enforcement is weak but sufficiently cheap to still have both courts and regulators (Garoupa and Gomez, 2004).

9.4 Conclusion

The previous chapters discuss several reasons why Becker's (1968) recommendation to impose maximal sanctions may not be efficient. This chapter covers an additional reason, namely that maximal sanctions can lack credibility because the enforcement authority is unable to commit itself to imposing severe sanctions on violations that are regarded as minor offenses. Moreover, this chapter offers an explanation for why maximal sanctions, even if they are efficient, are not observed in practice: Law enforcers may have no incentive to minimize budgets and/or impose maximal sanctions because their objectives are better served with other enforcement policies.

Since law enforcers are also human beings, they do not necessarily seek efficient enforcement, but may have different, even counterproductive, objectives. This may, of course, harm social welfare. However, this chapter argues that one-sided objectives of law enforcers may be beneficial for two reasons. First, a social welfare-maximizing law enforcer may be unable to commit itself to an efficient enforcement policy, leading to the toleration of ex-ante inefficient, but ex-post efficient violations. Therefore, social welfare may be improved if law enforcers are instructed to enforce violations regardless of individual circumstances. Secondly, social welfare may be better served by a biased, political regulator than an independent judge, because the first is more inclined to invest in acquiring the necessary information in a given case.

10 The optimal structure of law enforcement

This chapter concludes on the theoretical analysis of the optimal enforcement method. The previous chapters analyze whether and how private or public law enforcement can induce optimal precautions. This chapter compares the alternative enforcement methods of private, criminal and administrative law enforcement. Section 10.1 starts with addressing several distinctions that can be made between these methods, namely the stage of intervention, the type of public law enforcement and the initiating party. In section 10.2 I examine the joint, simultaneous, use of private and public law enforcement. When and why is this joint use efficient? Should this use then be independent or interdependent? In section 10.3 I summarize the arguments for the choice between a compliance and a deterrence strategy. Section 10.4 summarizes the theoretical part by providing the conditions under which private, criminal and administrative law enforcement are effective and efficient in inducing precautions. This summary makes clear that for the problem of enforcing fire safety in horeca establishments a mix of enforcement methods is likely to be efficient. However, how the costs and benefits should be balanced and exactly which mix is optimal, requires an empirical analysis.

10.1 Analyzing and comparing enforcement methods

There are three basic ways in which an appropriate sanctioning risk can be realized: through private, through criminal, or through administrative law enforcement. A comparison of these alternatives implies comparing several distinct characteristics. Firstly, they differ in the stage of intervention. Secondly, there is a difference between the two types of public law enforcement: criminal or administrative. Thirdly, there is a difference between private and public law enforcement. These are discussed below.

10.1.1 Stage of intervention: prevention, act-based or harm-based sanctions?

The different stages of enforcement

Shavell (1993) discusses the fact that there are three basic stages of enforcement.

Firstly, intervention may take place before an act is committed by means of prevention of the act. This prevention rests on physical force or something close to it. A truck with dangerous substances may be prevented from driving through a tunnel at the toll-gate. A speed ramp may prevent automobilists from speeding. A fence may prohibit entering a forbidden territory. Denial or withdrawal of a license, or imprisonment may prevent firms or individuals from undertaking an activity and causing danger.

Enforcement relies on act-based sanctions if it takes place after an act has been committed (or refrained from), but before harm results, or independently of whether it does so. For example, a restaurant that is fined because it fails to maintain its emergency lighting and marks in working order (independently of whether there is a fire). Or an automobilist who is fined for speeding irrespective of whether he causes an accident. Enforcement actions are not aimed at preventing the restaurant from operating nor the automobilist from driving, but at guaranteeing that precautions are taken to prevent harm.

Enforcement takes place through harm-based sanctions if intervention takes place after harm has occurred. Examples are when a victim sues a restaurant owner for the damages that result from a fire in his establishment, or the automobilist for a car accident. Another example is when public enforcement is started after a murder is reported. Harm-based sanctions may deter undesirable behavior but do not prevent it.

Note that private law enforcement using liability rules is harm-based enforcement. Administrative law enforcement is often carried out by means of act-based enforcement (safety regulation) or prevention (withdrawal of license), although harm-based sanctions are also used sometimes (especially in environmental law for remediation of harm). Criminal law enforcement is used in all three stages, but is applied in the latest stage possible because of the severe consequences of criminal conviction.

Factors that determine the optimal stage of intervention[145]

The optimal stage of enforcement is determined by several factors:

1. *Enforcement costs*. In general, harm-based enforcement is relatively cheap compared to act-based enforcement, because enforcement costs are only realized after harm has been done. The costs of prevention vary. If prevention requires a once-only expenditure on a technical solution (such as a speed ramp or a fence), it can be relatively cheap. If, however, prevention requires the continued attention of enforcement agents (as in watching for any establishment without licence or in operating a toll-gate), the costs can be quite high. More generally, it can be argued that daily activities, ordinary and modifiable behavior can best be enforced in a later stage of enforcement, because these do not necessarily result in harm and require large enforcement costs. On the other hand, fixed physical objects such as safety devices can be enforced in an earlier stage because their presence

145 Shavell (1993). See also Shavell (1984a, 1984b).

or absence can be observed relatively easily and, as a consequence, evidence of violation is easily provided.[146]

2. The *magnitude of the sanctions*, in particular the ability to pay monetary sanctions. If the magnitude of the sanctions is too low to deter violators (efficiently) – for example because the level of wealth is limited – prevention must be used to control harmful activities. If the available magnitude of the sanctions is high, harm-based sanctions can be used. This argument is especially relevant for accidental harm, where harm-based sanctions are only carried out with the probability that an accident occurs, while the costs of precaution have to be paid in advance. To compensate for this probabilistic sanctioning, the magnitude of the harm-based sanction should be higher than in the case of an act-based sanction.
   In previous chapters I argue that, given the maximal magnitude of a possible sanction, the probability of detection should be raised high enough to deter. However, this is costly and it may be more efficient to switch to an earlier stage of enforcement to implement the efficient level of precautions. This is more likely to happen when the probability of detection or conviction is inherently small, for example when harm is widely dispersed and/or collective in nature, when there is no clear causal relationship between harm and injurer, or when injurers can invest in the non-detection of the harm.
3. The quantity and quality of *the information of the government* or courts about the costs and benefits of acts, in particular their harm, and hence the (un)desirability of offenses. The better the information, the better the allocation of enforcement costs. In general, this information becomes available (without effort) only after the occurrence of harm, which creates a tendency towards a later stage of enforcement. If a government or court is able to observe the harm done, there will be a preference for harm-based sanctions, because the government or court is able to impose an appropriate sanction likely to deter efficiently. Since the information about harm is limited ex-ante, act-based sanctions, and even more so preventive ones are more likely to be inflexible.

146 However, Innes (2004) shows that this argument only applies to situations in which the harm from accidents is freely observable, i.e. if no resources are needed for detecting and apprehending injurers, but only for prosecuting them (enforcement is characterized more by investigation than by monitoring, section 3.2). This might be the case with fire accidents, but not for food, drug and product safety nor for some types of environmental damage. In these cases, ex-ante regulation provides an inherent cost advantage to ex-post liability. The advantage of regulation is that it provides a stronger incentive to take precautions because it always sanctions negligent behavior rather than only sanctioning such conduct when an accident occurs (provided the same sanction is imposed). As a result, a lower probability of monitoring applies and enforcement costs can be reduced. Therefore, an ex-ante policy will be superior if the costs of an ex-ante inspection are not much higher than the costs of an ex-post inspection, and if the probability of an accident is relatively low.

4. The quantity and quality of *the information of injurers* about the acts. If injurers are well-informed about the expected harm and costs of precautions, they can be deterred efficiently through harm-based sanctions. If, however, injurers do not recognize how much harm they might cause, they will not be adequately deterred by harm-based sanctions, but might better be deterred by act-based ones. Act-based sanctions might force the injurer to take the efficient level of precautions, provided that two assumptions are met. First, the government (or court) must have superior information about the acts and be able to determine more efficient standards than the injurers themselves. This is the case, for instance, when information takes effort or special expertise to develop and/or the benefits of information development are non-exclusive. Secondly, the government must be unable to provide injurers with this information in any other way. If injurers learn of the expected harm (and compliance costs), harm-based sanctions can be used. However, people are limited in terms of the amount of time and effort they can invest in learning and are sometimes unable to handle technical or statistical information. Moreover, the government may have difficulty communicating the information because the (potential) injurers are hard to identify or too numerous.
   If injurers do not even recognize that certain levels of precaution will result in sanctions (because there is a large, and unavoidable, uncertainty about the standard or because they are unable to learn about efficient compliance techniques), prevention might be the optimal enforcement method.

10.1.2 The optimal method of public law enforcement: criminal or administrative

This subsection analyzes whether public law enforcement should be carried out through criminal or administrative law enforcement. This comparison is different from the one discussed above because the distinction between criminal and administrative enforcement does not depend on the question whether enforcement should be harm-based or act-based. It is also not related to the question whether detection should take place through monitoring or investigation. Moreover, although administrative law enforcement predominantly makes use of compliance strategies, the enforcement style is not necessarily related to administrative or criminal enforcement.

Criminal or administrative sanctions?

What factors determine the choice between criminal and administrative sanctions, hence criminal and administrative law enforcement?[147] The main characteristic of criminal law enforcement is that it allows imposing imprisonment. In addition, criminal sanctions lead to stigma on the individual or firm. Therefore, criminal sanctions are considered more severe than administrative sanctions. This implies that procedural requirements are higher for criminal sanctions.[148] Therefore, the explanation of the choice between criminal and administrative sanctions, should explain when higher sanctions are needed or beneficial, or when lower procedural requirements are needed or beneficial.

1. If *imprisonment* is required to impose a sufficiently high sanction, criminal law enforcement is required (Shavell, 1993). The need for imprisonment is more likely to arise if the level of wealth is relatively low, the costs of compliance are relatively high, the probability of sanctions is relatively small, the expected harm from an offense is relatively high, or the informal sanctions are relatively weak.
2. Administrative enforcement includes the possibility of *prevention* of the activity by withdrawing a license or closing a plant. Prevention can be beneficial if the offense cannot be deterred. In addition, administrative sanctions are aimed at remediation. Note, however, that where imprisonment implies incapacitation, this also produces a preventive effect.
3. The *stigma* that criminal law enforcement imposes is helpful in creating deterrence, especially when an offense is morally undesirable, rather than the by-product of an otherwise beneficial activity.
4. Administrative law enforcement (often) implies *law-making powers* for the enforcement authority. A criminal sanction can only be imposed if the activity is explicitly forbidden by law (*principle of legality*). In cases where considerable discretion is required, especially due to underinclusion, administrative law enforcement is more favorable. For undesirable activities, which are always morally reprehensible, criminal law enforcement

147 It is striking that there is (as far as I know) no literature on this point, except for Shavell (1993). The literature building on Shavell (1993) in essence provides arguments in favor of criminal law enforcement (i.e. when is imprisonment needed?) and no positive arguments in favor of administrative law enforcement. Moreover, the given explanations often refer to the distinction between harm-based and act-based enforcement and/or the enforcement style (cooperative versus deterrence). Although this has some intuitive appealing, it is incorrect since administrative law enforcement can also be harm-based (as when damages have to be remedied), while criminal law enforcement is often act-based (as in the case of a prohibition on gun possession).

148 See for example Ogus and Abbot (2002).

can be used because law enforcers do not have to determine whether an offense is (un)desirable.[149] Because of the considerable discretion, administrative law enforcement sanctions are relatively small to limit the potential abuse of this discretion.

5. Due to the severe sanctions that can be imposed by criminal law enforcement, individuals or firms should be able to *know which actions will be punished*. Therefore, criminal law enforcement applies to activities of which everyone knows or ought to know that they are forbidden. On the other hand, administrative law enforcement is most suitable in case of technically complex regulations, especially when the law enforcement authority also has the task of informing parties about regulation and compliance.
6. Since under administrative enforcement, the *infringements on private rights* are limited and stigma is absent, the authority is allowed to inspect firms even if there is no presumption of violation. Therefore, administrative law enforcement is more likely to be optimal if there is little information concerning an offense. Criminal law enforcement is especially suitable in cases where an offense, or a potential offense, is known (for example because of reports by victims), while the offender is unknown and must be traced using public resources. For example, it is known that a murder has been committed or that oil has been spilled into a lake. Due to these resources for apprehension, it is beneficial to save enforcement costs by imposing severe sanctions and minimizing apprehension. On the other hand, administrative law enforcement is especially suitable in cases where the potential offenders are known, but it is not known whether an offense has been committed. For example, it is known that the restaurants in a municipality may create fire accidents, but not which particular restaurants use inappropriate decorations; or it is known which firms may spill oil but not whether oil has been spilled.
7. For similar reasons, administrative law enforcement is most suitable for cases where there is *no clearly responsible* individual or firm. Especially with respect to safety regulation, and for larger firms, it may be difficult to identify the role of individual employees, while it may also be impossible to argue that the firm as a whole is guilty of violation, especially if firms cannot easily monitor the behavior of their employees. In such cases, administrative law enforcement is most appropriate. In the most extreme case, if firms cannot monitor their employees' behavior, the fines imposed on firms will create no deterrent effect, while enforcement costs will be made. Social welfare might be improved if (lower) administrative sanctions such as warnings are used. Offenses for which individuals can be made responsible (such as fraud and similar offenses) are enforced criminally.

149 Except for clear cases of self-defense or circumstances beyond one's control for which the burden of proof lies with the injurer.

10.1.3 Private or public enforcement?

Whether enforcement should be initiated by private parties or public agents depends on the method which most efficiently results in identification, apprehension and conviction of the injurer.[150]

If (potential) victims naturally possess information about the identity of injurers, it is socially desirable that enforcement be private, because it would be a waste of resources if public resources were spent on uncovering offenses. On the other hand, if effort must be expended to identify or apprehend injurers, enforcement by public agents may be required. Apprehension requires coordination because several individuals (victims) benefit, and, in some cases, this requires an investment in information systems. The incentive for victims to take action is high if they are provided with a monetary reward for taking action, as under liability where they receive compensation for the damage suffered. However, if the harm is widely dispersed or even collective in nature, if it takes a long time before the harm manifests itself, or if the victims are unable to prove a causal relationship between harm and injurer, the incentive to take action might be insufficient.

Furthermore the choice between private and public enforcement depends on the available information. Optimal sanctions generally require the ability to observe the expected or actual harm and precautions. There is no general argument for deciding whether the efficient level of precautions can be more easily determined in private proceedings (by courts) or in public proceedings (inspectors, regulators, administrative bodies, courts). Moreover, public enforcement might be required to inform injurers about the relationship between taking precautions and the expected harm.

Finally, public criminal law enforcement can be beneficial if a high sanction is required to create a deterrent effect, because it allows imprisonment. This is especially helpful if the wealth of the injurer is limited making him unable to pay large damages. Note, however, that criminal sanctions raise the burden of proof, so that the probability of conviction is lowered.

Two related questions must be raised at this point:

A. As discussed in section 2.3.2, insurance companies may be required to monitor the behavior of injurers to control for moral hazard. If injurers are insured, an additional relevant question is whether the injurer's behavior should be monitored and controlled by insurance companies or by government agencies. No general statement can be made on this point. Insurance companies might use the threat of not paying out. Sometimes, they might be more experienced due to economies of scale. On the other hand, the government can use imprisonment as a threat, as well as the preven-

150 Shavell (1984a; 1984b; 1993; 2004). Some of the arguments are similar for the use of harm-based and act-based sanctions because private enforcement is by definition harm-based, while public enforcement is not. I try to provide arguments that especially draw on the distinction private – public (given harm-based or act-based sanctions).

tion of the activity. It may also have more opportunities to interfere with private behavior, although this governmental interference is restricted by legal constraints (more than that of private organizations).

B. The analysis above restricts private enforcement to harm-based enforcement by liability rules. However, if it is optimal to have some act-based enforcement, such enforcement can also be provided by private parties. The government might impose the obligation to have an horeca establishment or an automobile inspected by some certified private company. Alternatively, it might at least use certification in granting and inspecting licenses. For example, horeca proprietors have to provide private certificates about the inspection of fire extinguishers. The relevant question is whether private organizations can monitor behavior more efficiently than public institutions. Due to competition, private organizations may be cheaper in providing the same level of monitoring, while public institutions may suffer from enforcement officials who are not maximizing social welfare (section 9.2). Moreover, economies of scale might be involved if a private organization can offer to immediately repair shortcomings. In addition, since the injurer must pay for the services of private organizations, he will be confronted with the full costs of his activities.

10.2 Private and public enforcement

None of the enforcement methods always dominates the others in effectiveness and efficiency. The question is therefore whether different enforcement methods can be used together, so that the advantages of each system can be combined and their weaknesses compensated.

10.2.1 Criminal law enforcement as a supplement

Criminal law enforcement allows for high sanctions, possibly imprisonment. This allows for the incarceration of people who refuse to pay (civil or public) monetary sanctions, or who appear to be notorious and incorrigible offenders. Criminal law enforcement can be used as an ultimate sanction (*ultimum remedium*) to enforce private or administrative sanctions. The threat of criminal sanctions is then the big stick which guarantees that compensation or fines are really paid.

Private law enforcement and criminal law enforcement can be used together in cases involving both monetary and non-monetary loss. Liability is used for monetary loss. Imprisonment is used to deter the non-monetary part of damages to prevent over-compensation.

A different combination involves regulation and administrative law enforcement being used as act-based enforcement against a firm. It is possible that individual employees might not be sufficiently induced to take the efficient level of precautions, because they do not (fully) take into account the sanctions imposed on the firm (section 2.3.6). The threat of personal prosecution and possibly imprisonment can deter at least the most serious offenses in which an individually assignable employee was clearly and/or deliberately negligent in complying with the regulations. Similarly, compensation claims in private litigation may improve employees' behavior.

10.2.2 *The simultaneous use of ex-ante regulation and ex-post liability*

The use of regulation as act-based enforcement is widespread. However, often, the use of regulation does not imply that there is no harm-based enforcement or that victims of accidents are unable to sue the injurer for damages. The combined use of regulation and (negligence) liability is explored in the literature.

The use of liability alone may fail to deter because injurers are judgment-proof or difficult to identify as the authors of harm (Shavell, 1984b and 2004). Regulation might be useful in forcing these injurers to take at least some precautions that are subjected to regulation. On the other hand, regulation alone may be inadequate because the government does not possess the perfect information concerning an injurer's behavior which would allow it to determine individually efficient standards. The use of liability may induce some of the injurers to take more precautions than the regulations prescribe.

There are two relevant points to consider. First, should there be a compliance defense, which states that compliance with the regulatory standard relieves the injurer from liability? Secondly, should there be a per se rule, which states that non-compliance with the regulatory standard automatically results in negligence? In fact, victims can actually become the enforcers of the public regulation by claiming damages when the proprietor does not comply with the regulation. Under the compliance defense, the regulatory standard is at least as high as the court standard. Under the per se rule, the court standard is higher than the regulatory standard. This leads to four different cases to be considered (Rose-Ackerman, 1991; Shavell, 1984a; Burrows, 1999).[151]

Independent use of liability and regulation: no relationship between standards

First, consider the case in which liability and regulation are used together but in an independent way, so that there is no relationship between the regulatory and the court standard. This implies, in fact, that neither the compliance

151 In fact, Rose-Ackerman (1991) distinguishes three different complementarities between liability and regulation. Shavell (1984a) only discusses the answers to the two defenses independently. Burrows (1991) analyzes whether the standards in liability and regulation should be dependent or independent. Taken together, we can identify four cases, which correspond to the two questions with two possible answers for each. An overview of the discussion can be found in Faure (2007). See also Ogus (2007).

defense nor the per se rule applies. The injurer can be punished twice, first by sanctions from the government and then by paying damages to victims. If the standards are adequately enforced, this implies that the injurer will choose the tightest standard (Burrows, 1999). If both standards are lower than the efficient standard or enforced below the efficient level, then the simultaneous use of both will be welfare-improving.[152] A compliance defense and/or per se rule will not be efficient, but only eliminate this benefit. On the other hand, if both standards are higher than the efficient level, the simultaneous use of liability and regulation can increase the danger of excessive precautions, because the injurer faces two penalties.

A related issue concerns the joint use of regulation and strict liability.[153] If the regulatory standard is set at the efficient level, and the damages are set as equal to the harm done, regulation and liability need not conflict.[154] If the firm complies with the regulatory standard, it will not have to pay a fine. The possibility of a liability payment provides an additional incentive for choosing the efficient level of precautions. The benefits of strict liability are that victims might be compensated for their damage and that the injurer has an incentive to choose efficient activities (if the regulatory standard only applies to taking care). Innes (2004) shows that the joint use of ex-ante regulation and ex-post strict liability (civil or criminal) can be beneficial if some accidents are freely detected and some are not. If the expected liability is sufficiently low (because there is a judgment proof problem) supplemental ex-ante government observations and the enforcement of precautions can reduce underdeterrence.[155]

The use of both compliance defense and per se rule: equal regulatory and court standards
Consider the situation in which both the compliance defense and the per se rule apply, i.e. compliance with the regulatory standard implies non-negligence and non-compliance with the regulatory standard implies negligence. In fact, the court will follow the regulatory standard in determining negligence (Burrows, 1999). Such a joint use of liability and regulation is beneficial if liability functions as a stopgap which applies in the absence of more stringent regulation (Rose-Ackerman, 1991). Since legislation is necessarily a slow process, regulation will fail to establish a regulatory program in all

152 In the absence of enforcement costs. In the analysis in this section, I do not discuss the impact on enforcement costs. Of course, using more enforcement methods together will increase enforcement costs. The right question is whether the increase in enforcement efforts from a simultaneous use of enforcement methods balances the increase in deterrence that is gained, compared to the use of one method alone. The text only investigates whether simultaneous use can increase deterrence, which is also the focus in the literature (with Innes, 2004, as a notable exception).

153 Under strict liability, there is no court standard of due care, so that neither a compliance defense nor a per se rule apply.

154 Rose-Ackerman (1991).

155 Whether ex-post or ex-ante supplemental government activity is optimal depends on the factors identified in Innes (2004) and described in section 10.1.1.

areas in which regulation is superior. Liability is then required as a stopgap for cases for which regulation fails to provide. Courts will take the regulatory standard as the level of due care. Liability is only required as a supplement in cases where regulation is missing, so that courts have to determine due care themselves. Priority is given to the regulatory standard.

Similarly, Shavell (2004) argues that regulation may fail to deter because the government is unable to capture in regulations – no matter how detailed – actual daily behavior, such as, for instance, keeping beer-crates away from emergency exits or limiting the number of people inside a building. Liability will be used to enforce precautions that concern modifiable behavior, while regulation concerns easily observable safety devices such as sprinkler systems and fire extinguishers.

Finally, the joint use of regulation and liability can be beneficial if regulation is used as an advisory function through which the injurer is informed about the expected harm (and hence sanctions) and/or the least costly method of compliance. Regulation alone might be unattractive because of the high costs of frequent inspections. If a periodic regulation is used to provide the injurer with information, it is possible to further rely on liability to induce precautions. This might minimize the costs of inspections. In such cases, it seems logic that the court follows the regulatory standard.

Interdependent use: the court standard exceeds the regulatory standard
Consider the situation in which the combined use of liability and regulation is interdependent because non-compliance with the regulatory standard implies negligence (per se rule). Assume, however, that compliance with the regulatory standard is not sufficient to escape liability (no compliance defense). In fact, regulatory standards are intended as minima that can be supplemented by more stringent court standards. There are several arguments against a compliance defense, i.e. reasons why the regulatory standard should be lower than the court standard. If the court standard equals efficient precautions, regulation can only be beneficial if some conditions for an effective use of liability are not met. If liability is failing, regulation may be beneficial.

1. Liability alone may fail to deter because it is possible for injurers to escape litigation (and there are no punitive damages; Schmitz, 2000). Regulatory requirements are generally based on imperfect information and do not perfectly describe standards for individual firms (Shavell, 1984a, 1984b, 2004).[156] The government may be unable to describe all behavior in (detailed) regulations, while courts (ex-post) may be able to evaluate this

156 Note, however, the remark in Hiriart et al. (2004) that ex-ante transfers may induce firms to provide information to the government. The joint use of ex-ante regulation and ex-post liability ensures that the firm has an incentive to reveal its information because a firm that provides incorrect information runs the risk of being caught ex-post if a lawsuit is started and the true harm and compliance costs are revealed.

behavior.[157] Hence, regulation alone may underdeter, and the use of liability may induce some of the injurers to take more precautions than the regulations prescribe, which is socially beneficial. If regulatory standards equal the minimal level of precautions that every injurer ought to take, a compliance defense will be unfavorable, because (some) injurers ought to do more than the regulatory standard. If, however, due to limited information, regulation is too excessive, this would provide an argument in favor of a compliance defense and against the per se rule.

2. If there is uncertainty about the level of due care that is handled by the court, negligence may lead to underdeterrence (section 2.3.5). If this uncertainty is high, a higher level of precautions results in a slightly reduced likelihood of being held liable, thereby paying for the victim's losses (Kolstad et al., 1990). If the marginal costs of precaution around the legal standard are high, undercompliance will substantially lower the injurer's costs of precautions but only slightly reduce his expected liability costs. Kolstad et al. (1990) argue that using regulation (which has no uncertainty) as a supplement to liability will limit the problem of underdeterrence. Under the per se rule, the regulatory standard will inform the injurer that the court standard cannot be lower. Therefore, no injurer will choose a level of precautions that is lower than the regulatory standard. As a result, if it is efficient to use regulation as a supplement, the regulatory standard will be lower than the regulatory standard used alone, and lower than the efficient precautions level, i.e. the average court standard.
   These conclusions, however, depend crucially on the assumption that there is Bayesian up-dating about the legal standard, where only the belief that the court standard falls below the regulatory standard is affected. Other assumptions can also be made. For example, the regulatory standard may affect the belief that the court standard is substantially higher than the regulatory standard.[158]
3. Whether regulation or liability is the optimal enforcement method depends (among other things) on the wealth of the injurer. If wealth is sufficiently large, liability is favorable, if wealth is limited, regulation is. However, the joint use of liability and regulation can be beneficial if wealth is limited and varies among injurers (Schmitz, 2000). In such a situation, regulation might be preferred for low-wealth injurers, while liability might be preferred for high-wealth injurers. Joint use of liability and regulation can increase social welfare by making low-wealth injurers stick to the regulatory standard and by inducing high-wealth injurers to take optimal, or at least higher, precautions. In that case, the optimal regulatory standard is strictly lower than the standard used under regulation alone, which is based on efficient average precautions.

157 Whether courts are indeed better able to estimate the benefits and costs of a firm's behavior is disputed by for instance Bergkamp (2001).

158 For a more extensive discussion, see Burrows (1999) and Ewerhart and Schmitz (1998).

Interdependent use: the regulatory standard exceeds the court standard
There is hardly any literature providing a rationale for situations in which there is a compliance defense but no per se rule, so that injurers who fail to comply with the regulatory standard are not necessarily negligent. Shavell (1984a) argues that a per se rule is not necessarily efficient because it induces some injurers, undesirably, to comply with the regulatory standard when they would not otherwise have done so. There will be some injurers (a) who ought not to meet the regulatory standard because they face higher than usual costs of precautions or lower than normal expected harm, and (b) who will not have been forced to satisfy regulatory standards due to flaws in or probabilistic methods of enforcement. By allowing these parties to escape liability in view of their circumstances, the possibility that they would still be led to take wasteful precautions can be avoided. However, similar arguments explain why a compliance defense is not favorable, so these arguments do not explain why there is a compliance defense but no per se rule, just why there should be no per se rule.

An argument for applying a compliance defense without a per se rule is that regulation is 'excessive' because it is biased to favor high precautions due to private interest theories (section 9.2) or one-sided information (see – although in a different context – Calcott and Hutton, 2006). However, this argument only explains the existence of regulatory standards that exceed the court standard, not how this existence can be efficient (as is the case above for the reverse).

Another reason why non-compliance with the rules should not automatically lead to negligence is that negligence should only be allowed if there is a causal relationship between non-compliance with the regulatory standard and the occurrence of an accident. A related point is that not using the per se rule strictly can be efficient in providing victims (in case of bilateral accidents) or other injurers (in case of multiple injurers) with sufficient incentives to take precautions.

10.2.3 Liability and prevention

The former analysis focuses on regulation as enforcing regulatory standards for taking precautions through the use of act-based sanctions. However, government regulation can also be carried out by means of prevention, a situation discussed by Calcott and Hutton (2006). The combination of prevention by licenses and liability is frequently observed.

An activity (or product) should be allowed if the benefits of the activity are larger than the expected harm and other external costs. In ideal circumstances, either regulation or liability will implement efficiency. Liability could be set in such a way that firms will incur the full social costs of their activities and hence will only undertake socially beneficial activities. In addition, a regulator might withhold approval from activities whenever the social costs exceed the benefits.[159] However, liability may, due to the reasons identified before, underdeter. Moreover, Calcott and Hutton (2006) argue that liability may overdeter if firms underestimate the social benefits of new activities (e.g. consumer surplus, employment). Under strict liability, a firm will bear the damages, so the expected net benefit from activities will be lower than under negligence-based liability and so will the incentive to propose activities. However, if firms under-appraise activities, negligence-based liability is more favorable to stimulate proposals.

If there is a perfect regulator, regulation alone will also be efficient. In general, a regulator will be imperfectly informed and may also be biased. If the regulator is not perfectly informed about the private benefits of the activity to the firm (but he is unbiased), there is a danger of permitting activities that are not socially beneficial. In such cases, strict liability will be a more effective constraint on proposals than negligence. Harsher regimes are more desirable if the expected harm is higher. Therefore, for low harm risks, the sole use of negligence is at least as efficient as the alternatives. For intermediate risks, either supplementing negligence with regulation or the sole use of strict liability becomes beneficial. For high risks, the joint use of strict liability and regulation is efficient.

There can also be failures in regulatory enforcement because the regulator is biased (Calcott and Hutton, 2006; see also section 9.2). First, regulators might be biased *against* proposed activities because they give more attention to possible harm than to the benefits of proposed activities, for example because criticism is higher for problems with approved activities than with foregone activities. Similarly, there may be a regulatory culture which emphasizes caution, because it bases its decisions on 'upper-bounds' on accident risk, rather than on best estimates. Alternatively, regulators might be biased toward activities because of regulatory capture or an internal conflict of interest.

If the regulator is biased against activities, strict liability might be a better supplement to regulation than negligence. Under strict liability, the firm is more constrained in proposing activities. Since the regulator knows that under strict liability, the expected benefits of proposed activities are likely to be higher, it will approve more projects. On the other hand, negligence-based liability encourages a firm to apply more often, but discourages the regulator from approving so many proposals. Because none of the regimes are unambiguously more restrictive, there is no general reason to prefer strict or negligence-based liability when the firm under- or over-appraises. Moreover, because the regulator is biased, it may be more efficient to refrain from regulation and only use liability. On the other hand, it is not necessarily optimal to have negligence-based liability in addition to a regulator who is biased against activities ("excessive regulation") because this will not induce the

159 Note that in the case described, the expected harm from an activity is exogenous and does not depend on taking precautions. The only relevant question is whether firms have an adequate incentive to undertake only socially beneficial activities.

regulator to approve more projects. A regulatory bias against proposals does not provide an argument in favor of softer liability regimes.

10.3 Deterrence versus compliance strategies

In public law enforcement, both administrative and criminal law enforcement, the efficient enforcement policy also constitutes the choice for an enforcement strategy or style. In chapter 8 I describe the distinction between a deterrence and a compliance strategy. I discuss the need for and the benefits of a compliance strategy from an economic point of view. This section tries to conclude by describing whether and when a deterrence strategy or a compliance strategy should be used.

10.3.1 *The appropriate distinction*

It is important to limit the distinction between deterrence and compliance strategies to a distinction between the enforcement *styles*. The style describes what an enforcement authority will do when it detects a violation. A deterrence style is a style in which every violation is immediately punished and the enforcement authority bases its decisions solely on its own information. A compliance style is a cooperative style in which the enforcement authority might forego some violations (for some time) when this induces the firm to reveal important information necessary for enforcement.

The enforcement style should not be confused with what might be called the objective of enforcement. In regulatory enforcement of firms maximizing social welfare may include several intermediate (not mutually exclusive) enforcement objectives: deterrence, remediation, advice, education, and completion. It is important to start with identifying the objective of enforcement and subsequently examining which enforcement strategy can best achieve this objective.

10.3.2 *Deterrence*

Under deterrence (the focus of most economic literature) the objective is to induce compliance by making non-compliance unattractive. The threat of sanctions signals that non-compliance does not pay. If the objective is to deter offenses, the primary enforcement style is a deterrence strategy in which offenses are immediately punished. The deterrent effect of enforcement relies critically on the expectation by potential offenders that a detected offense will consistently be prosecuted (Fenn and Veljanovski, 1988). Under a compliance strategy, the firm is able to delay compliance or to negotiate a lower level of compliance.

Failures of deterrence strategy

However, a deterrence strategy assumes that enforcement authorities are granted sufficient resources to enforce compliance (chapter 5). In practice, they are usually not, so that there is serious underdeterrence and the maximum achievable compliance rate is a partial compliance rate. It might then not be optimal to directly impose severe sanctions because such sanctions might – on balance – increase underdeterrence. For instance because they lead to a waste of resources on imposing sanctions, because they crowd out the intrinsic motivation to comply or because such sanctions do not sufficiently discriminate between the different harm from offenses. In short, there might be problems if imposing sanctions is sufficiently costly (see for a more extensive list chapter 8). These costs become relevant if there is underdeterrence.

Under a deterrence strategy, the enforcement authority strictly enforces the legal standard. If law is incomplete, a deterrence strategy that legalistically enforces the legal standard to the letter will not induce the first-best level of precautions (chapter 6). Note, however, that if the enforcement authority is perfectly informed and is granted sufficient discretionary powers it is able to discriminate between firms and to induce the first-best level of precautions by strictly enforcing this level for each firm. Moreover, a strict enforcement of the standards forces the lawmaker to avoid unnecessarily incomplete standards.

A final reason why a deterrence strategy may fail is that enforcement officials are insufficiently interested in pursuing deterrence and in consistently sanctioning every violation. This might be the result of credibility problems or may stem from a misalignment between public interests and the private interests of enforcement officials (chapter 9). The officials might be more interested in their short-term careers than in deterring future offences, or they might not want to spend sufficient time on (actual) enforcement.

In sum, a deterrence strategy fails to deter (i) if imposing sanctions is too costly, (ii) if regulation is too incomplete, or (iii) if the enforcement authority and its officials have insufficient interest in pursuing deterrence.

Alternatives to a deterrence strategy

If a deterrence strategy fails to deter, different other enforcement methods might be more efficient. If only a partial compliance level is achievable, it is optimal to target enforcement on some subgroup of the population. This will induce the target group to fully comply, while there is no sanctioning in the non-target group. Hence, the enforcement authority will avoid the costs of imposing sanctions. The best targeting scheme is one that depends on past compliance, for example by issuing warnings. Yet, the primary enforcement style is a deterrence strategy. The only point is that this strategy might forgo a part of the population or that the first enforcement action involves moving to the target group instead of a direct penalty.

If the legal standard cannot be enforced, it is generally optimal for the enforcement authority to adjust the standard downwards and to enforce some informal standard. Again, the costs of imposing sanctions are avoided. A partial compliance rate is realized by inducing every firm to partially comply with the standard. This is not a pure deterrence strategy because the enforcement authority should not legalistically enforce every breach of the standard but instead forgo minor violations. However, it is still a deterrence strategy in enforcing the adjusted standard.

If, due to the incompleteness of the law, there is much two-sided uncertainty about the standards and the compliance methods, a cooperative, flexible enforcement style can be efficient. The enforcement authority can achieve higher levels of safety against lower enforcement costs by offering mutually beneficial deals to the firm. This implies that the firm will be required to reduce the most important safety problems in exchange for reduced sanctions for minor violations.

10.3.3 Optimal strategies for other objectives

Remediation

Remediation implies that enforcement is aimed at restoring compliance. For example, repairing a technical device that causes non-compliance or cleaning-up the environmental harm resulting from some emission (chapter 4).

Only in cases of purely accidental offenses where it is impossible to make enforcement dependent on the level of remediation will immediate punishment of violations have no merit, and a strategy of warnings be efficient. If, however, it is possible to sanction insufficient remediation, a deterrence strategy that imposes sanctions dependent on the level of remediation is efficient, as this is most likely to induce the firm to choose remediation (given the general conditions for a deterrence strategy discussed above). Similarly, if offenses also occur deliberately, there is no need for lowering deterrence.

Advice

Advice implies that the objective of enforcement is to ensure that the firm is sufficiently informed about all the benefits and costs of (non)compliance, the standards that apply and the least costly method to create compliance.

If the firm is able to acquire the relevant information at a reasonable effort, advice is not needed. Imposing efficient deterrent sanctions provides firms with the efficient incentive to inform themselves. If the firm is unable to do so (especially under technically complicated regulation), advice might be relevant. If all firms lack the same kind of information, general information campaigns will be sufficient to inform them. If the law is (very) incomplete, so that an individual firm cannot learn from others, it might be efficient for public inspection to have advice as its objective. This requires a more compliant strategy, in which inspection is characterized by speech, explanation and cooperation.

Education

Enforcement may also have as objective the stimulation of norms and values that enhance compliance. Public inspection may be needed to maintain, strengthen or even create norms and to realize the costs and benefits of norms (chapter 7).

If the costs of norm-violation are only realized once an offence is detected – which is especially true for informal social sanctions – public inspection might be necessary to detect offences. The magnitude of sanctions may be relatively low, but generally violations must be punished, and immediately so.

If the costs of norm-violation depend on the level of compliance in the relevant population, public enforcement through sufficient deterrence is needed to prevent a breakdown in compliance and norm-subscription. If norm formation occurs by observing that non-compliance is punished, a more deterrent enforcement style is appropriate. If norms are stimulated by being inspected if in compliance because this signals that compliance is important and valued, a more cooperative style, aimed at compliers, is appropriate.

Another problem is that public enforcement may destroy norms, because imposing sanctions crowds out the intrinsic motivation to comply. Then deterrence might fail and preventing the breakdown of norms might require a compliance strategy.

Completion

Another objective of public enforcement is to supplement private enforcement. There are several reasons for this complementary function. First, public enforcement might be used to provide the firm with the information it needs in order to take efficient precautions. Public enforcement is especially needed to reduce the uncertainty concerning the levels of due care in private litigation claims. Moreover, public enforcement is needed when firms and their employees are not, or not fully, aware of the existence and consequences of private enforcement because they suffer from cognitive limitations. For these cases, public enforcement has advice as its objective, as specified above. In addition to the arguments above, information about standards might only be credible if the public standard is actually enforced. Therefore, in general, public enforcement should apply a deterrence strategy.

Finally, public enforcement might be used to create deterrence in cases where private enforcement fails to deter efficiently. Deterrence through private enforcement may be hindered by for instance wealth constraints or litigation barriers for victims. Here, public enforcement has the objective of deterrence. In general, a strict enforcement policy is needed to induce the privately underdeterred firms to take precautions.

10.4 Effective and efficient enforcement

10.4.1 *Conditions for the enforcement methods*

Part I of this thesis deals with the conditions under which private, criminal or administrative law enforcement provide an efficient incentive for firms to comply with safety standards. If an appropriate expected sanction is imposed, the expected harm that the firm inflicts will be internalized into the decision of the firm. Let me summarize the most important conditions identified.[160]

Private law enforcement

Characteristic for private enforcement is the fact that enforcement is left to the victims. The sanction consists of compensation of the victims for the damage they suffered. Therefore, in order to realize an expected sanction that equals the expected harm, the probability of sanctions must be 100%. If enforcement costs are negligible (for instance because liability fully deters or because a settlement is arranged) an efficient outcome results. If, however, the firm can sometimes escape liability, the incentive to take precautions will be insufficient. Therefore, private enforcement is effective under the following conditions:

1. Most of the harm done consists of monetary damages, or, if there are significant non-monetary damages, courts should be able and willing to award adequate compensation for these damages.
2. The firm has sufficient wealth to pay for the damages, or is fully insured for liability. If it is insured, insurance companies should be able to easily monitor and enforce precautions.
3. Courts and firms have sufficient information about the possible accidents so that courts are able to enforce socially optimal levels of precaution and firms are able to anticipate on courts' decisions.
4. Victims have sufficient access to the legal system so that the probability of sanctions is sufficiently high. Therefore victims should be able to identify the injurer and should suffer sufficient harm to find it worthwhile to file a claim.

Criminal law enforcement

For those failures to take precautions that fall under criminal law, enforcement takes place through a public prosecutor. Generally, the injurer must be traced, so that public resources are required to realize the probability of sanctions. Due to these costs, it is usually not efficient (if possible at all) to realize a probability of conviction of 100%. As a consequence, to maintain the expected sanction, the magnitude of the sanction should be correspondingly higher. If monetary sanctions are insufficient, imprisonment may be used. Due to the severe sanctions that may be imposed, and because conviction also raises the

160 See Shavell (1984a; 1993; and 2004), Ogus (2007) or Faure (2007) for an overview.

(moral) question of guilt and stigma, the criminal law enforcement process must satisfy important requirements of due process. This implies a high burden of proof, especially with respect to the relationship between the behavior of the injurer and the harm inflicted. Criminal law enforcement is especially suitable to enforce precautions in the following situations:

1. The injurer is unknown and has to be traced using public resources.
2. Imprisonment is required to impose a sufficiently high sanction.
3. The harm can be related to a specific injurer and the magnitude of the harm is fairly well known. There is be 'intent' and 'responsibility' of the firm.
4. The firm has sufficient knowledge about the possibilities for preventing harm, because otherwise there is no 'guilt' and even extremely high sanctions will not deter.
5. The regulations refer to norms that are widely held in society, so that it is certain that offenses are regarded as immoral and undesirable, and enforcement actions are always desirable.

Administrative law enforcement

If public intervention is required, administrative law enforcement may be an attractive alternative. This is especially true in cases where it is uncertain who is responsible for the harm caused, what the magnitude of the harm is or when it comes about. If the expected harm is a by-product of an otherwise desirable activity, it may be inefficient to completely ban the activity and the question of guilt is less prominent. Since penalties are small, the procedural requirements are low and the enforcement authority has much discretion. Administrative law enforcement is appropriate in the following situations:

1. Enforcement is aimed at repairing technical devices (remediation) that occur accidentally.
2. Offenses are not associated with guilt, for example because an offense is not deemed immoral for lack of causality.
3. Compliance requires much technical expertise that is not freely available. In such cases, warnings might be helpful especially when intrinsic motivations to comply are relevant.
4. Discretion is required to specify the laws for individual cases and to determine whether and which enforcement action is appropriate.

Joint use of enforcement methods

Observation of enforcement in the real world shows that in many cases all three enforcement methods are employed. This may be justified by the fact that criminal law enforcement is used as an *ultimum remedium* to deter severe offenses and clear neglect, or to impose public sanctions against an individual employee. Private enforcement is used because it is a relatively cheap way of providing an incentive to take precautions and offering victims compensation for their monetary loss. Additional act-based enforcement, especially by means of administrative enforcement, is required because some firms are wealth-constrained and/or have difficulty in predicting the level of due care and/or are unaware of the best compliance techniques. In these cases, it is likely that the court standard will exceed the regulatory standard: compliance with the regulatory standard will not exclude from liability. Moreover, public enforcement is needed as a supplement to private enforcement if damages are difficult to determine and if prevention of non-monetary loss by private enforcement is insufficient.

Enforcement costs
Besides the factors mentioned, the enforcement costs are relevant for determining which enforcement method is socially optimal. The theoretical part repeatedly argued that the efficient enforcement policy balances the costs of enforcement with the benefits of a higher level of compliance (i.e. a lower level of expected damage minus the costs of compliance). There is no general argument that enforcement costs are lowest for one of the enforcement methods. A joint use will increase enforcement costs but this might be justified by the higher level of deterrence gained. The level of enforcement costs and its effects on (non)compliance is, in the end, an empirical issue. The empirical part of this thesis attempts to estimate the level of enforcement costs and to evaluate whether this level is socially optimal for the field of enforcing fire safety regulation in horeca establishments.

10.4.2 The case of fire safety in horeca establishments

For the case of fire safety in horeca establishments chapter 2 argues that it is likely that some enforcement activity is needed. The analysis thereafter analyzes whether in general enforcement by private, administrative or criminal law enforcement is effective and efficient in inducing compliance with the regulation. It has to be examined whether and to what extent these arguments apply to the field of enforcing fire safety in horeca establishments. Such an investigation is not possible without filling in the analysis with the available data on this field.

The relevance of the conditions identified
A quick survey of the fire safety case reveals that almost all conditions stated above matter for fire safety in horeca establishments. In fire damage, non-monetary damages are an important part of the harm to victims, but not of the compensation award. Many proprietors will be able to pay damages, but it is unlikely that all proprietors have sufficient wealth to pay for all the harm a disaster like the one in Volendam causes. As argued in chapter 2, insurance companies might be able to monitor technical devices, but not daily behavior of the proprietor. Whether courts and proprietors have sufficient information about the accident problem can not be decided in general. Knowing due care in advance might be problematic for proprietors especially when courts devi-

ate from the regulatory standards in determining negligence. Given the clear presence of harm in the fire safety case, access to the legal system seems to be sufficiently guaranteed in the Netherlands (as in most Western European countries), but proving that the proprietor was negligent may require technically complex investigations and coordination by the victims. So, in fact, all named barriers for private enforcement are present. Still, private enforcement might be beneficial because of the relatively low enforcement costs and the incentive for victims to pursue a case.

Criminal law enforcement is also not unambiguously effective or ineffective. In the fire safety case, the injurer is generally known. But public resources may be helpful in conducting technical investigations into the negligence of proprietors and in unraveling enterprise structures. And imprisonment might be needed for those proprietors that are unable to pay for a high sanction. On the other hand, the remaining conditions are not necessarily satisfied. If for instance a fire is caused by third parties or is caused by an (unpredictable) technical failure the harm can not be related to a specific injurer. If the proprietor can not prevent the accident, criminal sanctions are inefficient. In such cases there may be no 'guilt'. As argued in chapter 2, proprietors are not well informed about expected harm so that they might not be able to understand which precautions should be taken. While some requirements concern acts of which everyone knows or ought to know that they are dangerous (like – possibly – a beer crate in front of an exit, far to many visitors inside), other requirements might not be viewed as immoral by society (perhaps locking an emergency exit because of prevention of burglary or maintaining fire extinguishers).

Administrative law enforcement can be an effective method for enforcing fire safety in horeca establishments, especially for requirements concerning technical devices (like for instance repairing broken escape-route signs or fire extinguishers). These requirements are primarily aimed at guaranteeing that the devices are adequately maintained and repaired. Violations of these requirements are primarily viewed as the by-product of an otherwise socially beneficial activity and therefore not regarded as immoral. Compliance requires knowledge of technically complicated devices. However, for other requirements these conditions do not hold, like for instance blocking emergency exits or allowing too many visitors. And the disadvantage of administrative enforcement is the lack of an incentive to prevent technical failures. In the end, the (expected) harm is the result of the level of precautions of the proprietor, including the adequate maintenance of technical devices, and the proprietor might thus be viewed as the causer of an accident. Moreover, a gross or deliberate level of negligence might be viewed as immoral. An argument in favor of administrative enforcement is the discretionary powers of municipalities. Because exploiting a horeca establishment is a socially desirable activity, fire safety regulations should be applied with wisdom. The primary requirement is that an establishment is fire proof, not that the regulations are complied with to the letter.

The need of further information

As should be clear from the description above there is, as yet, no unambiguous conclusion on the optimal enforcement method for enforcing fire safety in horeca establishments. None of the enforcement methods is superior to the other ones. The optimal enforcement policy is likely to constitute a mix of private, criminal and administrative enforcement. The real question is not whether a condition is satisfied or not. As described, almost all of the conditions are likely to be of relevance for the case of fire safety in horeca establishments and none of the conditions is perfectly satisfied in this case. The real question is the intensity of the problems: is it a real barrier for enforcement? This can hardly be decided in general, without filling in the analysis with actual data for this specific problem.

Take for instance the condition on an adequate compensation of non-monetary damages for private enforcement. It is generally known that in the Netherlands the compensation for non-monetary damages is small (in light of the preventive effect of liability). But the question is to which extent this affects the incentive to take precautions given that the proprietor will have to compensate an important part of the monetary damages and given that under negligence a less-than-full compensation might be sufficient. This can only be concluded by filling in the analysis with information on the actual compensation the proprietor has to pay and on his costs of precautions. Likewise, saying that fire accidents are an undesired by-product of an otherwise socially beneficial activity, does not deny that gross and deliberate negligence may occur at times and require criminal law enforcement. Or consider the fact that many fire safety regulations concern technical devices. Therefore, non-compliance may be the result of technical failures instead of the intention of the proprietor. But does administrative enforcement provide a sufficient incentive to repair and maintain the fire safety devices?

Similar uncertainty remains concerning the discussion on the optimal enforcement style. The previous section argues that neither a compliance nor a deterrence strategy is always superior to the other. Which enforcement style should be applied, depends on the circumstances. On first view, the fire safety problem seems to be an outstanding example of the benefits of a compliance strategy. Remediation, advice, education and completion can all be relevant objectives. Cooperation with proprietors can be expected to increase the willingness to comply more than strict punishment of every violation. At least, the actual enforcement policy does not look like the description of Becker's (1968) trade-off with high sanctions. Moreover, in many cases only administrative repair sanctions are available. But doubt on a compliance strategy is at place. The different goals might also be achieved by a deterrence strategy. The lack of deterrence allows the proprietors to postpone compliance. The arguments for a compliance strategy might seem plausible arguments, but the question is whether they really make a reasonable case for fire safety in horeca establishments. Especially, to what extent do they apply to the case of fire safety in horeca establishments? In the end, the optimal enforcement policy concerns a mix, a balance of deterrence and compliance strategies. Does the actual enforcement policy apply the right mix? The arguments in favor of a compliance strategy do not necessarily justify the present intensity with which a compliance strategy is applied.

As a final example of the need of further information, consider the question which stage of enforcement is optimal for fire safety in horeca establishments. The presence of limited wealth and information problems point at act-based enforcement, but the large enforcement costs that are involved with act-based enforcement are a clear disadvantage. So it is important to understand how large these costs are and how severe the information problems are.

So, although one might fairly say that enforcing fire safety in horeca establishments requires a mix of enforcement methods and styles, this is a rather empty phrase if this mix is not further specified. Such specification requires that the analysis is filled in with data on the relevant cost functions. These should decide how intense one enforcement method can be used and to what extent it fails to induce precautions. This part explains how, according to the economic analysis of law enforcement, the costs and benefits of enforcement should be weighed at an abstract, general level. To determine the optimal level of enforcement for fire safety in horeca establishments, one has to specify the benefits and costs in the concrete.

The analysis in part I describes the optimal enforcement policy of safety standards for firms in general. A next question is whether actual enforcement policies correspond to this recommended policy and follow the arguments provided in the literature. Such a judgment can not be made in general but only by considering specific enforcement problems. Moreover, such a judgment is only valid when it is based on the available data for this specific problem. Therefore the next part analyzes whether the actual enforcement policy for fire safety in horeca establishment adequately applies the theoretical arguments of the literature and is in line with the description of the optimal enforcement policy.

Part II

Empirical analysis

Part I of this thesis provides a theoretical analysis of the conditions under which some type of enforcement will be effective and demonstrates that the costs and benefits of enforcement must be balanced. Further conclusions can only be drawn once the analysis is filled in with specific, empirical data. This is done in this part where I attempt to analyze the enforcement of fire safety regulation in horeca establishments in the Netherlands. This mainly concerns the enforcement policies that are adopted following the Volendam disaster, in the period 2001-2007.
First, in chapter 11, I describe this case and the broader context of the Volendam disaster and its consequences. Secondly, in chapter 12, based on the interviews with enforcement officials, I interpret the enforcement policy in light of the economic literature described in part I. Is this policy effective? Particular attention is given to the evaluation of the use of a compliance strategy. Thirdly, in chapter 13, the question is whether the enforcement actions following the Volendam disaster are efficient. Do they pass a cost-benefit analysis? A further step in chapter 14 is to consider whether private actions by victims, such as those of the Volendam disaster, may, through filing and settling damage claims be expected to be more effective.
In chapter 15, I take a different approach. I consider the case of a hypothetical but representative Dutch municipality and examine which enforcement policy is most welfare-enhancing, assuming that a horeca proprietor acts according to the rational choice framework of the economic model. This simulation allows me to examine the effects of alternative enforcement policies and to examine whether the conditions formulated in part II for private, administrative and criminal law enforcement are satisfied. The final chapter (chapter 16) concludes on the socially optimal enforcement policy, based on both the theoretical and the empirical part of this thesis.

11 Fire safety regulation in horeca establishments and policy developments

In this chapter I introduce the problem of enforcing fire safety regulation in horeca establishments in the period 2001-2007. This period begins with the Volendam disaster and ends with the introduction of the Use Decree and the abolishment of use licenses, now planned for mid 2008. First, I give a brief description of the horeca sector. Then, I describe the fire safety regulation and the importance of and the response to the Volendam disaster. Finally, I describe how this response to the Volendam disaster fits to an increasing attention to enforcement problems in general. This builds a bridge to the investigation in 13 municipalities in the next chapter.

11.1 The horeca sector

I use the term horeca to indicate a group of businesses, aimed at serving food and/or drinks that are consumed at the establishment itself, i.e. restaurants, cafés, bars, pubs, discothèques, party centers, etc.

11.1.1 *A general description of horeca establishments*

Tables 11.1 and 11.2 provide some summary statistics concerning the horeca in the Netherlands in 2003, in order to give a short impression of the characteristics of horeca businesses. There are about 7.4 establishments in the drinking sector (cafés/bars, disco, party centre) per 10,000 inhabitants. These establishments vary greatly. Some of the neighborhood cafés or brown cafés are located within a particular neighborhood, are relatively small and attract a rather constant number of regular customers. Instances on which these proprietors might allow more people than the maximum prescribed are for example soccer matches and similar events.

Other cafés, as well as discothèques, are located more in the centre of town and aim at day-trippers, tourists and/or the café-frequenting public in the weekends. They may allow more visitors than the maximum prescribed almost every weekend. Over-attendance may also occur when the establishment is hired out (for private parties). Sometimes, these establishments request an entrance fee, which makes easily accessible emergency exits problematic. Drinking is usually quite severe, which makes people more careless and difficult to handle.

Table 11.1 Characteristics of the catering industry in the Netherlands[a]

		Number		Sales area (m^2)[b]		Exploitation in (years)[c]
		Total	Per 10.000 inhabit.	Total	Average	Average
Café, bar:		10.900	6,7	1.174.000	110	5,6
Of which	*Neighbourh., brown*	*7.010*	*4,3*			
	Pub, diner	*2.240*	*1,4*			
	Grand café	*260*	*0,2*			
	Music café	*420*	*0,3*			
	Theme, target group	*970*	*0,6*			
Disco		380	0,2	144.400	380	10,8
Party centre		880	0,5	768.100	880	11,5
Fastfoodrestaurant		310	0,2	58.200	190	6,0
Restaurant sector:		10.220	6,3	1.563.800	150	
Of which	*Bistro*	*1.420*	*0,9*	*115.300*	*80*	*8,7*
	Restaurant	*5.910*	*3,6*	*800.500*	*140*	*8,2*
	Café-restaurant	*2.790*	*1,7*	*616.500*	*220*	*11,9*
	Road-house	*110*	*0,1*	*31.400*	*300*	*13,9*
Of which[d]	*Dutch / French*	*5.850*	*3,6*			
	Mediterranean	*1.890*	*1,2*			
	Chinese, Indian	*2.210*	*1,4*			
	Other Asean	*360*	*0,2*			
	Other Foreign	*1.490*	*0,9*			
Total		22.690	14,0			

a. Source: *Horeca in Figures 2003*. Set day: July 1, 2003
b. Sales area: the number m^2 that is fitted up as consumption on the spot, excluding the area for the kitchen, bar etc.
c. Duration of exploitation: the average number of years that a firm is registered on name of one proprietor. The figures in the table are for existing firms, not for those who stopped business.
d. Figures for classification into kitchen includes hotel-restaurants and hotel-restaurant-cafés, therefore the sum does not fit the total of the restaurant sector.

Party centers are relatively large, because they usually have several rooms. They earn their living from letting rooms for private parties. This implies that there are a frequent and rather unpredictable number of instances on which the maximum number of visitors may be violated. When the number of guests at a party is uncertain in advance (e.g. a wedding party), it may be difficult for the proprietor to enforce the number of people inside.

Table 11.2 Employees and turnover in the catering industry in the Netherlands[a]

	% of firms with employees on the pay-roll	Average number of employees per company (with employees)	Average turnover (excl. VAT)
Drinking sector	69	8,0	€ 155.000
Restaurant sector	90	10,5	€ 390.000

a. Source: *Horeca in Figures 2003*. Set day for number of employees: September 30, 2002. See Suurmond (2004, appendix 2).

There is a group of establishments, such as pubs/diners, grand cafés and café-restaurants, which serves food as well as drinks. Extra fire danger is present because of the existence of a kitchen. Since people want to have some space at their table, the maximum number of visitors is not easily violated. What may happen is that in order to create more tables, some of them are placed in front of emergency exits or on an escape-route. In addition, since there is an apparent peak in the stir (eating times), the danger of neglect of fire safety by the personnel at such times is high. The same holds for regular restaurants. Differences in fire damage do not seem to be related to the type of kitchen, but more to the target group. The more luxurious, special restaurants will not have problems with the maximum number of visitors.

In 2000, 92% of the Dutch people between the ages of 12 and 74 years (about 12.7 million people) had a drink at an horeca establishment. They did so on average 36 times, amounting to more than 400 million visits. In 2002, average consumption varied from €10 (in cafés) to €20 (in discothèques). Similarly, 88% attended a horeca establishment to eat something, on average 21 times, which makes almost 230 million visits. Average consumption varied from €6 in fast-food restaurants to €14 in Chinese/Indian restaurants and €28 in Dutch/French restaurants.[161]

11.1.2 Competition and ownership

The horeca sector is characterized by monopolistic competition. Within a neighborhood, there are usually various horeca establishments or similar alternatives for eating and drinking, so that price elasticity is quite high. However, not every horeca establishment is the same. There is heterogeneity in the atmosphere, quality and target group. Note also that the horeca sector is aimed at the local market. Individuals do not tend to drive to another city in order to visit a café or restaurant (with the exception of weekend nights by youngsters).

That the market is quite competitive is reflected in the relatively high number of new firms. In 2002, for example, 3.300 of the 37.100 firms in the

161 For more details, see Suurmond (2004*) in which I combine data from various sources.

whole recreation and horeca sector were closed and 1.200 new firms were established.[162] In this sector, bankruptcy and change of ownership is a common phenomenon. Frequently, a new horeca firm is set up in an existing horeca establishment. Sometimes, only the owner of an establishment changes, keeping the same names and brands. In these cases, the new horeca firm also takes over the structural engineering state of the establishment, leading to the same level of fire safety and the same fire safety prescriptions.

In many cases, a horeca firm will consist of (only) one establishment with a single proprietor who is both the owner and the manager of the firm, and who is able to monitor the behavior of his (limited number of) personnel. Often, the horeca entrepreneur will rent the establishment from a brewery, investment holding or otherwise, or he will own the establishment himself. Sometimes, one horeca entrepreneur owns several establishments in one or more cities. In such cases, every establishment is quite independent and will have its own target group. In any case, the decisions about compliance with fire safety regulations are taken at establishment level since fire safety prescriptions are relevant at that level (and not at firm level). All these horeca firms are characterized by a rather simple organization with few hierarchy levels. The manager is usually the person who receives the profits of exploiting the horeca establishment. There are a few larger chains, notably in fast-food restaurants, hotel-restaurants and roadhouses (e.g. McDonald's), which are characterized by a professional organization with possibly special departments for purchase, promotion and dealing with legal-administrative issues.

11.1.3 Regulation

The horeca sector belongs to one of the most highly regulated sectors. Sixteen national laws and regulations have been identified (usually each with their own Inspectorate), in addition to several local bylaws, covenants for copyright and the like.[163] To name the most important ones: the Drink and Horeca Act (for serving alcohol), the Environmental Protection Act, the Act of Area Planning, the Housing Act, the Food and Drugs Act, the Act on Working Conditions and the Building Bylaw. In addition, the Act of Monuments and Tax Laws are also (sometimes) relevant.

Given the large number of regulations and Inspectorates, the importance of fire safety can clash with other safety interests. Horeca proprietors often complain about contradictory requirements. Upon further examination, however (see section 11.3.1), the number of real contradictions appears to be very small. The main problems are either the diverse interpretations and inconsistency of the Inspectorates, or the high costs that may be at stake in applying a

162 Source: CBS, Statline. The year 2002 is the most recent year for which definitive data are available.

163 See Bekkers et.al. (2003*).

satisfying solution. In any case, for horeca proprietors, the fire prescriptions are just one type of prescription out of many.

11.2 Fire safety regulation

I analyze the enforcement of the requirements that concern a fire-proof *use* of an establishment. Let me describe these requirements and the enforcement of these requirements in the past.

11.2.1 *Fire safety prescriptions*

The care for fire prevention is delegated to municipalities in the Fire Services Act 1985.[164] A brief summary of the historical development of fire safety regulation is given in the report of the Committee of Inquiry Café Fire New Year's Eve 2001.[165]

General history of regulation

The Fire Services Act of 1952 reflected the tendency for decentralization following World War II. Municipalities were made responsible for fire safety and therefore adopted municipal bylaws which regulated fire safety. In the seventies, fire safety was professionalized. In particular, there was an increase in the knowledge and quality of technical aspects of fire safety. The Association of Dutch Municipalities developed a model fire protection bylaw in order to structure the many municipal bylaws. In this model bylaw, a use license system was introduced.

A further stimulus to fire safety was provided in the eighties. In 1985, a new Fire Services Act appeared. A real revolution in fire safety regulations came in 1992, when a major revision of the Housing Act took place. In this revision, the Building Decree was introduced, in which all technical fire safety requirements were harmonized into one national Building Decree, which is updated on regular basis. The Housing Act also stipulates that every municipality must lay down a Building Bylaw which must contain (among other things) prescriptions with respect to fire safety. Although these Building Bylaws may in principle differ among municipalities, all municipalities tend to a large extent to follow the model Building Bylaw of the Association of

164 A more extensive description of fire safety regulation can be found in Zoomers (2001*). See also Weerkamp (2004*). Helsloot et al. (2007*) is a general handbook on fire safety and fire safety organization in the Netherlands. Helsoot (2006*) is a short introduction into the enforcement of fire safety regulation.

165 Also known as the Alders Committee (2001*). The historical description is found in Part A, p.60-63 of which I give a summary here.

Dutch Municipalities.[166] As far as I know, all municipalities have adopted the use license system from the model Building Bylaw. According to this system, the fire safety prescriptions can be divided into three parts: (1) construction, (2) installations, and (3) use.

Construction requirements
In order to prevent fires and to limit their spreading, certain structural requirements are laid down in the (national) Building Decree 2003. These technical prescriptions concern for instance the creation of fire and smoking compartments, the use of inflammable materials and the presence of emergency exits and escape-routes. A building license is required in order to supervise whether a building satisfies these prescriptions. The owner must apply for this building license at the municipality, which is charged with the inspection and enforcement of the Building Decree regulations.

Installation requirements
In order to detect and fight fires, buildings must be supplied with certain equipment, such as sprinklers, a fire alarm system, an evacuation alarm system, escape-route marks and emergency lighting. Especially the latter two are relevant for horeca establishments. These prescriptions are laid down in the municipal Building Bylaw. Together with the construction requirements, the installation requirements are specified in the necessary building license and enforced by the municipality. The Act of Occupational Health and the Environmental Protection Act contain regulations with respect to fire extinguishers, evacuation plans and instructions, as well as safety measures with respect to dangerous substances.

Use requirements
The regulations for the fire-safe use of a building are also included in the municipal Building Bylaw. Hence, it is the municipality which is charged with the enforcement of these regulations. Appendix three and four of the model Building Bylaw contain the regulations a user must satisfy. These regulations have a general effect on all the exploiters of a building. In addition to these, Article 6.1.1 (1) of the model Building Bylaw stipulates that structures that satisfy one of the following characteristics must possess a use license:

- fifty or more people can be present at once;
- there is storage of fire dangerous substances;
- night residence is given to more than ten people, commercial or related to care;
- day residence is given to more than ten children or disabled persons.[167]

166 In Dutch: Vereniging van Nederlandse gemeenten (VNG). For the model Building Bylaw, see VNG (2002*).

167 Municipalities differ on the numbers in these requirements (e.g. twenty-five instead of fifty people present) but all municipalities have chosen for a system of use licenses.

The granting of a license (with an unlimited period of validity) guarantees that the municipality investigates whether the structure satisfies the general requirements of appendices three and four. Beside these, the use license may also stipulate the maximum number of people allowed in (a part of) the building. In addition, other building-specific regulations may be included. However, the Board of Mayor and Aldermen is not allowed to stipulate further regulations on the construction or the installations, such as prescribing an extra escape-route. The structural state of the building determines how many people can safely use it. The use requirements stipulate that this number of people may not be exceeded and that the structural state of the building must be well maintained. If a building does not possess a use license, the maximum number of people allowed is in principle restricted to fifty since operating with more people present without a use license is illegal.

I analyze the enforcement of the use requirements, whether or not laid down in a license, *given* the structural state of the building and the present installations. The most important use requirements (for horeca establishments) can be summarized in the following seven points:

1. The (emergency) exits and escape-routes should be kept free. This means for example that emergency exits can be opened immediately if necessary, and that escape-routes may not be blocked by trash-bins, (beer) crates, etc.
2. Furnishings (such as curtains and drapes) and decorations should be correctly hung and impregnated.
3. Escape-route signs (called transparencies) should be clearly visible and burn when people are present, also in case of voltage or current failures. The emergency lighting (if required) must light up in case of voltage or current failure within 15 seconds and keep on burning for at least 60 minutes. Periodical inspections, and if necessary servicing, are required to check whether these installations satisfy the requirements.
4. Fire extinguishers must be indicated clearly (with the correct pictures) and must be inspected annually (and if necessary serviced or replaced) by a professional company.
5. Candle lights, trash, ash-trays etc. must be used safely.
6. The maximum number of persons allowed in (a part of) the building may not be exceeded.
7. Remaining use requirements must be obeyed, such as those in the domain of evacuation plans and personnel instructions, the periodical inspection and service of required fire alarm or evacuation alarm systems, etc.

   Note that parts of these regulations require periodical attention, while other require continued attention by the user (during use). The former type implies that if the user regularly (say yearly) inspects whether he complies with these regulations, he will do so for some period (say a year). Fire safety is then stable in that period. The latter type implies that complying with these requirements is a daily (or even hourly) decision and hence expected fire damage differs from day to day.

The enforcement set-up
Municipalities control compliance with the use requirements. Four types of inspections can be distinguished:

A. On application for/granting of a licence.
B. Periodical inspection.
C. During 'events' (Christmas, Carnival or local events), also called actions.
D. Repressive (after a complaint or report, or after a false fire alarm).

For violations detected during inspections of type A, possible sanctions include not granting a license. For violations detected during other types of inspection, as well as for operating without a license, several administrative sanctions can be imposed:

- (duty under) penal sum;
- administrative coercion (execution by the municipality on the firm's costs);
- (partial or temporary) closing, which is the (ultimate) consequence of the withdrawal of a licence.

Besides these, relevant enforcement tools are: toleration or legalization of violations, warnings, re-inspections, deliberation, etc. The municipality can also use sanctions such as naming-and-shaming, less willingness to cooperate in other respects (e.g. renovation) or, for example, no more municipal parties or receptions allowed in the building.

Offenses can also be enforced through criminal law by the Public Prosecutor, especially when the offenses create clear and immediate danger.[168] The maximum sanction is a fine of €4500 or four months of imprisonment.

11.2.2 *Short history of enforcement and the Volendam disaster*

The granting, inspection and enforcement of building licenses is usually assigned to a department of the municipality which is charged with building and housing supervision. The fire brigade gives advice on the granting of these licenses. In the nineties, (municipal) building supervision became more professional, also due to the pressure from the Ministry of Housing, Spatial Planning and the Environment. New buildings are almost always provided with the necessary building license. Enforcement problems are usually caused by relatively small-scale renovations for which no permission is asked, so that current situations do not necessarily reflect the original building license. As regards the problem of fire safety, these renovations might imply that the maximum number of people that is allowed is smaller than the number stipulated in the original license.

With respect to the use license, most municipalities did not grant, inspect

168 According to article 12.1 of the model Building Bylaw (VNG, 2002*) violation of the use requirements is a criminal act and is punishable by imprisonment for a term not exceeding four months or by a fine of the third category.

and enforce these licenses during a substantial period. Since 1992, the Ministry of Internal Affairs, the Association of Dutch Municipalities and national, regional and local Fire Brigades have often pointed to the importance of a municipal fire safety policy. Only as late as the late nineties did several municipalities start programs for granting use licenses. Priority was given to sleeping and care institutions. What happened in many municipalities was that from time to time, inspections were carried out, especially in horeca establishments in periods preceding Christmas or Carnival. Enforcement in cases of violation hardly ever occurred. Administrative attention for fire safety was low.

The Volendam disaster and its consequences

These problems came to the fore after the Café Fire in Volendam at New Year's Eve 2001. What happened that night in café "'t Hemeltje" in Volendam has been described in the report of the investigating committee:

> Like any year, groups of youngsters celebrated New Year on the dyke. 't Hemeltje was still in the Christmas mode.(...) Shortly after midnight, 't Hemeltje was completely packed...[A]round midnight, the proprietor had opened the doors to some three hundred people in total who were only able to shuffle around. (...)
> Half past one. A boy took a packet of sparklers and lit the packaging with a burst of flame as the result. He got a fright and held the packet upwards consequently setting fire to the Christmas decorations. Panic set in. The fire spread through the dry branches in no time and burning branches dropped onto the floor which people trying to stamp out. Clothing was set alight.
> The temparature in 't Hemeltje quickly rose to some 500 degrees...In less than three minutes the fire had smothered itself.
> Everyone was trying to get away. But where to? People...got stuck in the crowd on the stairs and in the bar itself. There was panic there also. The ...exit has inwards-opening doors....'t Hemeltje had an emergency exit with an emergency lock. The lock, however, was fitted upside down and covered by a beam. Therefore it took quite some time before anyone was able to open it and people could escape using the concrete outside stairs. Meanwhile, people were trying to break the windows on the front of 't Hemeltje. The windows, however, are bolted and have bars in front of them. People were pressed against them. (...)
> But [on the flat roof at the back] there were no stairs leading down off the flat roof. People jumped down. In the mean time, people were fainting because of the lack of oxygen and lay – sometimes in piles of four or five – on the floor on top of each other....The youngsters who were in the toilets...vainly attempted to break the windows of the door on the small hallway.
> (Committee of Inquiry Café Fire New Year's Eve 2001 (or: Alders Committee), English Summary, p.4-6)

In the end, fourteen youngsters died in the fire and approximately 300 people were wounded. It appeared that fire safety in the café was far below standard. The building was not in accordance with the building license that had been granted. The owner had consciously created a situation with inadequate (difficult to use) emergency exits. He had hung fire-dangerous decorations, and he was allowing entry to too many visitors. The owner did not possess the prescribed use license.

However, the municipality of Edam-Volendam was also to blame, because they were consciously not granting the use licenses. Only at the end of 1999 did they employ someone to be responsible for fire prevention. Their only feat of arms was a letter to horeca entrepreneurs on November 30, 2000 to warn against flammable Christmas decorations. The owner of 't Hemeltje did nothing about it, and neither did the municipality.

Enforcement practice

It appeared that such practices, although somewhat extreme, were not exceptional, but representative for many more municipalities. A quick scan by order of the investigating committee showed that:

- Seventy percent of the municipalities did not have an active use licence policy;
- Eighty percent of the municipalities had great backlogs in granting use licences, especially for existing buildings. On 1 January 2002, the Public Order and Safety Inspectorate (IOOV, 2002*) ascertained that in the meantime, almost every municipality had adopted a policy of granting use licences, but that only 29% of the buildings under obligation in fact possessed a use licence, and hence 71% did not. Without further action, this backlog will only be caught up on after 13 years.
- This backlog was accompanied by a personnel shortage, both for granting and inspecting of licences. AEF (2002*) estimated this shortage for the fire safety policy in case of application of the so called prevap-norms[169] at 465 fte. More generally (and importantly), the investigations ascertained a clear lack of qualified personnel.
- In almost no municipality was there a systematic and adequate inspection of the licences. If there were inspections, they usually did not take place at a time the building was thoroughly used. As an indication: horeca proprietors mentioned that in the five-years period of 1996-2001, 20% of them were never inspected, and 75% only once. This period includes the first months after the Volendam disaster, when many municipalities carried out inspections in horeca establishments. Eighty percent of the municipalities did not pay attention to the number of visitors during inspections.
- In most municipalities, a good foundation for the enforcement policy was lacking. Generally, norms and guidelines were lacking on the basis of which licences might be granted and enforcement executed. The IOOV (2002*) gave only 65% of the municipalities a pass for the structural policy with regard to the granting of licences; for enforcement, this was not more

169 The Prevention Action Plan, published by the Ministry of Internal Affairs (1997*), gives detailed guidelines for the inspection of fire safety in buildings. These guidelines give each type of building a priority. Depending on the priority provided in the guidelines an indication is given of the number of hours required for the granting and inspection of a license, as well as an inspection frequency.

than 37%.[170] The Fire Brigade, to which the task of fire safety was usually mandated, had little legal knowledge and was not backed by the legal departments of the municipality. Real enforcement was therefore exceptional.

These observations caused a major increase in municipal effort in fire safety enforcement, especially relating to horeca establishments. Municipalities began to eliminate the backlogs in use licenses and to regularly inspect the licensed buildings. Due to the Volendam disaster, horeca establishments were labeled as having the highest priority.

11.3 The enforcement problems and actions of recent years

The Alders Committee (2001*), which investigated the Volendam disaster, concluded that the enforcement of the use requirements was far below standard. Many municipalities launched new programs to grant licenses and enforce fire safety regulations. These conclusions and actions do not stand alone. They fit into a more general description of the state of administrative enforcement around 2000.

11.3.1 The 'enforcement deficit'

In 1998, the Michiels Committee concluded that the government was seriously lacking in the enforcement of its adopted rules.[171] This "enforcement deficit" appeared to exist in many fields, from environmental to fire safety, from European subsidies to school inspections. For fire safety, the committee concluded that there was a serious backlog in granting (use) licenses and that the inspection of compliance with the rules was not as required. The work of the Michiels Committee led in 2000 to the Project 'High Quality Enforcement', aimed at supporting lower governments in their enforcement efforts and eliminating the enforcement deficits that had been detected.

Several disasters alerted the general public to these problems. The two most important ones are the fire in café "'t Hemeltje" in Volendam on 1 January 2001, and the fireworks disaster in Enschede. On 13 May 2000, a fireworks trade company exploded in the middle of a residential area in the municipality of Enschede, leading to 22 dead, approximately 950 wounded, and much material damage. In both cases, the primary responsibility rested with the

170 See also the references in note 174.

171 Michiels Committee (1998*), officially the Committee Administrative and Private Law Enforcement. This committee followed on the work of several prior committees. Van de Peppel (2006*) provides a retrospective on the Michiels Committee and the notion of enforcement deficits.

entrepreneur, but the government was also to blame for failing to grant and/ or inspect the necessary licenses. Since several governmental agencies had failed to supervise the licenses, the entrepreneurs did not feel the responsibility to take the regulations seriously and accidents could therefore end in disasters.

The reports by the Oosting (Enschede) and Alders (Volendam) investigating committees initiated much attention for the problems regarding both regulation and enforcement, and especially for fire safety. DHV (2002*) reported that regulation is complex, in content as well as form, and is experienced as inaccessible by the business community as well as the executing and enforcing institutions. In addition, the coordination between the departments involved is insufficient. There are no real contradictions in the system or the text of the regulations: the large number of complaints about the conflicts of requirements usually concern conflicts in interests (for instance a horeca establishment in a monumental building).[172]

Moreover, studies by the Ministry of Justice have shown that the number of regulations is rather high,[173] forcing the authorities to set priorities. However, reports also revealed that in most municipalities, a good foundation for the policy is lacking. Norms and guidelines are usually lacking concerning which licenses are granted and when enforcement is executed.[174] These problems are accompanied by a quantitative and qualitative personnel shortage.

11.3.2 *The response*

As a consequence, the attention for administrative enforcement has strongly increased. In response to the Oosting and Alders Committees, large-scale action programs have been adopted to reform both regulation and enforcement. The level of resources for enforcement has increased, at least in relation to other budget alternatives. The detection of enforcement problems has led to different kinds of policy initiatives. At the national level, the work of the Michiels Committee led to the project 'High Quality Enforcement', followed by 'Rich in Enforcement' and 'Enforcement with Effect'. There is an 'Expert Centre Law Enforcement', a 'Service Centre Enforcement', a 'Taskforce Succeeding in Safety', and a 'Complaints Office Conflicting Rules', usually supported by their own websites. The idea of 'programmed' and 'integral' enforcement is stimulated.[175] Instruments such as the 'Table-of-eleven' and

---

172 See also reports by the Ministry of Economic Affairs, *Contradictory Regulations in Practice*, in 's Hertogenbosch, Maastricht en Arnhem.

173 De Jong and Herweijer (2004*) conclude that the number of valid laws has increased since 1980 from 1100 to 1800. Other regulations have also strongly increased. 'High Quality Enforcement' argues, in accordance with the report of AEF (2005*) that a municipality has approximately 500 enforcement tasks.

174 Netherlands Court of Audit (1996*, 2002*, 2005*), IOOV (2002*) and AEF (2002*).

175 See Expert Centre Law Enforcement (2006*).

the 'Risk Matrix' (Ministry of Justice, 2004 and 2002*) have been developed to support municipalities in setting priorities.

These actions have had effect. In many municipalities, plans have been or are being developed to intensify and reorganize enforcement. By now, almost all municipalities use 'programmed and integral enforcement'.[176] Programmed enforcement implies that enforcement proceeds according to plan and not to incidents. The enforcement program includes the following parts: (i) the formulation of a view on enforcement, (ii) an inventory of the size of the enforcement tasks, (iii) a risk analysis (using the Risk Matrix), (iv) the establishment of priorities, and (v) an implementation plan for targets, methods and costs. This must guarantee that the enforcement process becomes cyclical (including a regular evaluation), democratic (the plan must be approved by the council), integral and cooperative (it concerns all enforcement tasks, and internal and external officials are expected to work together), and transparent (priorities become visible).

11.3.3 Evaluation of these policies

Unfortunately, all these investigations, plans and instruments provide little information about the costs, and even less about the benefits of the additional enforcement efforts. The Netherlands Court of Audit (2005*) concludes that enforcement has indeed become more professional over the past years, but that this professionalization has not as yet had much result.[177] In three of seven fields, the court still detected important enforcement deficits due to failures both in regulation and in enforcement (capacity). Choices and priorities often remain implicit. Enforcement officials usually do not know to which extent their enforcement efforts contribute to compliance with the regulations and to solutions for underlying problems. The risks to which citizens are exposed at the chosen level of enforcement remain unknown, and it is unclear how and at what price risks can and should be reduced.

In the academic field, several studies have also investigated enforcement policies, but they merely describe these policies and/or evaluate them in terms of their judicial merits. Sometimes, the level of compliance is discussed and the effects of the policy are described, but the effectiveness or efficiency of the policy are hardly ever examined or discussed.[178] The same is true of

---

176 See Expert Centre Law Enforcement (2006*). The Service Centre Law Enforcement estimates that in early 2007, 85% of the municipalities had an enforcement program. See also Boek and Michiels (2006*) and Boek (2006*) for a description of 'High Quality Enforcement' and programmed enforcement.

177 See also Leeuw and Willemsen (2006*). See, furthermore, the Michiels Committee (1998*) and the Netherlands Court of Audit (1996*, 2002*, 2005*). Similarly, Voermans (2007*) questions the effectivenessis of the project 'High Quality Enforcement'.

178 See for example Huisman (2001*), Hoitink and Michiels (1993*), Blomberg and Michiels (1997*), Van der Tak (1988*), Wiering (1999*). See Huisman and Beukelman (2007*) for an extensive overview of national and international literature.

the enforcement style (compliance or deterrence strategies). Both the national and international literature examine whether these strategies are used.[179] However, whether they are effective, or which is most effective, and under which circumstances, is hardly ever tested.

Recently, Leeuw and Willemsen (2006*) have collected evidence on enforcement at the national level. They show that national enforcement agencies hardly provide any information about the (social) consequences of their actions. Since it is very fashionable to be transparent and informative, many of these agencies do provide information about their performances: how many schools have been visited, how many sanctions have been imposed, how many licenses have been withdrawn, etc. They also investigate customer satisfaction and quality management. However, none of these measures guarantees that inspection and enforcement is effective, let alone efficient. Their conclusions match those of the Netherlands Court of Audit (2005*). Many enforcement agencies have professionalized their organization, but whether this has yielded effects for compliance is unknown. Enforcement agencies do provide more or less information on their *output*. However, what we want to know is how this output affects the level of compliance, and even more, how compliance affects the expected damage and compliance costs (*outcome*). There are, as yet, no data on this relationship.

These conclusions are not typical for the Netherlands. For instance, in the UK, a program on Better Regulation has been launched over the past years. Reviews by Macrory (2006a, 2006b) conclude that enforcement authorities publish on their output, but not on the effectiveness of their enforcement policy. "Tangible data is absent in this area" (Macrory, 2006b, p. 24). He recommends that "regulators and government departments should make every effort to identify and measure regulatory outcomes" (p. 32).

11.3.4 The analysis in the next chapters

In general, there is no knowledge whatsoever about the effectiveness, let alone the efficiency of the increased enforcement efforts. I therefore focus on a specific problem in the hope that this knowledge may be available at a lower level. I investigate the enforcement policies of the use requirements for fire safety in horeca establishments which are enforced at the local level.

The next chapters all deal with a part of the main question: what is an effective and efficient enforcement policy for enforcing fire safety in horeca establishments? Given the lack of direct available data on enforcement, compliance and expected damage, several alternative methods are employed to collect and analyze the enforcement policy. Chapter 12 analyzes the current administrative enforcement policy. I describe how I obtained information about the enforcement policy of municipalities by interviewing enforcement

179 See the references in note 128.

officials and collecting the available policy documents. I investigate to what extent this policy is in line with the description in part I of an optimal enforcement policy. In chapter 13 I use the data available in national reports and supplemented by experts to examine whether the policy of municipalities following the Volendam disaster passes a cost-benefit-analysis. In chapter 14 I search parliamentary documents, newspapers and information of relevant organizations to collect data on the settlement of the damage claims of the victims of the Volendam disaster. I analyze this settlement process to learn whether private enforcement by victims is effective in inducing proprietors to take precautions against fire damage. Finally, in chapter 15 I integrate all the available information that I can find to simulate enforcement policies for a representative municipality. This simulation investigates which enforcement policy is efficient.

12 Analyzing current municipal enforcement policies: Which strategy is effective?

In this and the next chapter I analyze the municipal enforcement policies as they were implemented in response to the Volendam disaster. This chapter evaluates the enforcement policies in 13 municipalities in light of the economic literature discussed in part I. In section 12.1 I first describe how I obtained information on these municipalities. In section 12.2 I give a general description of the enforcement policy in these municipalities. In the following sections I analyze this policy, from the perspective of the issues covered in part I of this thesis. Section 12.3 discusses whether the observed policy confirms the Harrington Paradox. Section 12.4 analyzes whether the observed use of a compliance strategy is justified. Section 12.5 discusses the legal constraints applying to the fire safety enforcement officials. Section 12.6 returns to the general policy developments described in the previous chapter: to what extent have these led to an improvement in enforcement actions for fire safety regulation in horeca establishments? Section 12.7 concludes.[180] The next chapter considers the benefits and costs of these policies.

12.1 Data collection for municipalities' enforcement policies

The analysis of the effectiveness and efficiency of the enforcement policies requires information on these policies. Several national reports exist that will be used in chapter 13 in a nationwide cost-benefit analysis (e.g. IOOV, 2002*, AEF, 2002*, and AVD, 2004*) as well as on websites reported in section 11.3.2. However, these suffer from the same problem as most enforcement information: these sources mostly provide information concerning output, and hardly any information concerning outcome.

In order to obtain more information on enforcement policies, I try to obtain information at the local level, from municipalities. In 13 municipalities I investigate current enforcement policies and interviewed local officials.[181] This investigation serves two goals. First, together with the national reports, it should provide a description of the actual enforcement actions of municipalities. Secondly, information and expertise at the local level should yield insight into the effectiveness of these enforcement actions. By analyzing the

180 The analysis in this chapter, especially sections 12.1 to 12.4, is also presented (in a shorter version) in Suurmond (2007).

181 Appendix 1 provides a list of persons interviewed, the municipalities they belong to and some more background on these interviews.

enforcement policies of different municipalities and/or in different years, a judgment about the most effective and efficient policy should be possible.

This research focuses on the current administrative enforcement policy following the Volendam disaster.

12.1.1 *The relevant factors and the required information*

What are the relevant enforcement factors? How is compliance by horeca proprietors explained?

First, we need information on the quantity and level of enforcement actions as well as the number of violations. How do horeca proprietors pass through the enforcement chain? How many inspections are performed? How many violations are found? Which actions are undertaken against these violations?

In the second place, we must focus on the characteristics of the Fire Department, as well as the other relevant departments and their officials. Part I of this thesis reveals that the level of compliance does not only depend on the expected level of the sanctioning risk and the quantity of the enforcement actions, but also on the manner in which they are executed, in other words, 'the quality'. The most important factors are the enforcement style applied by the enforcement officials (a compliance or deterrence strategy) and how they use their discretion or what they want to achieve by their enforcement efforts (the objective). It is also important to know to which extent the enforcement officials are informed about the costs of compliance and the expected damage.

Finally, we want to know whether the effect of enforcement depends on the characteristics of horeca proprietors. First of all, the level of compliance can be explained by the level of the costs of compliance. Which factors explain differences in compliance costs? And what are the other relevant constraints on compliance and/or characteristics which explain differences in the level of compliance? How relevant are other compliance motives, such as concerns for one's reputation or (intrinsic) concern for customers? And to what extent are the proprietors informed about fire safety rules and the enforcement policy?

12.1.2 *Sources of information*

I investigate three possible sources of information. First, I study the written material including policy documents and plans, (annual or quarterly) reports concerning fire safety, information brochures, websites, etc. I could find this information on the municipality's website myself, or asked the municipality to send me the documents. Secondly, I investigate whether the municipality and Fire Brigade possessed useful databases of enforcement actions, compliance levels etc. Finally, I conduct interviews with enforcement officials and proprietors to obtain relevant information that is not available in a written source.

Sixteen municipalities were selected beforehand, from four Fire Department Regions, in order to create a representative selection of all Dutch municipalities with more than 15,000 and less than 250,000 inhabitants and at least 20 horeca establishments. Regions were selected in such a way that the geographical spread included rural, urban and touristy municipalities, as well as those celebrating Carnival.

12.1.3 *Response and available information*

Thirteen municipalities were willing to cooperate. In these municipalities, I studied the available policy documents and interviewed one or more officials. No databases were available which contained detailed information on all enforcement steps. Some municipalities do have (computer) databases, but these only store information concerning the granting of licenses, and occasionally on when (official) inspections are held and whether formal sanctions are imposed. The (more important) process around each inspection and the findings of these inspections are not systematically collected.

Therefore the major information has to come from interviews with local officials. In each municipality I interviewed officials with three types of functions (which are sometimes represented by one and the same person, so that less than three interviews were conducted): (i) executive officials, managers, who are responsible for policy plans, priorities, finance etc.; (ii) inspectors, usually from the Fire Brigade, responsible for the actual inspection of horeca establishments; (iii) officials concerned with legal actions (including sanctions).[182] The interviews were conducted with a basically unstructured list of questions.

The interviews with horeca proprietors were not successful and/or failed to provide relevant information. Their self-reports concerning compliance and enforcement actions were too general, unreliable or even invalid. In a first try-out, I had the impression that the proprietors were giving 'desirable' answers (despite my assurance that I was not in any way connected to the municipality), especially concerning their compliance behavior. They all assured me that they fully complied ("of course", "never something wrong"). They only provided very general answers about the costs of compliance ("takes much time", "fire safety is exaggerated"). If asked about the costs of compliance, they most often referred to the costs they made in the course of obtaining a use license. Finally, they were unable to provide more precise information about the inspections (frequency, duration etc.) of the fire safety department. I have therefore not invested further in collecting information from horeca pro-

182 Sometimes the officials are employed at different departments, like the Fire Department (inspection) and the Building Inspection (licenses). In addition, back-up is often provided by the Legal Department or the Department of General Affairs. I use the term fire safety department to mean the department or departments that are charged of granting and inspecting use licenses and imposing sanctions.

prietors. They did not provide information about the relationship between enforcement activities and their own behavior, or about the costs and benefits of compliance. It was my estimation that the most reliable information about enforcement actions came from the municipalities themselves.

12.1.4 *Interviews and interpretation of the results*

Given the lack of detailed, written, objective information on compliance levels, the main sources of information are interviews with the enforcement officials. This implies that an evidence-based conclusion about the effectiveness or efficiency of enforcing fire safety is not possible on the basis of these data.[183] Besides the information available in the written material, the data consist of answers, impressions and opinions of the enforcement officials.

That does not imply that these data are useless. On the contrary. Given the lack of quantitative data, interviews are an important tool to obtain more detailed (qualitative) information about enforcement actions, compliance and the interaction between enforcement officials and proprietors. Based on the interviews, the additional written material and the national reports, it is possible to offer a picture of the enforcement policies and actions of fire safety departments. That has merit on its own, because we need to know more about actual enforcement policies. Moreover, the interviews and the policy documents enable a critical evaluation of this policy in the light of the literature discussed in part I. I try to explain the observed policy, its effects, and the observed level of compliance. At first sight, the observed enforcement policy confirms the Harrington Paradox: despite the low expected sanctions, the level of compliance seems to be pretty high. Moreover, the observed enforcement policy fits the description of a compliance strategy. The following sections analyze these two observations after describing the policy itself.

12.2 Description of current municipal enforcement policies

Let me try to summarize the most important observations concerning the common enforcement policy and actions in the 13 municipalities. What this research brought to light confirms the information from other sources describing the national and local approach to fire safety.[184] It appears that the enforcement actions of municipalities do not seem to differ greatly, at least

183 A related problem is that there is too little variance (in time, between municipalities) to be able to compare different strategies. There is no 'control group'. All municipalities began intensifying enforcement following 'Volendam' and the accompanying investigating report and action program. There is no research for the period prior to Volendam.

184 See for instance AEF (2002*) and AVD (2004*), newspaper articles and municipal policy documents. See also the more general enforcement studies and websites discussed in section 11.3.

not when it comes to actual actions and impressions of compliance by proprietors. If relevant I describe the differences. Otherwise I speak of the municipal enforcement policy that might be considered as the general policy that all municipalities apply.

Granting use licenses and history prior to Volendam

Prior to the years 2000/2001 (the Volendam disaster), most municipalities did not have a structured supervision of fire safety in horeca establishments, although a number of municipalities had launched some kind of use license-policy. They carried out only repressive inspections, and did so scarcely, as well as an occasional action during Carnival.

- Two municipalities did, in 2004, not as yet have a policy, although they were aware that they should start one, and their supervision of the use of buildings was uncoordinated.
- One municipality did not grant licences, but inspected cafés every year at Carnival.
- Four municipalities had (recently) adopted a policy plan, that sometimes had also been launched but was far from completed. These plans focused especially on granting licences (rather than on inspection).
- Two municipalities began granting licences a longer time ago but had not as yet finished and did not always have a full inspection and sanction policy.
- Four municipalities had mostly licensed horeca establishments, although the licences were in some cases outdated. Only one municipality had a policy of periodical inspections, once per two years. Two others carried out inspections every year at Carnival.

The municipalities differ in their opinion concerning the disaster in Volendam and the catching-up required. Some argue that 'Volendam' might have happened anywhere while others argue that what happened there was exceptional and should in principle not be possible in their municipality. There is, however, no relationship between these comments and the extent to which the municipalities in question have or had an active enforcement policy. Almost every municipality indicates that, over the past years, the willingness and cooperation of horeca proprietors has increased, that horeca proprietors are better informed about the rules, and that fire safety has therefore improved. On the other hand, five municipalities warn against an 'exaggerated' reaction ("witch hunt") following the Volendam disaster and two others warn that actions should not lead to 'paper safety'.

Aims of fire safety department, political involvement and organization

The aim of most fire safety departments is either seeking maximal or sufficient fire safety, or achieving the legal standards (as a minimum safety level or just because it is legally required). Most officials argue that they do take the costs and 'reasonability' into account, but do not accept deviations from the legal obligations. The flexibility lies in the time that is allowed for taking the necessary precautions. There is also some flexibility in applying equivalent solutions, but this only applies to granting licenses and construction or installation requirements.

The fire safety department functions quite independently. Although the political attention for fire safety is (much) stronger than before, the political involvement with fire safety policy is low. Since Volendam (2001), almost every municipality has some policy, but the political administration merely approves the plans made by the fire safety department. The only active involvement consists of requiring that such plans are made and of setting priorities for granting licenses. The daily activities are not strictly monitored by political powers. The financing of fire safety departments with respect to enforcement consists of a lump sum. The granting of licenses is financed through legal fees.

Objective and method of inspections

All municipalities stress the importance of informing the horeca proprietors about the regulation and about compliance. They argue that information and explanation (1) create an understanding of the rules and of the importance of complying with the rules, and (2) alerts horeca proprietors to their own responsibility, and thus leads to cooperation and eventually compliance. Fire safety is something all proprietors have to be reminded of now and then. In addition, there are always some troublemakers who must be kept on a tight leash.

Therefore, most municipalities say they begin with a cooperative attitude, aimed at informing, advising and persuading the proprietor to comply. Inspection is only a random indication, in which the personal responsibility of the proprietor is emphasized (personal responsibility is mentioned by all officials, directly in 8 municipalities, indirectly in the others). However, if the proprietor is not going to comply, a more stringent policy is adopted which may eventually lead to sanctions (more 'police-like enforcement'). External, private consultancy firms are judged to be more 'police-like'. Moreover several officials stress that they do try to be kind and cooperative, but simultaneously try to keep distance in order to show that enforcement is taken seriously. This allows them to easily switch to a more formal approach if necessary. For example, some argue that they do not always accept coffee or other drinks, so that they do not owe the proprietor anything.

View on horeca proprietors

Most municipalities are quite satisfied with the cooperation of horeca proprietors, although there is always a group of unwilling proprietors. Most municipalities indicate that this group is maximally 20%. As mentioned above, all municipalities argue that the willingness and cooperation of horeca proprietors has improved over the last years, because they are better informed and more willing to comply, because they are more conscious of the importance of compliance, and/or because they fear their own liability in case of a fire. Three municipalities complain (more extensively) about cooperation. They argue that horeca proprietors are only willing to comply as long as it is not too expensive. This is particularly problematic in cases involving 'contradictory requirements' (requiring expensive solutions) or fear of bankruptcy (so that proprietors do not want to spend money on fire safety).

Two municipalities indicate that the relationship between the fire safety department and the horeca proprietors is reminiscent of an 'old-boys-network'. Most of the other municipalities say that they are familiar with the proprietors they are dealing with at a general level.

The extent to which proprietors are informed about the rules seems to differ. Two municipalities argue that proprietors are very well informed about the rules. Most others argue that they are generally informed about the rules, while one states that they are only moderately informed. Most argue that the horeca proprietors themselves do not realize the risk of fire safety but only do so because of the actions of the fire safety department.

Concerning other characteristics, the reports do not include many systematic and consistent differences between different types of horeca establishments. The only important thing mentioned by several officials is that (small) bars or restaurants that have recently been established and/or are struggling to survive, are less willing to comply, and that the larger establishments are more accustomed to regulation and enforcement. Other points are reported by one or only a few officials, such as the fact that non-Dutch proprietors cause somewhat more trouble because communication is difficult. However, it is also argued that this only makes a difference during the period of re-inspections. The "Mediterranean" (Greek, Turkish, Moroccan, etc.) and the "Asian" proprietors do try to negotiate more on the necessary improvements ("this way good too?"). On the other hand, a uniform has more authority for these proprietors. The Mediterranean proprietors are often not members of the sector organization Royal Horeca Netherlands (RHN) plays a double role. On the one hand, it is an important discussion partner which can be helpful in passing on information, contacting the target group, and increasing cooperation. On the other hand, RHN also requests 'sweeping actions' and a more stringent approach, because this will harm the establishments not connected to the organization. The smaller bars in the neighborhood do seem to be less aware or frightened from the possibility of a fire ("nothing will happen here"). The often older proprietors were not raised with fire safety regulations and often either are or play at being ignorant. Anyhow, for characteristics like these the interviewees do mention differences between proprietors or between establishments, but it remains unknown how these differences affect compliance behavior.

Inspections

a. *Periodical.* In fact, all municipalities inspect, or try to inspect horeca establishments periodically once a year. All municipalities but one usually carry out their inspections during the day, most often either announced or by appointment. However, not all municipalities are yet ready to begin a structured inspection of licenced horeca establishments.
   Most periodical inspections are performed by one inspector. The time

per inspection varies from 20-30 minutes to 45-60 minutes. This excludes travel time and administration.

b. *Action.* Most municipalities carry out some kind of action around either Christmas or Carnival. In the last years, the horeca sector was often a priority. However, some municipalities indicate that there will in future not always be a horeca action, but that other themes may be chosen (e.g. care institutes, churches). These inspections usually consist of a quick sight spot that lasts about 10-20 minutes and is done by 1 or 2 inspectors, dependending among other things on whether it is a joint (integral) action with other governmental agencies. Such actions focus on the establishments located in the centre or other important areas in which many establishments are grouped together.
   In addition to these actions, municipalities also carry out inspections during special events or feasts. Especially the smaller villages have 1 or 2 local events per year. Preceding or during these events, the fire safety department inspects the fire safety. This is especially important when special tents are put up.

c. *Repressive.* Every municipality takes reports seriously, but they are not judged to be important. No municipality has experience with reports leading to sanctions. The violations found during repressive inspections are not deviant from those encountered during regular inspections.

Violations

Almost every municipality indicates that they are able to detect at least one violation in almost every establishment (more than 90%). What's more, even if the inspectors are not over-zealous, most establishments must improve compliance on some points. The number of troublemakers who are not willing to cooperate and/or inclined to comply, is estimated at about 10-20%.

With respect to the actions, hardly any violations were found that are specifically related to Christmas or Carnival (such as decorations). With respect to the specific violations the following can be observed:

- Failures in escape-route signs are most common. This violation is mentioned as the most frequent violation. However, municipalities differ in the evaluation of the seriousness of this violation and in the way they respond to it.
- Blocking the emergency exits is seen as one of the most serious violations. However, since most inspections take place during the day and announced, it is not known how this rule is complied to during full use. This also holds for the maximum number of visitors allowed.
- Emergency exits are regularly observed to be locked. Discotheques wish in this way to prevent youngsters from entering without payment. Other establishments wish to be sufficiently burglar-proof. In some cases, a proprietor may have no time or inclination to unlock the exits every time someone is present.
- Decorations have improved over the past years. In reaction to the Volen-

dam disaster, the inspection of decorations has increased. New decorations is almost always impregnated according to the norms. The problems arise with old decorations. Some municipalities say that due to the trouble of strict inspections, proprietors have decided to reduce the (temporary) decorations in their establishment (only one municipality states that decorations are still a recurring problem).

- Since fire extinguishers are certified by private companies and most establishments now conclude service-included contracts with these companies (sometimes following the advice or pressure from the fire safety department), compliance has improved and is no longer a major problem. However, some report that problems arise because the private companies fail to fullfil their obligations.

Re-inspections and sanctions

The cooperative proprietors in particular tend to restore compliance following an inspection. The troublemakers often try to stretch the process. Some municipalities choose to always (or almost always) re-inspect. Others choose to re-inspect only in cases of more serious violations and/or if proprietors are judged to be uncooperative. In these cases too, re-inspection occurs in more than 50% of the cases.

Hardly any formal sanctions are imposed, except at the stage of granting use licenses. If municipalities were to impose such sanctions, they would use penal sums and the temporary closing of the establishment (until the necessary improvements had been carried out). Several municipalities complain about the ineffectiveness of administrative sanctions and procedure. Horeca proprietors are able to stretch the process. Troublemakers are often smaller establishments that have recently been founded and/or are likely to go bankrupt soon. Many Fire Departments are afraid of the legal requirements they must fulfill in order to impose (administrative) sanctions. Furthermore, these departments report (especially in cases involving licenses) that they often are uncertain as to who the sanction should apply to: the owner or the tenant. Moreover, once they are finally ready to execute a sanction, the ownership-structure is sometimes changed so that they have to start the whole enforcement process anew.

Some municipalities indicate that they have difficulty enforcing the number of visitors. First, it is difficult to determine the number of people inside. Unless the fire safety department uses costly tools, it can only give a general estimation of the number of people present and the proprietor can contest this estimation. Secondly, sanctioning is difficult because of the fear of violation of public order. If an establishment is suddenly closed, it creates a lot of trouble due to the sudden large number of people on the street. The Mayor is often not inclined to carry out such closings, nor is the police (if they are present at all). Therefore, the general solution is to ensure that no one may any longer enter the establishment, only exit (flowing off). Note however that many municipalities do not inspect the number of visitors at all, so that these discussions are partially theoretical. Moreover, several municipalities report that the number of visitors is not easily violated. There is usually enough opportunity to fill up the horeca establishment. Horeca proprietors are glad if the establishment is completely filled, but their own maximum is often also the maximum number that is stipulated in the use license.

Criminal law enforcement is possible but does not occur (except for one Christmas action). Public Prosecutors and Police do not have (or make) time for it.

Many problems arise when several municipal departments are forced to cooperate, for example because the department granting licenses (e.g. Building Inspection), the one responsible for inspections (e.g. Fire Department) and the one that imposes sanctions (e.g. Legal Department) or registers the collection of sanctions (e.g. Finance), all use different computer systems. Moreover, the other departments (may) give fire safety less priority than the Fire Department, so that files remain open.

Miscellaneous

The following interesting points also arose:

- Municipalities themselves are often very slow in installing fire safety devices in their own buildings (and/or in applying for use licences).
- The image of external, private firms which help to grant licenses, and in the future possibly also to inspect them is mixed. On the one hand, some argue that these firms are objective and obey procedures, while the fire safety department is not evaluated by anyone and may have a double hat. On the other hand, it is argued that external, private firms are too strict in applying rules and are unable to provide made-to-measure solutions. Moreover, the quality of their work is poorer because they are often only jugded on the number of licenses that is granted and because of a personnel shortage. Due to the increase in enforcement following Volendam, "masses of people have taken a course", so that the quality of inspectors diminishes and experienced personnel is scarce.
- Almost no municipalities advocate a system in which establishments receive an identifying mark that expresses the safety of their building. An inspection only provides a random indication and should therefore not be followed by a mark with long-term validity.
- One of the interviewees mentioned that "there are thousands of rules that you should enforce. Rotterdam once began counting, but stopped at 4000. I have heard the number 7000 mentioned. All things with which a municipality is concerned and that you might potentially enforce. But should you? A research bureau by order of Justice comes to 300-400 enforcement tasks that you can reasonably pick up. For a relatively small organisation that is still quite substantial." And another one: "For an information afternoon for the horeca I counted how many governmental organizations a horeca company is involved with. I came up with 26 departments (of which fire safety is one)." Yet another: "16 departments are concerned with the enforcement consultation".

12.3 A HARRINGTON PARADOX?

Several studies have identified a Harrington Paradox (section 8.1), which implies that there is a puzzling paradox between the level of compliance and the expected sanction. At least the first two observations of Harrington (1988) are confirmed here. The frequency of inspection is low and predictable, and sanctions are small, almost absent. The level of compliance is more complicated. Enforcement officials indicate that they do find many violations during inspections, yet they argue that the level of fire safety is in general sufficiently high. In this section, I investigate whether we can indeed observe a Harrington Paradox. I argue that the observed enforcement policy fits well with the general rational choice explanation. The level of compliance corresponds to the level of the expected sanction.

12.3.1 Low inspection rates?

Most bars and restaurants are inspected once a year (besides re-inspections). Bars and sometimes restaurants in the centre of the city often face an additional inspection during events or feast periods. Given the low penalties, this might seem to provide insufficient incentive to comply. However, a distinction should be made between two types of requirements. Some requirements concern continued daily attention on the part of the proprietor, such as keeping escape-routes free, carefully handling candle lights and limiting the number of visitors. For these types of requirements, an annual inspection is insufficient; even weekly inspections might be ineffective. Other requirements, however, concern the maintenance of technical devices requiring an annual check up, such as the escape-route signs, the fire extinguishers and the decorations. If the municipality carries out one inspection a year, it can induce the proprietor to carry out this type of check-up and guarantee that he is in compliance for the rest of the year.

Moreover, there are several reasons why a low inspection rate does not necessarily imply that serious non-compliance remains undetected. If the proprietor severely violates the rules, this fact is likely to be observed and reported to the fire safety department: visitors, who are not generally afraid of insufficient precautions, will be concerned about serious threats to life and body, or at least one of them will. In addition, other municipal departments are very likely to take notice and inform the fire safety department. For example, a bar or restaurant is also inspected for hygiene, serving liquor, public order, etc. These inspectors might not notice minor violations, but will most certainly report severe violations or a broad non-compliance record concerning all regulations. Finally, a bar or restaurant that has been shown to commit serious violations will sometimes be inspected more often and at least more thoroughly and more strictly.

12.3.2 *The threat of sanctions*

Formal sanctions, and particularly criminal sanctions, are hardly ever carried out. However, this does not imply that there is no *threat* of sanctions. Again, we have to distinguish between the serious violations and the minor ones.

For serious violations, the threat of sanctions suffices to induce proprietors to comply. For these violations, criminal prosecution is a real possibility. Moreover, if a proprietor commits such a serious violation, the municipality can close the establishment until the violation has ended. The use of these sanctions is credible. If the fire safety department can show that there is real danger, the municipality will not want to risk ignoring its responsibility and being blamed for inertia if something happens. The threat of these sanctions seems to be sufficiently high in relation to the costs of compliance, so that almost all proprietors comply. As a consequence, the real imposition of these sanctions is hardly ever observed.

Minor violations are a different story. They do not involve the threat of criminal prosecution or immediate closure. On the contrary, should the inspectors detect non-compliance the proprietor will receive a (written) notification that he must restore compliance. The fire safety department will almost always carry out a re-inspection. If the proprietor has adequately responded to this warning, no further action is undertaken. If the proprietor ignores the warning, the threat of sanctions becomes increasingly more severe. Usually, after 2 or 3 re-inspections, the municipality will issue a formal warning that the proprietor must restore compliance failing which he will have to forfeit a penal sum or his establishment will be closed. These formal sanctions do not have to be imposed very often, and the actual operation of the threat is never used. The threat of a legal procedure and of formal repair sanctions is sufficient to enforce compliance, even for uncooperative proprietors. The administrative sanctions are expensive for proprietors. If they are actually imposed, the financial costs can be quite high. Moreover, just the time and procedure involved in (re-)inspections is expensive for proprietors, especially if the proprietor knows that in the end he will only be able to pacify the fire safety department by complying.

In fact, the policy in cases of minor violations is a variant of the state-dependent, targeting enforcement policy proposed by Harrington (1988) and others.[185] Harrington (1988) demonstrates that it might be effective to use a system of warnings in which firms are in first instance not sanctioned but moved to a target group. If the firm repeats its violation in this target group, it is sanctioned. This is what happens when following a first inspection the proprietor is given a warning. It appears that this almost always suffices to induce compliance in the target group. As a consequence, sanctioning in the non-target group is absent and observed compliance is high. Proprietors have no incentive to choose compliance from the beginning but can await inspection and see whether they are moved to the target group.

185 See section 5.3.

There is one important problem reported by several municipalities. Proprietors who do not do well and fear bankruptcy, seem not to care much about the (financial) consequences of sanctions. They already assume that they will go bankrupt and do not care whether this happens as a result of administrative sanctions or for other reasons. These proprietors take every opportunity to postpone or even escape current expenses on fire safety. They are therefore the ones who cause most problems and for whom the fire safety department is most strict and inflexible.

12.3.3 *The level of compliance*

Although there are no hard data concerning the level of compliance, there is at least an impression from the enforcement officials and some evidence concerning the actions they undertake. They do not detect many situations in which the technical state of the establishment requires immediate action. Only one of the thirteen municipalities reported such a situation over the past years. Given the expected threat of sanctions, this is in accordance with the general economic model. The enforcement officials also failed to find serious violations of the regulations that require continued attention of the proprietor. However, given that they generally carry out announced inspections in the daytime, it is not unexpected that they should fail to detect many of these violations. Moreover, if they do observe for instance that an escape-route is blocked, this might be, and often is immediately corrected, so that no further action is taken.

Minor violations are very often detected. Enforcement officials report that there is always something wrong. In less than 10% of inspections does the report state that there were no failures at all. In more than 50% of the cases, the enforcement officials announce a re-inspection. Apparently, in these cases, there were enough violations to warrant a re-inspection.

With respect to the different type of requirements, the following observations can be made:

- The maximum number of visitors is problematic for only some of the establishments. Some proprietors do not want to exceed the maximum themselves because they want to have sufficient space for their visitors. Other establishments do not tend to attract such numbers of visitors that the maximum number would be violated.
- Easily accessible emergency exits are especially problematic for discotheques, because these establishments want to prevent youngsters from entering without payment. For other establishments, the problem is sometimes caused by the fact that emergency exits are not sufficiently burglar-proof. Alternatively, in some cases the proprietor has to unlock the exits every time someone is present (for which he may have no time or inclination).
- Temporarily blocked emergency exits (because of a table or beer crate)

can hardly be prevented through administrative repair sanctions. Such violations would never lead to an official sanction. If an inspector detects the violation, the proprietor can easily remove the object blocking the exit. Since he is then in compliance, no sanction can be imposed. Therefore, especially for this type of violation, fire safety departments plead in favor of an administrative fine for repeated violators.

- Since it is quite easy to repair a failing escape-route sign, the proprietor usually responds to the instruction to repair this type of failure. However, proprietors usually wait with repairing until after they have been inspected. Therefore, these violations are very often observed on a first inspection.
- Repairing fire extinguishers is more difficult because this must be done by a professional company. Due to the periodical inspections of the fire safety department, the proprietor knows that sooner or later he will have to certify the fire extinguishers. Therefore, many proprietors have concluded contracts with these certifying companies. This is probably cheaper than irregular certification.
- Violations of the decorations are not often detected. Since this was a major problem in the Volendam disaster, fire safety departments have been very tough with respect to this issue. In response, certification has improved and most proprietors only buy certified decorations or simply choose not to decorate anymore. However, most fire safety departments inspect the decorations especially during Christmas or Carnival. It is unknown whether the decorations during soccer matches or other events throughout the year also satisfy the requirements.

12.3.4 A Harrington Paradox?

The level of compliance is not unexpectedly high relative to the expected sanction. Enforcement officials do find many minor violations (during a first inspection) because proprietors are not punished, but only warned that sanctions might follow in future. This enforcement policy is as described by Harrington (1988). In the first instance, violators are not punished but moved to a target group in which they are more closely monitored by means of re-inspections until they comply. In response, proprietors choose to delay compliance until they are inspected. The only important point is that they should not willingly and deliberately create a fire-hazard situation. There is a credible threat of sufficiently high sanctions that proprietors are deterred from these major violations. As a consequence, actual imposition of sanctions is hardly ever observed.

Recently, a similar point has been made by Nyborg and Telle (2006). They conclude that there is no Harrington Paradox in the compliance with environmental regulation in Norway. For major violations, the threat of punishment suffices to induce compliance. For minor violations, enforcement is weak, but these violations flourish. Based on the evidence available in the literature, they conclude (p. 14): "Thus, until additional empirical evidence is available

– for Norway or for other countries – the claims that firms comply with environmental regulations to a surprisingly high degree must be regarded as a yet unconfirmed myth, rather than as an established fact." With the additional evidence presented here for fire safety regulation in the Netherlands the claim remains a myth.

12.4 Is a compliance strategy effective? The defenses analyzed

The main puzzle to the economic model is not that the level of compliance does not correspond to the expected sanction, but why enforcement officials choose a cooperative enforcement policy (a compliance strategy) that is different from the standard economic model of strict enforcement of offenses (a deterrence strategy). Enforcement officials defend their policy by arguing that advice, persuasion and warnings are more effective than direct punishment. They start with a cooperative attitude with no intention to immediately prosecute an offense. The remainder of this chapter is concerned with analyzing the use of a compliance strategy. This is an important element of the question which policy is effective and efficient for enforcing fire safety regulation. In section 12.7 I conclude by describing the case of enforcing fire safety regulation in light of the general description of section 10.3. Several steps will lead to this conclusion. First, in this section, I analyze six general claims of the enforcement officials with respect to the desirability of advice, persuasion and warnings. Secondly, in section 12.5, I analyze whether the conditions for a failure of a deterrence strategy are satisfied. Finally, in section 12.6, I argue that a major problem is the fact that both a compliance and a deterrence strategy fail if paperwork and policy development dominate actual enforcement actions.

This section analyzes the arguments of the enforcement officials that were interviewed in favor of a compliance strategy. These arguments correspond to the academic arguments (see chapter 8). In the interviews of the enforcement officials six general claims can be detected that defend a compliance strategy. These are analyzed below.

1. Most proprietors are insufficiently aware of the regulations, of compliance methods and of (the importance of) fire safety. Inspection is required to explain to the proprietor his failures and the danger of non-compliance. Enforcement is characterized by advice.
2. Proprietors can not always help being in non-compliance. Inspection is aimed at restoring compliance and thereby fire safety.
3. Most proprietors are willing to comply. There should therefore be no witch-hunt on non-compliance. Inspection is needed to persuade the proprietor that compliance is socially and morally desirable.
4. The problem with imposing sanctions is that many strict legal requirements have to be satisfied, which costs much time and effort. Inspectors achieve more compliance, and they achieve it sooner, through explanation and cooperation.

5. As long as the proprietor appears to be cooperative, the enforcement officials are cooperative too. However, should the proprietor turn out to have no intention of complying, the officials begin a strict, deterrent enforcement strategy. Some say that they do not prosecute every offense. Their primary question is: is this establishment fire-proof? Others disagree and regard the legal requirements as a minimum that must be satisfied. One should not compromise on this issue.
6. Fire safety is the personal responsibility of the proprietor. Inspection is only a random indication to stimulate proprietors to take this responsibility.

12.4.1 *Information and advice – non-compliance by ignorance*

The enforcement officials argue that inspection is needed in order to explain to the proprietor the requirements and the importance of fire safety, and to show him which shortcomings they observe and how he can deal with these shortcomings in the future. Most proprietors are insufficiently aware of the regulations. A relaxed, cooperative enforcement style, in which officials take time to explain relevant matters to the proprietor, is more effective than immediately punishing violations. Therefore, inspectors make appointments for inspections and take plenty of time for their inspections.

The theoretical defense

If a proprietor is insufficiently informed about the relevant costs and benefits, he is unable to adequately choose a level of compliance (section 6.1). It is likely that proprietors do not naturally possess information about expected fire damage as this is relatively expert, technically complicated information. How fires occur and spread, how to prevent harm to persons present, etc. is not common knowledge. It is not a natural by-product of exploiting a bar or restaurant. If the proprietor underestimates the probability of a fire, he will fail to sufficiently comply if he is sanctioned for the occurrence of a fire. If the proprietor does not know which standards apply, because they are too numerous or too vaguely formulated, he may choose either excessive or too limited precautions. For example, the fire safety regulations contain requirements about the maximum number of seconds before emergency lighting should burn in case of a power failure, or the minimum number of minutes that escape-route signs have to burn. Proprietors cannot be expected to know the regulations in such detail.

Moreover, it may be inefficient for every proprietor to incur the costs of fully acquainting himself with the law and the least costly method of compliance. If the costs of public inspections are small relative to those of self-audits by firms, the social costs will be lower when local firemen, who naturally possess this information as part of their job, visit bars and restaurants and share this information with the proprietors (section 6.1).

Evaluating the argument

However, none of this provides a sufficient argument in favor of advice and a compliance strategy. The problem of insufficient information about expected fire damage might be solved by enforcing in an 'earlier' stage (Shavell, 1993; section 10.1.1). If proprietors are unable to predict the expected damage (and therefore the sanctions), the solution might be to choose for act-based enforcement, in which the proprietors are inspected and sanctioned irrespective of the resulting damage. This is precisely what has been observed in the field of fire safety in bars and restaurants. Act-based enforcement only requires the proprietor to be aware of the act-based standards and of the sanctions that might follow.

It is argued that proprietors cannot be expected to know the act-based regulations in detail. Actually, the rules are not that difficult. The detailed rules about seconds and minutes are captured in other rules that are quite easy to remember. For example, replacing the lamps in the escape-route signs yearly, checking the emergency lighting on a yearly basis and yearly taking care of re-certification of the fire extinguishers, all these guarantee that the proprietor will satisfy the detailed requirements on minutes and seconds. The rules summarized in section 11.2.1 are not difficult to remember and it is clear how proprietors can comply with them. In fact, most fire departments have a brochure that quite simply summarizes the fire safety rules. There is no reason why public inspection is required in addition to this type of general information campaigns.

Moreover, it is not clear why advice should be a continued objective of inspections. If a proprietor is informed about the law and compliance methods, it is expected that he will have this information for quite some time, up to several years. There is no fast technological or institutional change making information outdated within a year. Enforcement might consist of two stages. In the first stage, for example during the granting of the fire safety license, the proprietor is informed about the relevant regulations. After that, a deterrence strategy can be used in which offenses are adequately sanctioned.

The argument in favor of advice is usually sustained by referring to the fact that inspectors do find many violations during inspections. Many inspectors have the impression that the proprietors do not know or simply are not aware of their violations, as in the case of placing a table in front of an emergency exit. However, I believe that this type of behavior is primarily the result of a relaxed enforcement strategy. Since violations are not sanctioned (at least not immediately), proprietors have no incentive to acquire information about their compliance status. If a strict punishment policy were adopted, the blocking of emergency exits would soon be over. Proprietors would certainly be aware of it.

In summary, fire safety is a technically complex issue. It certainly requires professional, expensive information by fire experts. However, I would argue that a proprietor does not have to be an expert in order to comply. Apparently, the complicated technical requirements (especially concerning technical devices) are and can be translated into a number of easy requirements for proprietors' behavior. The proprietor would not have to be an expert himself. He can comply by periodically hiring an expert to inspect the technical devices. There is indeed a need to inform proprietors about the importance of fire safety, the relevant regulation and compliance. However, this is primarily achieved through general information campaigns and a deterrence strategy, not a compliance strategy.

12.4.2 *Remediation – non-compliance by accident*

Another claim is that proprietors cannot always help their failing to comply. Failure to comply is not always the fault of the proprietor. Non-compliance can often be the result of defects in technical devices, such as the sudden break-down of a lamp in the escape-route sign. Moreover, for many technical devices, the proprietor is dependent on supplying companies. It is argued to be ineffective and unfair or disproportionate to immediately punish such violations. A deterrence strategy would lead to an exaggerated 'witch hunt' which would not lead to more compliance. Inspections are aimed at restoring compliance and thereby fire safety (not deterrence).

The theoretical defense
As argued in section 4.2, if the enforcement authority detects a violation, it can and must take action to induce compliance as quickly as possible. In the field of fire safety, remediation especially implies repairing broken technical devices. As explained in Livernois and McKenna (1999), if offenses occur accidentally, the optimal enforcement policy might be to issue a warning to the proprietor. First, he is given the opportunity to restore compliance. Only if he fails to comply before some specified future date is he sanctioned. This is what happens when an inspector who detects a violation issues the proprietor with a warning to repair the installation and be in compliance within a period of – say – 6 to 8 weeks. The only difference is that Livernois and McKenna assume that firms self-report their violations, while in our case the warning is only given if the inspector detects a violation.[186]

Evaluating the argument
However, upon further examination, it turns out that offenses are not 'accidental'. The probability of technical failures is to a large extent under the control of the proprietor. The standards are precisely aimed at inducing the proprietor to take sufficient care so that technical failures in the equipment

186 There are no benefits of self-reporting for proprietors. If they self-report non-compliance, the fire safety department will say that the proprietor must correct the failures. This is also what happens when the establishment is inspected by the fire safety department. It is therefore better for the proprietor to await inspection instead of self-reporting violations.

are reduced to a minimum. For example, replacing the lamps in the escape-route signs on a yearly basis should prevent them from suddenly breaking down (only in exceptional cases they do burn for less than a year). Moreover, the enforcement officials can sanction the fact that proprietors fail to immediately replace failing escape-route signs (Innes, 1999b; section 4.2). The level of repair is quite easily observable for the fire safety departments, even more so since proprietors have had to register the repairing they carried out on their installations. In such circumstances, the policy of Livernois and McKenna (1999) is only efficient if less than full compliance is desired. This is not the case, since there is no reason for delaying the replacement of failing lamps. The current policy allows the proprietors to delay compliance. A deterrence strategy that imposes sanctions dependent on the level of remediation would deter the underinvestment in preventing technical failures.

12.4.3 Persuasion – voluntary compliance

The enforcement officials argue that most proprietors are willing to comply and/or to restore compliance. Persuading the proprietors to comply is more effective than punishing every violation. The latter will break down their cooperation. Enforcement is much easier if it builds on this willingness to comply. Therefore enforcement is not aimed at punishing non-compliance, but at stimulating voluntary compliance. Inspection involves a moral appeal on the proprietor. Enforcement officials try to persuade the proprietor that compliance is socially and morally desirable. Proprietors must be taught how to behave as good and responsible citizens. Once again, this requires enforcement officials to take much time for inspection and to make an appointment.

The theoretical defense

As discussed in chapter 7, proprietors not only consider the direct costs of compliance (in terms of time, effort and money) and the expected formal sanction by the government, but also the costs of informal sanctions and the benefits and costs of the intrinsic motivation to comply.

The existence of intrinsic motivation to comply justifies both a more deterrent and a more compliant enforcement strategy. If enforcement is or becomes weak, proprietors might lose their belief in the norm and no longer subscribe to it, especially if they observe that others can violate the norm unhindered. Enforcement must be strict enough so that proprietors keep on believing that the norm is actually important and should not be violated.

On the other hand, strict enforcement can also be counterproductive. Immediate punishment of every violation, even if minor, might induce proprietors to believe that the only reason for complying is the sanction that is the consequence of violation. Sanctions might crowd out the intrinsic motivation to comply and/or the sanction may be regarded as a price the proprietor can pay to buy non-compliance. Therefore, it might be better to forgo a strict enforcement policy of immediate punishment and choose instead a policy of warnings, negotiation and cooperation, in which it is assumed that compliance is voluntary, based on the intention to comply.

Evaluating the argument

The influence on proprietors of informal sanctions is very limited in the case of fire safety. Disapproval of the proprietors' behavior by customers, neighbors or colleagues is practically absent, unless it actually results in a disaster. None of the interviewees mention that visitors acquire information about an establishment and/or report non-compliance. Customers lack the information to make their consumption dependent on the level of compliance. There is also no culture in the horeca sector such that proprietors would stimulate each other to comply. Only the reputation with the municipality is somewhat important since the municipality would be less ready to help a non-compliant proprietor if he needs a license (fire-related or otherwise) or some other favor.

It is likely that the intrinsic motivation to comply is more relevant, although this is more a claim made by enforcement officials and a question of general common sense than an evidence-based fact. Whether and to which extent these motivations are really relevant is unknown. In any case, this motivation does not provide enough indication of what the enforcement policy should look like. As discussed above, intrinsic motivation to comply requires a balance of compliance and deterrence strategies. It seems that in the current situation, in which violations are treated relatively mildly, the enforcement policy signals that violation does relatively little harm. This stimulates a culture in which non-compliance is not regarded as a problem. The lack of large norm subscription and accompanying feelings of guilt and regret raises the question whether proprietors really have an intrinsic motivation to comply. In order to induce compliance earlier, the balance should be shifted towards a more deterrent strategy. This will teach the proprietors that regulations and compliance with the regulations is something that must really be taken seriously. Of course, the enforcement policy should take into consideration the possible crowding out of motivation. This might be solved by first issuing a warning to the proprietor, and then by strictly enforcing the regulations after this warning instead of issuing several more warnings and undertaking re-inspections. Moreover, if sanctions can be made high enough, the problem of the crowding out of intrinsic motivation is of no interest (Lin and Yang, 2006). If the expected formal sanction exceeds the direct costs of compliance, the proprietor will always comply, irrespective of his moral concerns.

Irrespective of the precise enforcement style, this case study does not improve our understanding of how norms are created and shaped by public enforcement. Does inspection directly create norms or does it create norms by increasing the proportion of complying proprietors? Are inspections needed for complying firms, since they will be confirmed in their belief in the norms through observing the punishment of non-complying firms? Or are inspections needed for non-complying firms, because punishment of the violation of norms leads to norm formation through guilt and regret and possibly shame? And what is the optimal number of inspections for norm formation? Why do moral intentions to comply not survive if public inspections are reduced? The knowledge about the working of norms in the case of fire safety in horeca establishments is too limited to answer these questions and draw any unambiguous conclusion.

In short, enforcement as morally educating the proprietor may be relevant. However, it is as yet unknown which enforcement policy serves this goal best. Education as an objective is not a sufficient argument for a compliance strategy. As with raising children, punishment is an essential tool too.

12.4.4 *Informal enforcement – tolerated non-compliance*

A claim that can be heard from almost all enforcement officials is that imposing sanctions is complicated. It costs a lot of time, effort and money. The legal requirements that have to be fulfilled are strict and high. The interviewees are afraid that their decision will not withstand the court's review because of these requirements, more than they are afraid that the judge will reverse their decision concerning content. More can be achieved, and achieved earlier by cooperation. "You don't impose a sanction if a single escape-route sign is not burning". A deterrence strategy is ineffective because it carries the threat that the whole budget will be spent on expensive legal procedures without improving fire safety. Therefore, the enforcement officials do not immediately impose formal warnings, but begin with informal notifications. The municipality will only begin an official enforcement procedure if it is really necessary, for instance because the proprietor does not respond to repeated informal warnings.

The theoretical defense

Sanctions are imposed by the municipality (the College of Mayor and Alderman). If a proprietor disagrees, he can file an objection to this decision. If this objection is not accepted, he can appeal the municipality's decision in an administrative court. Therefore, in the end, it is the judge who decides who is sanctioned. The municipal enforcement policy must conform to the court's ruling.

The proprietor might question the facts, the standard and/or the appropriateness of the sanction (section 6.2). An increase in the expected sanction may have an adverse effect on the level of compliance because it induces proprietors to use objection and appeal procedures more frequently. This effect is even stronger when such procedures are costly for the fire safety department, because the time spent on procedures cannot be spent detecting violations. In response, the enforcement authority might be better off enforcing lowered, informal standards (Khambu, 1989) and/or issuing a warning first (Nyborg and Telle, 2004; Fenn and Veljanovski, 1988).

Another problem is that since sanctioning is costly, the threat of sanctions is incredible (section 9.1). Therefore, at the supreme moment, the municipality (the College of Mayor and Alderman) might be more interested in minimizing current enforcement costs than in deterring future non-compliance. Of course, unwillingness to sanction might also result from a discrepancy between public interests and the private interests of enforcement officials (section 9.2).

Evaluating the argument

In general, fear for the court's review seems to be a problem of perception. Several studies have shown that there is no reason for municipalities to be afraid that judges reverse their decisions because of formal, procedural errors. A research by order of the Alders Committee into the judicial review of 30 fire safety decisions for public buildings showed that municipal decisions are not frequently corrected by courts. Moreover, in some cases, the court even sustained the decision despite formal failures by the municipality.[187] Similarly, Gundelach and Michiels (2003*) demonstrated for 430 decisions in environmental law, spatial planning and horeca law, that governments (usually municipalities) have no reason to fear judicial review as long as they 'simply' do their job (the title of their book is 'unfounded fear'). In only 10% of the cases in which a sanction was imposed was that decision not maintained by the court. The fear of legal procedures seems to stem from a lack of judicial expertise in fire safety departments. This is supported by the fact that several officials indicate that they would appreciate a better back-up from legal departments or even employing their own lawyer. In addition, the only department that does in fact employ a lawyer does not share this fear.

Moreover, it is hard to think of what might be contested by the proprietors. In general, the standards are quite clear (section 12.4.1 above and 12.4.5 next). If municipalities have granted proprietors time to restore compliance, they are entitled to impose sanctions. The facts also generally remain largely undisputed. The only violation that is difficult to establish is the number of visitors, since it is difficult to count the number of people present in a crowded bar. Generally, however, challenging the imposed sanction is of little use.

These conclusions do not deny that imposing sanctions requires time and effort and that the threat of a sanction might therefore be incredible. For example, enforcing the number of visitors is problematic because it can lead to problems of public order if people are sent out of a bar onto the street. Municipalities are unwilling to impose sanctions if the costs are not in proportion to the offense. There is no firm line for enforcing every single violation, for example when a single escape-route sign fails to light up. The College of Mayor and Alderman will often give priority to other, more severe, cases, such as

187 Alders Committee (2001; part 'Rechterlijke toetsing van brandveiligheidsbesluiten', pp. 265-288). See also Michiels Committee (1998*). Another relevant reference is Lomwel and Nelissen (2005*) which examines how enforcement actions are influenced by (the fear for a reaction from) citizens who decreasingly accept risks.

an environmental or building offense. The imposition of a sanction is often carried out by a different department (often a Legal Department or the Building Department). If the fire safety department eventually wishes to impose a sanction, the file begins at the bottom of the stack of the other department and possibly never reaches the top. Moreover, there are some indications that sanctioning is not always in the private interest of the enforcement officials, such as the sometimes poor cooperation between different municipal departments, and the fact that many inspectors prefer to spend time on advice rather than on sending letters. In particular, most fire safety departments have been shown to have more interest in granting licenses (for which fees are collected) than in enforcing these licenses.

If proprietors know that the enforcement officials may for some reason postpone or even refrain from the use of sanctions, they might choose to continue to non-comply, a fact which has been widely observed. Yet this does not mean that a compliance strategy is desirable. On the contrary, social welfare will be improved if the enforcement officials are forced to pursue a deterrence strategy of strictly imposing sanctions on every violation. The only real problem is caused by the requirement concerning the number of visitors. However, in the current situation, this requirement is also not really enforced since inspections are hardly ever unexpectedly carried out late at night.

12.4.5 *Cooperative enforcement – flexible compliance*

Enforcement officials are of the opinion that enforcement must be somewhat flexible, although there are different opinions on the form this flexibility should take. Some argue that the primary objective is not compliance, but fire safety. They do not investigate whether every single requirement is precisely obeyed, but ask (themselves and the proprietor): is this bar or restaurant fireproof? They negotiate about the precautions necessary to restore compliance provided that fire safety is guaranteed. Others argue that there should be no compromise on compliance with the regulations, as they describe a minimum. There is, however, flexibility concerning the possibilities to restore compliance. These officials do negotiate with the proprietor about the time allowed for restoring compliance, and whether the proprietor will benefit from precautions beyond compliance.

The theoretical defense

Generally, the law is incomplete (section 6.1.2), making it both under- and overinclusive.[188] This problem might be overcome by granting the enforce-

188 This is different from the uncertainty described in section 12.4.1 that proprietors might be ignorant about the law or make a mistake in choosing precautions. The problem here is that proprietors know, or ought to know, the standards, but that these standards are only partially applicable to their situation.

ment officials sufficient discretion to specify the law for individual firms and particular situations, in licenses and/or during inspections. Once specified, a deterrence strategy can induce the first-best level of precautions.

However, in general, a two-sided uncertainty arises. Enforcement officials have better knowledge of the applicable standards and expected damage than firms. Firms, on the other hand, have better knowledge about their production process and costs of compliance. In such a situation, both the enforcement authority and the firm may benefit from behaving more cooperatively (section 6.1.2). The authority will choose maximal enforcement if a firm has established a record of minimal compliance, and flexible enforcement if the firm has a record of flexible compliance.

Evaluating the argument

That fire safety regulations are incomplete is self-evident. There are very different types of establishments, from very specially situated restaurants and establishments in monumental property to regular neighborhood bars. The regulations themselves vary from very detailed requirements (such as the number of seconds before the emergency lighting should switch on) to very open-ended norms (such as being careful with candle lights). Part of this problem is solved by the obligation of carrying a license in which special requirements can be included.

It is more difficult to determine whether – in general – the regulations tend to be under- or overinclusive. As mentioned, there are two attitudes common among enforcement officials. Some say that the regulations represent a minimum to which everyone has to comply. For some proprietors, it is necessary to impose even stricter rules. This fits the description of the enforcement policy of Ricketts and Peacock (1996; section 6.1.2) in which the municipality tries to achieve as high as possible a level of precaution. Both the proprietor and the enforcement officials benefit from being cooperative and exchanging information. The proprietor benefits if he implements the required precautions, because compliance with this (adjusted) standard will relieve him from sanctions. If the enforcement officials learn more about the costs of the proprietor and his 'production process', they are able to bargain for higher precautions, which is beneficial if the law is underinclusive.

Other enforcement officials prefer to speak of a required level of fire safety. They argue that the basic question is whether a bar or restaurant is fireproof. This fits the description of Scholz (1991), where the municipality tries to achieve the highest safety gains. Both the proprietor and the enforcement officials benefit from being flexible. The enforcement officials overlook minor technical violations (resulting from overinclusion) in recognition of the firm's extra-legal safety efforts to reduce greater hazards not directly addressed in the regulations (resulting from underinclusion). The firm that chooses flexible compliance can tackle its worst health and safety hazards with the most efficient methods available rather than spending money on precisely complying with every standard. Hence, flexible compliance produces greater safety at lower costs.

In both cases, the enforcement officials benefit from cooperation because it allows them to achieve higher levels of safety. They have sufficient discretion to determine the precautions proprietors must take. However, cooperation is necessary because the enforcement officials lack the information required to determine and enforce first-best levels of precautions. Inspectors say that they are only moderately informed about the costs of compliance, but they do have an informational advantage concerning the expected damage and the regulations.

Another question is what makes the proprietor willing to cooperate. For the proprietor, there are three benefits from being cooperative and choosing compliance before the enforcement officials launch official procedures: (1) Re-inspections require time and effort and being cooperative implies fewer re-inspections; (2) Being uncooperative and being re-inspected implies that the fire safety department will choose strict terms for compliance, while the fire safety department may allow the flexible proprietor to choose compliance at a time that suits him better (for example with respect to supplying companies); (3) Being uncooperative may imply that the municipality will not be prepared to help the proprietor when needed (governmental reputation). However, it is unclear when proprietors choose a cooperative approach. Are they only behaving cooperatively after inspection or also in the absence of an inspection (throughout the year)? I suspect the former, because there are many violations found during first inspections, because proprietors are only generally informed about fire safety and must be informed over and over again, and because (as follows from the above discussion) being cooperative (uncooperative) is only rewarded (punished) following inspection.

There are sufficient arguments for stating that enforcement benefits from flexibility and cooperation. However, this does not imply that the current level of flexibility is optimal. There are several problems. First, cooperative enforcement implies that the inspectors are flexible for cooperative proprietors, but choose strict enforcement with uncooperative proprietors ("tit-for-tat"). The latter seems to be insufficient. Under the observed policy, uncooperative proprietors are able to a large extent to delay compliance, because it takes much time for official procedures to be started. The inspectors should immediately adopt a strict enforcement policy, as soon as they detect that a proprietor is unwilling to cooperate. Secondly, a cooperative enforcement strategy also includes the threat that enforcement officials might abuse their discretionary powers to maximize their own interests and apply unjustified discrimination between proprietors. The obligation to strictly enforce violations might lead to a better, although not first-best, result. It is not obvious that the regulations are so incomplete that strict enforcement will produce undesirable results. It may be that strictly enforcing a second-best standard for every proprietor would lead to a higher level of social welfare.

12.4.6 *Supplementary enforcement – compliance as personal responsibility*

A final comment often made by enforcement officials is that compliance is in the first place the responsibility of the proprietor himself. Inspection is only a random indication. Therefore, officials argue that they cannot be expected to enforce continued compliance at all times by arbitrarily punishing every violation. It is better to use inspection to persuade the proprietor that he has to guarantee compliance throughout the year.

The theoretical defense

In itself, an appeal on the personal responsibility of the proprietor seems to be a weak argument. Such an appeal can be made by every enforcement official, for every kind of regulation. However, in a somewhat different context, there might be reasons for stimulating the feeling of personal responsibility of the proprietor. In many situations, the importance of public enforcement lies not in inducing compliance on its own, but in supplementing private enforcement. Private enforcement may fail if the proprietor is ignorant of the (ex-ante) precautions he can take to prevent damages (ex-post). As discussed in section 12.4.1, public enforcement might be needed to provide the proprietor with the relevant information. There are three other relevant weaknesses of private enforcement.

First, public enforcement is needed to reduce uncertainty about the levels of due care in private litigation claims (section 10.2.2). Under such uncertainty, public regulatory standards imposed ex-ante might be helpful, at least if courts follow the regulatory standard in a predictable way. Proprietors learn how to escape liability.

A second – related – justification in favor of public enforcement as a supplement is that proprietors are short-sighted and do not think of the long-term consequences due to the severe competition or bounded rational perceptions (sections 1.4.2 and 3.1.4). In such cases, a proprietor might underestimate the low probability of a fire and remain unaware of the full consequences of liability. Inspection might alert these proprietors to the issue of potential liability damages and therefore prevent underdeterrence (or overdeterrence if they are too afraid of liability).

Finally, deterrence by private enforcement may be hindered by for instance wealth constraints or litigation barriers for victims. Public enforcement might be required in order to deter non-compliance by wealth-constrained firms, so that these firms take at least some precautions. Additional private enforcement induces those proprietors who have sufficient assets to take higher precautions.

As a result, there might be several reasons for having public enforcement in addition to private enforcement. Public enforcement attempts to persuade the proprietor that he is responsible for guaranteeing compliance throughout the year and that he bears the consequences of non-compliance.

Evaluating the argument

Despite stressing the personal responsibility of proprietors, none of the enforcement officials mentioned the supplemental function of public enforcement. Yet this function might be relevant. Let us examine the three arguments discussed.

It is quite possible that proprietors are uncertain about the evaluation of their behavior by courts, and that they are unable to obtain information about this evaluation at a reasonable cost. The probability of a fire is small. The probability that it leads to casualties is also small. And if there are victims who claim damages, most of these claims will be settled in relatively non-transparent ways. Therefore, it might be helpful to impose ex-ante regulatory standards that will determine negligence.

However, it is not immediately clear which enforcement style is most appropriate for these standards. Enforcing the regulatory standard might be characterized more by advice than by punishment, but as in section 12.4.1, we expect that public standards can better be communicated through general information campaigns than by means of inspections (and: how many inspections are required?). Moreover, communication about standards might possibly only be credible if the municipality really and effectively enforces the standards (by means of a deterrence strategy). Otherwise, the proprietor can argue in court that he believed that compliance with the requirements was not that important because the municipality was unwilling to allocate sufficient resources to it and really take action against violations.

The second problem discussed is that proprietors might be cognitively constrained so that they are unable to take optimal decisions. The interviews or other data do not investigate the rationality of proprietors and the information available to them. There is probably not much difference between proprietors and other human beings. Several interviewees observed that in the period following the Volendam disaster, proprietors were generally aware of potential liability claims. Once this memory vanishes, the awareness will disappear too. However, the period after the Volendam disaster is precisely the period in which public inspection was most frequent. This pattern does not fit in with the argument that public enforcement is needed at times when proprietors are insufficiently aware of fire safety. Apparently, public enforcement suffers from the same cognitive constraints. Moreover, the question is again how information can best be provided.

The third reason for public enforcement as a supplement does seem relevant. Although claiming damages is not an important problem (see chapter 15 next), there are wealth constraints, especially when several people are harmed at the same time. Moreover, the level of assets varies widely among proprietors. Some own several bars and restaurants and even other property, while others only rent the establishment from a brewery or investment holding. However, this is more an argument in favor of a deterrence strategy rather than a compliance strategy. In order to induce those firms with insufficient assets to choose at least some level of compliance, enforcement must be strict, not relaxed. Since public enforcement is already guaranteeing a minimal level of compliance, there is no space for negotiation.

12.4.7 *Conclusion*

This section analyzed different claims with respect to the optimal enforcement strategy in the field of fire safety regulation for bars and restaurants. These claims often imply that a deterrence strategy is ineffective. They largely correspond to the claims made by academic scholars concerning the effectiveness of a compliance strategy over a deterrence strategy (section 8.1). However, in the example of enforcing fire safety in bars and restaurants, we can conclude that the arguments in favor of a compliance strategy are usually invalid. On the contrary, most arguments imply that in the current situation, the balance must be shifted towards a more deterrent strategy. The arguments given fail for the same reason as the arguments in the academic literature fail: they do not distinguish between the 'objective' of enforcement and the 'enforcement style'. For instance, the fact that proprietors are willing to comply, or that they fail to comply due to mistakes or misinformation does not in itself provide an argument in favor of a compliance strategy.

12.5 THE LEGAL ENVIRONMENT – WHY DOES DETERRENCE FAIL?

Section 12.3 argued that in the case of minor violations, the expected sanction is initially small so that proprietors choose to delay compliance. In section 12.4, I examined the claims for the defense of this enforcement policy and concluded that a deterrence strategy of strictly enforcing offenses is likely to improve compliance. The question then is why the use of a compliance strategy is so widespread. Moreover, a deterrence strategy is limited by the legal framework in which the fire safety department operates. Currently, the fire safety department is not allowed to impose fines or other immediate sanctions, but only repair sanctions. If it is argued that the fire safety department should adopt a stricter, more penalizing, enforcement policy, this means that it should impose sanctions quicker than in the current situation, within the legal framework. Or that the legal options themselves should be expanded. This section analyzes why a deterrence strategy may fail and discusses the benefits of the observed compliance strategy.

12.5.1 *Legal constraints on the fire safety department*

Chapter 8 discusses that a compliance strategy is only favorable if there are severe enough limits on the sanctions and the enforcement budget so that deterrence fails. In the case of enforcement of fire safety in horeca establishments, these constraints are clearly visible.

The enforcement constraints

The municipality is not allowed to immediately impose sanctions. Only if there is serious and immediate danger the municipality can directly close an establishment. Enforcement policies rest mostly on administrative repair sanctions. Public Prosecutors give no priority to fire safety regulations.

Given these limited sanctioning opportunities, compliance can only be enforced by a high frequency of inspection and by quickly responding to offenses by immediately imposing these repair sanctions. This requires sufficient resources. However, the current budgets assume that the enforcement officials inspect an establishment once or twice a year and that some proportion of re-inspection is necessary. All other activities, including sanctioning, have to compete with other planned targets. Especially when the imposition of sanctions is carried out by the Legal Department, there is no direct budget for fire safety files and the Legal Department is not inclined to deal with a substantial number of fire safety files.

These limited powers explain that a deterrence strategy will possibly not be strict enough to really deter proprietors. As a consequence, proprietors will not be induced to take efficient precautions or to immediately restore compliance.

Alternative methods for strengthening deterrence

If a strict punishment of all offenses fails to induce full compliance, different enforcement methods may induce higher levels of compliance. A higher level of compliance can be achieved by (i) targeting enforcement or (ii) adjusting standards.[189] This would allow the enforcement officials to achieve the highest possible level of compliance.

Targeting enforcement is only effective if enforcement is sufficiently predictable. The enforcement officials use several forms of targeting enforcement resources. First of all, they target resources on different establishments. They inspect more in the centre town (the going out area), especially during actions at Christmas or Carnival. There is no reason why expected fire damage would vary per location, but the municipality can realize economies of scale by focusing on the centre. Since in the centre of the city establishments are situated closer together, many establishments can be inspected in one evening. Municipalities also save resources by rarely inspecting the smaller establishments that do not possess a use license (except during the actions just described).

Enforcement officials also target resources in time. Random inspection might be inappropriate to induce compliance but if the officials announce an inspection the proprietor will comply at least around the time of the inspection. Moreover, officials target resources on the persistent non-compliers by issuing warnings. This corresponds to the idea that the optimal targeting scheme depends on past compliance.

Enforcement officials also save resources by not enforcing all requirements

189 Moreover, I identified self-reporting, advice and cooperation/negotiation. These were already discussed in section 12.4.

(adjusting standards). Although they argue that there should be no compromise concerning legal requirements, they actually fail to enforce every standard to the same extent. Their focus lies mostly on technical devices. They rarely inspect the number of visitors, at least not on a Saturday late at night, when these violations are most likely. Enforcement officials, especially in the smaller municipalities, do not want to work at these times. Given the limited sanctions and inspections, enforcement officials can only induce compliance with the technical devices that need a regular (usually annual) check-up. The importance of the regulations that require continued attention of the proprietor is only explained during inspection but not actually enforced.

Why these constraints?

The legal opportunities of the fire safety department are limited. They try to maximize compliance and/or fire safety given these limits. But why are the legal opportunities limited? Why is there no higher budget? Why are severe fines not permitted? There may be several answers to these questions.

1. Full compliance may be inefficient. For high(er) levels of compliance, the costs of compliance and enforcement may exceed the benefits of a reduction in expected damage. Therefore, a limited budget is granted that forces the fire safety department to focus on the most severe problems of non-compliance.

   In the extreme case, compliance may be not desirable at all. The fire safety regulations and/or the enforcement of these regulations are a form of window-dressing. In order to satisfy the citizens and increase the feelings of safety, regulations are imposed and a small budget is allocated. However, it is not intended that proprietors really should comply. As in section 9.2.1 (point 6), we might expect stringent regulation and weak enforcement. More on this issue in section 13.4.
2. It may be important to have sufficient marginal deterrence. If other regulations such as those concerning the environment, occupational health, closing times, etc. are valued higher, there might be no opportunity for severe punishment of fire safety regulation violations. Since criminal fines must be reserved for clear and immediate danger, (small) violations of the use requirements might only be punished mildly.

   A problem of this argument is that it shifts the problem up to explaining why sanctions are limited for these other regulations. There seems to be room for an overall increase in the magnitude of sanctions. In any case, wealth problems are not great enough to explain the limitations of available sanctions.
3. Due to capacity constraints, the (marginal) costs of increasing enforcement efforts may be high. The increase in enforcement efforts following the Volendam disaster resulting in the current policy was accompanied by much training of personnel and/or the hiring of private companies. Questions about the quality of personnel, especially of these private companies, have already been raised. Higher budgets might possibly fail to lead to more inspectors as this personnel is simply not available.

Increasing powers
Given the limited powers, a deterrence strategy of prosecuting every violation is not likely to induce full compliance. However, this does not necessarily mean that a deterrence strategy would not work. There are two reasons for this. First, every optimal enforcement policy is a mix of a deterrence strategy and a compliance strategy. In the current situation, the share of deterrence is very small. The main claim of section 12.4 is that within the legal powers welfare would be improved by being stricter. Even if municipalities cannot impose fines or imprisonment, they can shorten the enforcement proceedings by imposing a penal sum or a closure of the establishment sooner, granting less re-inspections and spending less time on explanation and advice. This requires an improvement of the internal organization (especially the cooperation with the Legal Department) and reducing the fear of court procedures. Efficiency gains can be realized by a smarter allocation of personnel, especially in the evenings (for instance using firefighters' 'waiting time').

A second way to improve enforcement would be changing the legal environment as such, for example, by allowing officials to impose fines. However, a warning should be made here. Fines should only be introduced if this is done right. Given the current budget for inspections, a fine of €50 will probably not have much effect. Besides, such a policy should be clearly communicated. Otherwise, proprietors will have false expectations of the enforcement policy and continue to incur violations, which will require many resources for imposing these administrative fines.

Conclusion
The question is why the use of a compliance strategy is so widespread. One reason why a deterrence strategy fails is that the budget and legal sanctions are too small to induce (full) compliance. Therefore sanctioning might become a very costly process. A compliance strategy might be more effective and efficient in inducing compliance with the regulation. Nevertheless, if fire safety departments will become more strict when they find violations, this is likely to improve compliance. Moreover, allowing administrative fines might be considered.

There are two other potential failures of a deterrence strategy: incomplete law and private interests of enforcement officials. These are discussed next and provide another explanation for the widespread use of a compliance strategy.

12.5.2 *Explaining the level of discretion and private interests*

A different, competing, explanation for the observed policy is that there is a misalignment between the private interests of the enforcement officials and public interests (chapter 9). Enforcing fire safety in the way described is not in the interest of social welfare, but only in the interest of the fire safety department and its officials. For example, advice might be better for the officials' self-esteem than sanctioning. They want to show their competence. It is also possible that they do not like the personal effort required for sanctioning, but prefer the easy way of advice and cooperation. Alternatively, it may be in the interest of the fire safety department to have a large (but not too large) number of horeca proprietors incurring violations so that they can demonstrate the relevance of their work. Since enforcement officials are specialized, they develop their own expectations. They tend to see their own employment as relatively important. They perceive fire safety and compliance with fire safety regulations and thereby their own employment as being of vital importance (Lipsky, 1980). Higher budgets and more status might be important, perhaps more so than, or in addition to, maximizing social welfare.

Such inefficient behavior is only possible if enforcement officials have sufficient discretion to apply the policy they want.

The level of discretion

In fact, the fire safety departments do have much discretion. They are not under close scrutiny from the College of Mayor and Aldermen or the general municipal organization. They function quite independently, choosing their own methods. They can determine which establishments and buildings require more attention and when action should be undertaken. Following the Volendam disaster, discretion was reduced in the sense that the municipality required that licenses be granted and inspections carried out. However, these mostly concerned quantitative requirements. Under which conditions a license is granted or what is done following an inspection is primarily up to the judgment of the inspectors. There are no instructions from above concerning enforcement. Moreover, the resources granted for the required policy largely depends on information from the fire safety department and consist of a lump sum.

The benefits of discretion

The enforcement of fire safety regulations necessarily requires much discretion. The enforcement officials are experts who determine which actions will be taken. Relative autonomy from the College of Mayor and Aldermen and the municipal organization is generally required because only the inspectors themselves can determine which establishments are dangerous and how resources should be allocated. The College of Mayor and Aldermen or the municipal organization in general is not really able to evaluate the benefits of fire safety. The frequency of fires and casualties is too low to learn the importance of fire safety.

Even if there are uniform inspection lists and/or an obligation to strictly enforce the regulations, the decision on site (whether or not to take action against an offense) is difficult to control. It is difficult to write a manual that precisely describes which action should be performed in each possible situation. Moreover, even if this was possible, the costs of monitoring inspectors would be high.[190] Moreover, it is difficult to summarize the optimal enforcement policy in simple rules. Maximizing social welfare is not easily captured in a single rule. Public targets are necessarily ambiguous. What is sure is that the number of inspections cannot be the primary goal. Nor does the observation that inspectors prosecute in 50% of the cases say anything. It is therefore difficult to determine which parameters should be used to evaluate enforcement officials.

This is a different way of arguing that fire safety regulation is incomplete and that enforcement officials have to decide how it applies in specific circumstances. Enforcement should preferably respond to specific situations. A cooperative compliance strategy is desired. Such an enforcement strategy requires sufficient discretion on the part of enforcement officials. Moreover, since the budget and sanctions are limited, enforcement officials must set priorities.

The problem of discretion

It might be hard to reduce discretion because the fire safety department and its officials may not want to. They enjoy the power discretion provides them, so they try to demonstrate that work pressure is high. Discretion can be problematic if enforcement officials use their discretionary powers for purposes other than maximizing social welfare. The observation that fire safety departments use a compliance strategy can also be explained by arguing that enforcement officials themselves benefit more from a compliance strategy than a deterrence strategy.

That enforcement officials do have their own interests is unavoidable and is not necessarily a problem. If the enforcement officials are only generally informed about the costs of compliance, and hence unable to determine whether offenses are efficient, they should simply try to maximize compliance, given the budget that society wishes to spend on enforcement. If the efficient level of compliance and enforcement is unachievable, resulting in serious underdeterrence, the objective of the enforcement officials should be to obtain as much compliance as possible.

However, the private interests of enforcement officials are a problem if these interests cannot be brought in line with maximizing social welfare. There are generally two ways to deal with this problem. The first is to limit the discretion of enforcement officials. As discussed, this is difficult to realize in the case of fire safety enforcement officials. The only other way to protect public interests is then to limit the available budget and not allow officials to impose severe sanctions such as fines. If the enforcement officials inefficiently aim at maximizing compliance or case loads, society can be protected by granting them limited powers. However, this does not save society from

190 A related point is that discretion might be needed to provide the enforcement officials with sufficient enthusiasm for their work. They may derive pleasure from decision-making which would disappear if they were strictly following rules and being monitored.

inertia, nor from a fire safety department that applies a compliance strategy instead of a deterrence strategy.

Behavior of the fire safety department and its officials

The argument that the observed enforcement policy is inefficient because enforcement officials are more concerned with maximizing their own interests than social welfare is supported by several observations.[191]

- Following the Volendam disaster, no fire safety departments seem to have questioned and informed decision-makers (politicians) about the need for enforcement. On the contrary, they seem to have seized the opportunity to increase budgets by making claims for fire prevention, despite some statements warning against a witch hunt. The fire safety department has a vested interest in keeping work pressure high. This provides them with discretion to determine priorities, and therefore power.
- There are many problems of cooperation between the different departments within municipalities. This problem might partly be related to the fact that inspectors do not wish to transfer files to other departments.
- Municipal buildings are the last to be licensed, probably because the municipality does not wish to spend money on fire safety measures, and the fire safety department does not want to cause 'internal' trouble.[192]
- There is no incentive for fire safety officials to complete an inspection as quickly as possible. They would then simply be instructed to inspect another building. Officials always work at full capacity. They therefore take much time for inspections, for example by extensively advising on compliance and fire safety.
- Moreover, fire safety departments focus on the tasks on which they are directly reckoned, in this case especially the number of licences granted and the number of inspections. Other tasks that are less visible and easily measured are given less priority. This is acknowledged by several enforcement officials themselves and follows from observing the activities that fire safety officials carry out (reported in both the interviews and the policy reports). Besides, the fire safety departments (as well as the municipal organization as a whole) are opposed to the new Use Decree in which licences are abolished for most buildings. This opposition seems to stem primarily from the fact that they no longer receive legal fees for granting licences but have to finance inspections themselves.
- The fact that there are no inspections late at night (on Saturday) is also due to the fact that officials only want a 9-to-5 job. Firefighters who are on duty at this time apparently prefer to remain at the station.

191 For completeness sake, these arguments are not always sufficient. The observations might be a signal of both efficient and inefficient policies.

192 A different explanation might be that for municipal buildings, sanctioning is problematic because of the high social costs. A school cannot be closed. There is therefore an important credibility problem.

Conclusion: Explaining the enforcement constraints
The question is why the use of a compliance strategy is so widespread. The previous subsection discusses the fact that an efficient enforcement policy might require a limited budget and limited sanctions for enforcing fire safety regulation in horeca establishments. Given these limits a compliance strategy might be efficient. This subsection discusses an alternative explanation. An alternative hypothesis is that the budget and sanctions might also be limited in order to protect society against a fire safety department that is not concerned with maximizing social welfare, but instead tries to maximize its own importance. Fire safety enforcement officials are a typical example of what Lipsky (1980) calls 'street-level bureaucrats'. Street-level bureaucrats are bureaucrats who carry out a policy and whose position is characterized by relatively high degrees of discretion and relative autonomy from organizational authority. For this type of officials, the budget is always pressing. They (consciously or unconsciously) increase their own work pressure. I can not find unambiguous evidence that refutes this hypothesis.

12.6 Policy-making and enforcement deficits

Social welfare would benefit from a shift to a more deterrent enforcement strategy, but such a shift may be difficult to realize because of the resistance of executive enforcement officials. Moreover, the higher level officials might not be interested in investing in such a shift. However, more important than the precise enforcement style is the fact that enforcement must in fact be carried out. Compliance is induced by actual enforcement actions, not by deskwork.

12.6.1 *Problems with policy plans*

Almost all municipalities I visited have policy documents, plans, yearly priority programs, integral enforcement documents and meetings, etc. They especially began investing in these actions after the Volendam disaster. In this respect the national campaigns to improve (lower level) enforcement (section 11.3) have been successful. Of course, enforcement will improve if municipalities were to work more professionally. Moreover such initiatives may alert the whole municipal organization to the fact that enforcement is important and that imposing regulations is not sufficient.

Municipalities base enforcement on a carefully thought-out plan in which the effects of enforcement are considered. However, it is doubtful whether the effects are really weighed in the enforcement plans. My impression is that most (municipal) enforcement programs are nothing more than an inventory of the existing activities with the addition of some indicators and formation. A real choice of priorities is rare. This is logical because a school attendance officer cannot just like that switch to performing fire safety inspections.

Occupational resettlement requires training and hence money. Furthermore, as discussed before and confirmed in my own analysis, there is hardly any knowledge about the effects of enforcement. Municipalities are therefore simply unable to formulate well-thought-out policy plans in which the effects of enforcement are considered. If they do formulate plans, or are forced to do so, municipalities can do nothing more than simply base themselves on the output of previous years, on the highest perceived risks, etc. As long as adequate information on the effects of municipal enforcement is missing, we cannot force municipalities to base their policies on the effects of enforcement.

The problem is not only lack of information, but also the fact that several of the available instruments fail to induce municipalities to make a proper balance. The Risk Matrix (Ministry of Justice, 2002*), for example (which is part of programmed enforcement), is an instrument which helps municipalities to calculate several risks (such as physical safety, public health, social environment, etc.) into a (total) number. Unfortunately, it ignores the effectivity. It therefore only helps municipalities to list the various risks, but enforcement should do more than that and take into consideration whether it is possible to *reduce* risk. Maybe a risk of 10% cannot be lowered any further, while a risk of 1% can be halved. Therefore, the Risk Matrix is unsuitable for establishing priorities. A better instrument is the Table-of-eleven (Ministry of Justice, 2004). This table is a checklist by means of which municipalities and other governmental bodies can estimate the workability of and compliance with regulations. Following this estimate, the table can be useful in determining the necessary enforcement effort.

It is recommendable for municipalities and for fire safety departments to list their enforcement task in more detail instead of simply responding to incidents. However, the danger is that all documents and programs turn out to be nothing more than paperwork, only intended to please the national government and the citizens. Enforcement then becomes deskwork to keep the bureaucrats busy, rather than a policy that is actively carried out. In any case, I did not observe any differences between municipalities which have much policy development and those that only have a little (often smaller municipalities). What they actually end up doing is generally speaking the same. This implies that most administration and policy development around fire safety is unnecessary and does nothing more than increase the costs.

12.6.2 Organization – Separating granting and inspecting licenses

The policy development of the past years has also implied paying attention to the organization of the municipal departments. This again shifts the focus away from actual enforcement actions to policy-making. The national policy is that the department which formulates the rules and policy (such as granting licenses) should be different from the one that inspects licenses and carries out policies. Otherwise, a conflict of interest may arise. Some municipalities (especially the medium-sized or larger ones) have taken this idea very seriously, resulting in complete municipal reorganizations. At the very least, every municipality is expected in its policy plans to explain how they have taken this problem into account. However, in the practice of enforcing fire safety regulation, for the majority of municipalities, this is not a problem. In some municipalities, granting licenses has long been the responsibility of the Building Department, while the Fire Brigade carries out inspections. In other municipalities, the problem is solved pragmatically by dividing the work intelligently, so that an establishment is inspected by different individuals throughout the years. In yet other municipalities, especially smaller ones, the problem is simply ignored because there are too few individuals to divide the work. I do not think this is a real problem. Compensating for the danger of collusion, there is the real advantage of a single inspector for one establishment being able to build a long-term relationship with an individual proprietor. The fear of inconsistencies is one of the major complaints of the inspected proprietors and one of the major factors which may explain the failure of enforcement (May and Wood, 2003).

The separation of granting and inspecting licenses seems to be a toy for the national government and municipal managers to keep themselves busy. Directly spending the resources on inspection might possibly lead to higher levels of compliance. Similarly, the problem of imperfect cooperation between different municipal departments is not solved by meetings or documents. It is simply something that professional employees should do. The best way to promote cooperation is through joint actions, rather than investing effort in detailed plans and starting point notes.

12.6.3 Licenses

Following the Volendam disaster, most municipalities did throw themselves at granting licenses. However, the usefulness of this measure is questionable. All requirements of the Building Bylaw have direct effect, even if no license is granted. For horeca establishments and most other 'common' buildings, it is not necessary to formulate specific rules in a license.[193] It would probably have been much better if municipalities invested their time in inspections rather than licenses. This would imply spending more time on the streets and less at one's desk and therefore contribute to higher levels of safety than granting licenses. The problem is that inspections have to be financed by the municipality, while licenses are financed through legal fees. However, this should not be an argument for a municipality which is trying to maximize social welfare. In the end, it should not matter who pays for the enforcement costs.

193 The maximum number of visitors is the most important exception. However, as noted, this maximum is not always a problem and could possibly also be included in a general rule (for example maximum number of people per square meter or number of exits).

Just at the time when most municipalities have largely caught up on the backlog in use licenses, the government is about to abolish the system of use licenses and create a national Use Decree to replace the municipal bylaws. Of course, a fault confessed is half redressed, but it would have been better had the government considered the benefits of licenses before they began urging municipalities to catch up on the backlog.

12.6.4 *Conclusion: action program*

After the Volendam disaster, the investigating Alders Committee concluded that enforcing fire safety is something one simply must do. It is time that we finally began doing that which we have agreed upon. Tasks, responsibilities and requirements are all regulated; now it is time to implement the relevant measures. "This is not the first time. Will we never learn? Trust has been betrayed and – let us be honest – that is not the first time either. The Committee has reported, as carefully as possible, on the way tasks have been carried out and responsibilities have been assigned. But it did not stop there: it has also presented us with an action program: for right now, but also for tomorrow. Let's do it!" (Public version, p. 34). Enforcement of fire safety is only effective if enforcement officials actually go on the premises and see for themselves. This is not something that can be done sitting at a desk. As one of the interviewed official argued: "There are parties (external, private firms, but also municipalities), who are judged on the number of licenses. That is paper safety. From behind a desk. But you have to go and look, you have to go and inspect."

Such claims must be supported. What matters is how often an (horeca) establishment is inspected and which actions are taken when something is wrong. Maybe more important than the precise choice of enforcement style is the fact that enforcement must actually be carried out.

It is striking that it is precisely the action program of the Alders Committee, which stressed the importance of 'doing', which led to many policy documents being produced behind desks. The examples discussed above demonstrate that attention has been drawn away from actual enforcement actions to policy development, municipal (re-)organization, and granting licenses. The action program of the Alders Committee has induced many municipalities to focus on the defense of their policy and to granting licenses with the danger of creating only paper safety and much paperwork.

12.7 Conclusion

This chapter analyzes whether the current municipal enforcement policies are most effective in inducing compliance with fire safety regulations by horeca proprietors. The enforcement of fire safety in horeca establishments fits the model of the enforcement pyramid and of compliance strategies. It begins with advice, persuasion and warnings. If the proprietor is unwilling to comply, a more strict enforcement policy is adopted which leads in the end to sanctions being imposed. However, no municipality begins to enforce from the beginning, in the style of a policeman. They feel more comfortable in the role of consultant or relief worker. In conclusion, let me evaluate this policy in light of the summary in part I, section 10.3.

12.7.1 Deterrence

From an economic perspective the most important objective of enforcement is deterrence: making non-compliance unattractive by imposing sanctions. This objective is hardly mentioned by the interviewees and does not seem to be an important objective of enforcement actions. The cooperative, explanatory enforcement style in which sanctions are only imposed after several re-inspections allow proprietors to postpone compliance. I think it is wise that (general) deterrence should receive more attention.

Deterrence can be achieved by imposing sanctions. However, there are several reasons why immediately punishing every violation will be ineffective and inefficient.

Failures of a deterrence strategy

Criminal prosecution and closing of the establishment is saved for very fire-hazardous situations. Given the available budget and the repair sanctions the enforcement authority is not able to enforce full compliance with the fire safety regulation. Therefore, a deterrence strategy might – on balance – increase underdeterrence because it leads to high costs of imposing sanctions on the non-complying proprietors. Section 8.2.1 provides a list of reasons why imposing sanctions is costly.

1. Imposing sanctions on minor violations might reduce deterrence for serious offenses. However, for fire safety regulation, it seems possible to increase sanctions on all levels. Inability to pay sanctions is only problematic for a fraction of the proprietors (usually with smaller establishments) that have difficulty to survive and face bankruptcy. For them, the quick threat of closing the establishment will help.
2. Optimal sanctioning implies that the fire safety department treats existing firms with good compliance records more leniently than young companies or persistent violators. This difference is acknowledged by the fire safety departments. However, the analysis shows that good compliance records are hardly present, but that almost all proprietors somehow violate a rule. The difference especially applies to proprietors that respond to notifications of restoring compliance and those that do not. More strict enforcement will increase compliance by all firms. That does not exclude a somewhat more lenient approach to proprietors with good compliance

records. The fire safety department should use strict terms and less re-inspections when it detects a violation, to become more tough in the target group.

3. Imposing sanctions is costly if it reduces the incentive to self-report violations. This is irrelevant in my case study because self-reporting is absent.
4. Imposing sanctions is costly because there is a probability of wrongfully imposing these sanctions. However, these costs are not large for enforcing fire safety regulation. In general, it is quite clear whether a proprietor violates the rules, or not. This is supported by the fact that proprietors usually do not make objections to the observed violations, but only to the time granted to restore compliance and to the costs of the measures they have to take.
5. Imposing sanctions is costly if it leads to many procedures of objection and appeal against these sanctions. This effect of sanctions is feared by many enforcement officials. But existing evidence suggests this fear is unfounded.
6. Imposing sanctions on every minor violation might crowd out the intrinsic motivation to comply and thereby, on balance, reduce compliance. This is a potentially strong argument. However, crowding out is probably prevented by first issuing a warning where non-compliance is explained. After this warning strict enforcement should be applied. This is important to signal that non-compliance is a serious problem, so that the motivation to comply is sustained.
7. More strict enforcement is costly, when it implies that inspectors can perform less inspections. Given the budget, the number of violations exceeds the possible workload of inspectors. At least in the short run, more strict enforcement implies that inspectors have to spend time on prosecuting a violation. As a consequence, the expected sanction will be insufficient to deter. By starting with a warning they can deal with more violations and prevent more proprietors from becoming persistent violators. However, the problem is that inspectors can find time by reducing the duration of an inspection. Moreover, this argument justifies the use of a warning, not of several warnings before a sanction is imposed.

In summary, almost all arguments do not sufficiently explain why in the case of enforcing fire safety regulation in horeca establishments a deterrence strategy fails to induce compliance and a compliance strategy should be used. At least the arguments do not explain why sanctioning is such a slow process. The use of warnings can be defended, but there should be strict enforcement of the consequence of not responding to this warning.

Another reason for failure of a deterrence strategy is that fire safety regulation is incomplete. Some norms are open-ended and merely constitute a duty to care. Not every possible situation is regulated in detail. Therefore the fire safety department should treat the regulations with wisdom. It has sufficiently discretion to do so. It is not perfectly informed to be able to determine the first-best level of precautions for every proprietor, so that some cooperation between proprietors and enforcement officials is necessary. On the other hand, given the disadvantages of a compliance strategy, the question is whether strict enforcement of the regulation does not lead to a higher level of social welfare (although not first-best) than a compliance strategy.

A deterrence strategy also fails if imposing sanctions is incredible. This might only be true for enforcing the number of visitors. But this standard is also not enforced under the current compliance strategy. Moreover, a deterrence strategy will not work if the enforcement officials do not implement it because they do not like to spend time on actual enforcement actions.

Alternatives to a deterrence strategy

Several examples of targeting enforcement and adjusting standards downwards have been identified. For example, inspections are announced, inspections concentrate on dense horeca areas, inspections focus on a target group by re-inspections, inspections are at day-time, etc. Given the legal constraints, these measures help to achieve as much compliance as possible.

Enforcement should also be cooperative and flexible and focus on the most important safety problems. This is largely acknowledged by the enforcement officials. However, a problem is that such a flexible enforcement strategy should consist of a tit-for-tat-strategy, implying flexibility to the cooperative proprietor, but inflexibility to the non-cooperative proprietor. If a proprietor does not respond to a first warning, the fire safety department should choose for a strict, inflexible enforcement attitude.

12.7.2 Optimal strategies for the other objectives

Remediation

For fire safety regulation restoring compliance is an important objective of inspections, especially repairing failures of technical devices. Because proprietors are able to repair such failures immediately, the optimal enforcement policy is a deterrence strategy that imposes sanctions dependent on the level of repair. For example, the current enforcement policy induces proprietors to postpone the replacement of failing escape-route signs until they are inspected and instructed to replace them. This can be prevented by directly sanctioning such violations. The fire safety regulation is precisely aimed at preventing technical failures.

Advice

Proprietors are not naturally informed about expected fire damage. However, this is not necessary. It is important that they know the regulation they have to comply with and the methods to comply with the regulation. This requires some effort by the municipality to inform the proprietors. However, general information campaigns will serve this goal. There is no need that every inspection takes a long time to explain fire safety to the proprietor.

Education
Reputational concerns do not play an important role in complying with the fire safety regulation by proprietors. Whether intrinsic motivation is important, is questionable. There is no culture observed that stimulates compliance with the regulation. At least, the motivation to comply varies among proprietors. I judge that a strict enforcement of the regulation provides an important signal that compliance with the regulation is valued by society and therefore stimulates the motivation to comply.

Completion
The objective of completion is not advocated by the enforcement officials. However, public inspection and sanctions contribute to private enforcement in a number of ways. First, proprietors have difficulty with determining whether their behavior is negligent. This uncertainty can be reduced by public standards. Again, general information campaigns will suffice. Moreover, strict enforcement of the standards is necessary. Otherwise, proprietors can claim that they believed that compliance with the regulation was not considered desirable. Secondly, public enforcement is needed to impose sanctions in those cases where private enforcement fails because of wealth constraints. Again, this recommends strict enforcement of the regulation.

12.7.3 Conclusion

Every enforcement policy should be a smart combination of a deterrence and a compliance strategy. There is no abstract, optimal enforcement strategy. In the case I analyzed, for fire safety regulation in horeca establishment, society will benefit from a shift to a more deterrent enforcement strategy. Improving the share of a deterrence strategy, is desirable for all relevant objectives of enforcement and is therefore likely to improve the compliance of proprietors with fire safety regulations. I do not argue that there should only be immediate and severe sanctioning of all violations. But in the current situation the balance is too much to a compliance strategy. A more deterrent strategy can be obtained in two ways. First of all, the fire safety department should start with formal warnings at an earlier time. It should strictly enforce the contents of such warnings. Efficiency gains can be realized by reducing the number of re-inspections and by reducing the time spent on a (re-)inspection. Secondly, the legal opportunities of the fire safety department can be improved. The municipality should provide a better back up and follow up of the imposition of administrative (repair) sanctions. The national government can consider the possibility of administrative fines.

Three points should be made here. First of all, although conclusions are drawn from the specific example analyzed, the description of the enforcement of fire safety regulations in horeca establishments matches the policy observed in other studies. A quick scan by order of the Alders Committee (2001*, subparts A) showed that this practice was and probably is common to enforcing fire safety in general. Other research, described in section 11.3.3, showed that both at the local and at the national level, such enforcement policies are quite common.[194] Finally, scholars have found that in other countries, many enforcement agencies use cooperative rather than deterrent enforcement policies too.[195] The enforcement of fire safety regulations in horeca establishments is a classic example for the defense of advice, persuasion and warnings. Therefore, if the analysis of this example casts doubts on the benefits of a compliance strategy, it also urges for caution with respect to the use of compliance strategies in general.

Secondly, the analysis focused on the question whether enforcement can induce compliance. Due to a lack of data, it was not possible to draw any conclusions concerning the precise level of compliance which is or can be enforced. I also did not discuss in detail whether the costs of the enforcement policy are matched by the benefits of a higher level of compliance.

Thirdly, despite the current trend to base enforcement on well-thought plans rather than on incidents, the policy analyzed seems primarily to be a response to the incident of the Volendam disaster. Instead, the government(s) should have used a conscious balance of the social costs and benefits to decide whether it was desirable to intensify enforcement. This would have allowed the government to defend its need for more enforcement resources. The next chapter examines (ex-post) the costs and benefits of the observed intensifying of the enforcement policy. Despite the lack of knowledge concerning the effects of enforcement, a cost-benefit analysis appears to be very informative and useful. Such an analysis might also have been carried out ex-ante.

---

194 See the references in note 178 and Netherlands Court of Audit (1996*, 2002*, 2005*), Michiels Committee (1998*).

195 See references in note 128. The enforcement pyramid of 'responsive regulation' (Ayres and Braithwaite) is most popular. Most theories are based on an empirical investigation of occupational safety and health regulation or environmental regulation, usually at the national level in the United States, Australia or the United Kingdom.

13 Cost-benefit analysis of enforcement following the Volendam disaster

So far I have mainly evaluated the effectiveness of the increase in enforcement efforts for fire safety in horeca establishments following the Volendam disaster. The next step is to consider whether the observed enforcement policies are efficient at all. Have the changes in enforcement policies following the Volendam disaster improved social welfare? In order to measure the contribution of these policies to social welfare, we must conduct a cost-benefit analysis. Do the benefits of the additional enforcement efforts exceed the costs?[196]

13.1 The role of a cost-benefit analysis

A cost-benefit analysis aims to list all the effects of a policy intervention. Monetizing these effects allows us to compare them and to evaluate whether the intervention makes a positive or negative contribution to social welfare.[197] A careful analysis of the costs and benefits of an enforcement policy helps public decision-makers to choose their priorities and to resist unjustified claims to take action following an incident such as the Volendam disaster. This is not to argue that the cost-benefit analysis should replace public decision-making. Cost-benefit analyses are used to improve, not replace, decision-making by listing the consequences of a policy and as much as possible lumping them together. Public decision-makers have to balance the outcome of the cost-benefit analysis with the items that could not be monetized. These include at least the distributional consequences and the risks and uncertainties of the policy. Although a cost-benefit analysis can not replace decision-making, it might be helpful in making the balances explicit. Policy-making must often – consciously or unconsciously – take unto consideration how much society wants to invest in preventing deaths. If we reject a proposal of a €100 million investment in occupational safety which is estimated to save 10 people's lives, it means that we are not willing to give €10 million for a human life. If we accept a proposal of €50 million that is estimated to save 30 lives, we are willing to spend €1,67 million per life. A cost-benefit analysis makes these choices explicit. Numbers and values – such as €2 million for a life – are not used to make exact calculations which drive decision-making, so that a proposal that

196 This chapter is an update of the analysis in Suurmond and Van Velthoven (2003*).

197 Boardman et al. (2001) provide a general introduction into the methodology of cost-benefit analysis. An introduction to cost-benefit analysis in the field of criminal law enforcement can be found in Cohen (2000 and 2005) or Van Velthoven (2007*).

costs €1,999 per life is accepted and one that costs €2,001 is rejected. Instead we are given some idea of the order of effects.

Such cost-benefit analyses are common in other policy fields. In the Netherlands, national projects concerning the infrastructure are standard submitted to a cost-benefit analysis, along lines established after intensive deliberation by experts. Evaluations of this practice have led to the conclusion that a cost-benefit analysis contributes to a more professional decision-making because it offers a clear understanding of the effects of intended projects. The European Commission has proposed a policy agenda for traffic safety based on a simple cost-benefit-analysis-based rule of thumb. They computed that a policy measure that saves 1 casualty also saves 8 severely injured, 26 slightly injured and 200 material-damage-accidents, as well as leading to direct economic damage savings (medical costs, loss of productivity, material damage, transaction costs) of €1 million. Since smaller accidents tend to be underreported and there is also non-monetary damage, this €1 million is only a lower bound for the actual benefits of the measure. If a policy measure costs less than this €1 million, it goes without saying that it is socially profitable. The compulsory introduction of speed limiting devices for small vehicles and regulation and enforcement of alcohol use in traffic largely satisfy this requirement and therefore have the highest priorities.

This example shows that cost-benefit analysis can also be helpful when the data are too limited to conduct a full cost-benefit analysis as is the case in the enforcement of fire safety regulations. If there is no definitive balance of social benefits and costs, a careful analysis may provide upper or lower bounds. The cost-benefit analysis in this thesis does not provide a definitive judgment on the balance of benefits and costs, but it does provide a required reduction in expected fire damage. Such analyses can be helpful in guiding public decision-making and may force policymakers to think before they start on a new policy, especially resulting from an incident such as the Volendam disaster.

13.2 Costs and benefits of intensified enforcement policies

The analysis in this chapter focuses on the enforcement policies of all municipalities following the Volendam disaster. The data that are used have been gathered as much as possible from written national sources. In order to complete the data, I contacted several persons employed at other (national) organizations.[198] In addition, the interviews described in chapter 12 provide supportive information concerning inspections.

In the analysis I follow the definition of *horeca* given by Statistics Nether-

198 These persons are named in Annex 1.2. I would like to thank them for providing the information.

lands (the CBS). The total number of horeca establishments equals 22,600.[199] All values are in prices of 2003.

In order to evaluate the balance of social costs and benefits, we need to know whether the direct and indirect costs of the additional enforcement efforts are lower than the benefits in the form of the damage avoided. There are three major items to consider: (1) the costs of the government, (2) the costs of the horeca sector, and (3) damage from fire.

13.2.1 *Government costs from intensified enforcement*

The costs of the government relate to the granting of licenses, the inspection of licenses and, if necessary, the costs of imposing sanctions.

With respect to use licenses, IOOV (2002*) concluded that on 1 January 2002 there were 179,300 buildings that should possess a use license. However, 71% of the existing buildings did not possess a use license. There is no reason to assume that this backlog was initially different for horeca establishments. The sector organization Royal Horeca Netherlands (RHN) estimates that 80% of the horeca establishments need a use license. Therefore the backlog in the granting of use licenses at 1 January 2002 can be estimated at 12,800 establishments.[200] Granting use licenses for horeca establishments requires 15 to 35 hours working,[201] so that the backlog can be estimated at 154 to 359 fte[202], or

---

199 According to the classification of enterprises by the CBS (SBI'93) restaurants are category 553.01 and cafés and the like category 554. I use the number of establishments, not of enterprises. On January 1, 2002 the number of restaurants was 10,425 and the number of cafés and the like 12,150. So, the total number of establishments equal 22,575. The numbers are similar, albeit slightly higher, for other years. Source: CBS, Statline. See also the data of Bedrijfsschap Horeca en Catering [Associated Companies Hotel and Catering Industry], available at www.bhenc.nl .

200 22,575 * 80% * 71% = 12,823.

201 According to the 'prevapnorms' (see note 169) horeca establishments come in priority 3, 2 or 1, depending on the number of visitors, which requires 10 to 20 hours for granting a license. Most municipalities make their own norms based on these prevapnorms (AEF, 2002*, and the interviews). 15 to 20 hours is the most common number in the action plans of municipalities. Most persons involved indicate that these norms are lower bounds and that they have sharpened them. AVD (2004*) states that for all buildings, an average of 35 working hours is required. Horeca establishments are average buildings.

202 Assume that the net productive working hours equal 1250 per year (as in AEF, 2002*). The *Handleiding Overheidstarieven 2003* [Manual Government Tariff] of the Ministry of Finance mentions 1333 hours. AVD (2004*) argues that in practice 1200 hours is more realistic.

€10.5 to 24.5 million.[203] Making up for the backlog can be regarded as a once-only investment with an infinite duration, so that the social costs consist of the interest on the investment amount. If we take the real risk-free interest rate of 4% that is common in the Netherlands for government investments,[204] the annual costs equal €420,000 to €980,000.

In addition to catching up on the backlog, new horeca establishments, approximately 660 of them[205], also have to apply for a use license. Using the same numbers, this requires on a yearly basis 8 to 18 fte or €540,000 to €1,260,000.[206]

The use licenses must also be inspected regularly. There are no national data on the number of inspections. However, municipal policy documents, interviews and RHN indicate that before the Volendam disasters, there were hardly any regular inspections. Only at Christmas or Carnival did some municipalities carry out inspections. By now, most municipalities intend to inspect horeca establishments at least once a year in addition to the inspections at Christmas and Carnival, and during special events. This is also in line with the prevapnorms.[207] It therefore seems fair to assume that following the Volendam disaster, one additional inspection is carried out per year. If all 18,000 horeca establishments requiring a license (80%*22,500) are inspected

---

203 AEF (2002*) reports an average salary scale of 8 for granting use licenses. According to AVD (2004*), 90% of the activities require an intermediate and 10% a higher vocational education level, i.e. respectively salary scale 8 and 11. The following numbers are taken from the *Handleiding Overheidstarieven 2003* (Ministry of Finance):

Salary scale	*Salary costs per man-year (€)*	*Overhead (€)*	*Total costs (€)*	*Cost-effective tariff per hour (€)**
8	42,352	24,259	66,611	50
11	59,513	24,259	83,772	63
13	78,670	24,259	102,929	77

* Based on 1333 productive hours per year.

With this information the total costs can be calculated as 154 or 359 * (90%*66,611 + 10%*83,772) = 154 or 359 * €68,327 = €10,513,573 or €24,531,670.

204 See Eijgenraam (2000*, pp. 63-75) and *Kamerstukken II*, 2003/04, 29 352 nr. 1 (Letter of the Minister of Finance concerning evaluation of risks in public investment projects). This rate applied until March 2007.

205 In 2001, approximately 1200 new horeca enterprises were started, including hotels, snack bars, etc. In the period 1994-2005, the number of new enterprises varied from 900 to 1600, with an average of 1200. Given the current proportion of different types of establishments (22,575 establishments on 40.820 including hotels, snack bars, etc., in other words 55%), this implies that approximately 660 new restaurants, cafés, etc. have been set up.

206 660 * (15 to 35) / 1250 = 7.92 to 18.48 fte. With costs of €68,327 per fte this gives us €541,151 to €1,262,685.

207 According to the prevapnorms (note 169), horeca establishments (depending on the number of visitors) need to be inspected 1 or 2 times a year during 2 to 3 hours. Most municipalities base their plans on these norms. See further AEF (2002*) and AVD (2004*). The interviewees mention 40-60 minutes, but exclude travel and administration time. AVD (2004*) reports that 3.25 hours are required on average for all buildings, including re-inspections.

yearly for 2 to 3.25 hours this requires 29 to 47 fte annually.[208] The costs are €1,970,000 to €3,210,000 per year.[209] Note that I leave aside the inspection of establishments that do not need to possess a use license.

In addition to the inspections themselves, we would like to know something about the content and consequences of these inspections. To which extent are violations discovered? Are re-inspections needed?[210] Do inspections lead to sanctions? This information is rarely available. Interviewees mention that real imposition of sanctions is scarce. Appellation procedures almost always concern license fees. AVD (2004*) indicate that based on their experience there is sanctioning in 1% of the cases, which requires 20 hours work per case. Using this figure sanctioning requires 2.9 fte or €270,000.[211]

Before these intensified enforcement actions were started, there were many national and municipal preparations and overhead. Consider the steering committees, taskforces, research reports, risk inventories, policy documents, enforcement programs, etc., as well as the technical prescriptions and manuals that have been written, the information campaigns, etc. Suppose that in all 496 municipalities, one policy maker (scale 11, higher vocational education) put 2 months worth of effort in the initial project development and that we put 25% ahead for national overhead. Suppose further that 10% is put on the account of horeca establishments (18,000 of the 179,300 buildings that should possess a use license). Using the 4% interest rate as before, the annual costs can be estimated at €35,000.[212] Moreover, suppose that every year the same policy maker is busy for two weeks writing an enforcement priority program. These costs can be estimated at €220,000.[213]

13.2.2 *Compliance costs for horeca proprietors*

The intensified enforcement policies imply that horeca proprietors face higher costs of compliance with the regulations. An estimate of these costs has

208 18,060* 1 * (2 or 3.25) / 1250 = 28.9 or 47.0. 2 to 3.25 hours is based on the references in the previous note.

209 Inspection is carried out by the same kind of personnel as the personnel granting licenses (note 203). The costs are therefore 28.9 or 47.0 * €68,327 = €1,974,380 or €3,208,367.

210 Re-inspections are only included in the 3.25 hours of AVD (2004*).

211 1% * 18,060 * 20 = 3612 hours, or 2.89 fte. According to AVD (2004*), the activities require 50% higher vocational education level (scale 11) and 50% higher education level (scale 13). Using the data in note 203 this implies that the costs equal 2.89 * (50%*€83,772 + 50%*€102,929) = €269,746.

212 496 * 2/12 * €83,772 * 1.25 = €8,656,440. Of these 10% are for the horeca establishment, so that the costs equal 10% * €8,656,440 * 4% = 34,626.

213 496 * 1/24 * €83,772 * 1.25 * 10% = €216,411.

been made by the RHN.[214] They estimate the initial costs of catching up on the backlog at €28 to €52 million in the five years following the Volendam disaster. If we see these costs again as an investment with infinite duration and apply an interest rate of 4%, the annual costs are €1,120,000 to €2,080,000. The regular costs are estimated at €8.4 to €15.6 million per year.[215]

13.2.3 *Fire damage*

The objective of the intensified enforcement policies is to reduce the probability and/or severity of fire, thus: the expected harm. To acquire an impression of the possibly avoidable damage, tables 13.1 and 13.2 list the total damage of fire in horeca establishments in the years before the Volendam disaster. Table 13.1 shows that over the past years there were on average 348 fires per year. On average rather more than 2 people died in these fires, while slightly fewer than 25 were seriously wounded. The direct financial damage is calculated at €17.6 million per year on average. The other costs have to be estimated (table 13.2). For the costs of disposal I use the estimate of the costs of a turnout of the Fire Brigade. For the computation of the costs of casualties (medical costs, lost output, non-monetary damages) I use the best available estimates of the corresponding figures for traffic casualties (table 13.3). The non-monetary damage is estimated according to the statistical value of life as discussed in section 2.1.2. This leads to total costs of €2.39 million per fatality and of €0.23 million per wounded All in all the annual expected damage from fire accidents can be estimated at €30.4 million.

---

214 This includes the following costs: (i) time, administrative expenses, consultation; (ii) constitutional investments (building); (iii) investment in procedures (evacuation plans, exercises etc.); and (iv) additional costs for adjustments to the building and for new buildings as a consequence of fire regulations. Legal fees are left aside because they only have a redistributional effect between horeca proprietors and municipalities.

215 Of course an important question is whether this estimate (by the branche itself) is realistic. To compare, the administrative costs of granting a use license are estimated at €569 on average (SIRA Consulting, 2004*). If this also applies to horeca establishments, the administrative expenses alone of catching up on the backlog can be estimated to equal 18,060 * €569 = €10.3 million. The same report uses €50 as a standard for an hour work by an entrepreneur. If the regular annual costs only consist of time spent by the proprietor, this implies that he spends 9.3 to 17.3 hours per year on fire safety (€8.4 or €15.6 million / (18,060 * 50). That does not seem unreasonable. It also fits the general information of websites on the costs of technical devices.

Table 13.1 Number of fires in horeca establishments 1993-2001[a]

Year	Number of fires	Of which with damage	Damage (mln euros)[b]	Number of lost lives[c,d]	Number of wounded[c,d]
1993	400	351	20.5	3	5
1994	386	340	22.9	3	21
1995	353	300	15.2	-	19
1996	448	381	27.8	-	11
1997	408	346	15.6	-	18
1998	273	232	11.8	-	17
1999	259	231	16.1	-	15
2000	275	213	17.7	-	4
2001	329	276	10.6	14	113
Average per year	348	297	17.6	2.22	24.78
Average per fire	-	0.85	50,500	0.00639	0.07122

a. Source: CBS (according to statements of the municipal fire department).
b. In prices of 2003. Direct financial damage. Business damage (lost turnover) is excluded. Not in millions for average per fire.
c. Excluding fire men in function.
d. As dead are only registered the persons that have died within one day after the fire. The figure for 2001 is therefore in connection to the disaster in Volendam adjusted to 14 (in stead of 8). As wounded are registered the persons that are transported to or admitted at hospitals. This figure is not adjusted.

Table 13.2 Costs per year by fire in horeca establishments

Costs		In million euros
Direct financial damage [a]		17.6
Loss of profits		PM
Costs fire brigade (348 fires for € 5,100 per turn out)[b]		1.8
Monetary and non-monetary costs of victims:[c]		11.1
– medical costs	0.23	
– gross output loss	2.03	
– human losses	8.83	
Total		30.4

a. The average financial damage according to table 13.1.
b. According to table 13.1 there are on average 348 fires. The *Brandweerstatistiek 2005* [Fire Brigade Statistics] of the CBS mentions that the costs per turnout in 2003 equal € 5,059 (on average for all fires). Besides these there are other costs of disposal, like insurance, police-effort etc. Note that the real costs are higher because there are also in 85% of the cases false alarms. It is not clear how these should be ascribed to fire damage (they are partly a consequence of regulation and compliance).

c. The total costs per victim (TC) consist of medical costs M, the gross (discounted) lost output (GO) and the intangible loss (or: human losses, H) (De Blaeij, 2003). The human losses H can be estimated by the statistical value of life (VOSL). The VOSL however is generally assumed to consist of human losses as well as lost consumption (LC). Another general assumption is that wounded do not have consumption loss.
These data are not available for fire victims, so I use the data for victims of traffic accidents reported in table 13.3. According to table 13.1 the average number of fatalities is 2.22 and of wounded 24.78. Multiplying these numbers with the costs reported in table 13.3 yields the costs of victims mentioned above. For instance, the total costs are 2.22 * 2,394,942 + 24.78 * 232,653 = 11.1 million.

13.3 The balance of costs and benefits

Using the above-mentioned data we can try to compute the balance of costs and benefits of the intensified enforcement policies for fire safety in horeca establishments. Unfortunately, one crucial piece of information is missing, namely the relationship between the additional enforcement efforts and the probability and severity of fire damage. There is no source available which provides any useful information on this relationship. My own research did not provide this information either (see section 12.1). That means that it is not possible to give a clear indication of *the* balance of social benefits and costs of the enforcement policies following Volendam.

13.3.1 *A general picture of required effectivity*

We can, however, describe what the balance of benefits and costs would be for several hypothetical values of the effectiveness of enforcement. Table 13.4 provides three alternatives. If the intensified enforcement effort reduces the expected fire damage by 25%, the policy is inefficient no matter what.[216] If the measures reduce expected fire damage by 50% or 75%, the policy is only efficient if the costs are low. Under the higher estimate of the costs, the policy remains inefficient under an effectiveness of 75%.

Put differently, for a positive balance of benefits and costs of the additional enforcement actions, the required reduction in expected fire damage after Volendam should be at least 43%. This figure of 43% is the lower bound for the required effectiveness that applies if the costs of compliance and enforcement are limited to the lower estimates. If the real costs are higher, the minimum required effectiveness increases correspondingly, up to 78% under the highest estimate.

This computation suggests that the resources spent on fire damage in horeca establishments since Volendam have not been socially profitable. A reduction of the total fire damage with more than 40% seems rather high, although there are no data to determine whether this is the case.

216 Leaving aside the PM posts on both the benefit and costs side.

Table 13.3 Costs of victims[a]

Costs	Per fatality	Per wounded
Medical costs	6,400	8,800
Gross output loss	510,200	36,000
Human losses	1,878,300	187,800
Total	2,394,900	232,700

a. These costs are the estimates of AVV (2006*) for the costs of victims of traffic accidents for the year 2003.
The numbers for wounded are those that has been taken to a hospital (which corresponds to the reports of fire wounded). The intangible costs of a deadly casualty are computed with the estimated VOSL of De Blaeij (2003). Based on De Blaeij (2003), Wesemann et al. (2005*) advise to use a VOSL of €2.2 million ± 0.3 million (prices 2001). The advise for intangible cost is €1.8 million ± 0.3 million. The intangible costs of a wounded victim is (as usually) estimated to be 10% of those of a deadly victim.
To compare, Viscusi and Aldy (2003; values in dollars in 2000) have investigated the estimates in the international literature, most often based on labor market research. They find an average of $6.7 million. The most reliable estimates are in the range of $5 to $6.2 million. Furthermore they report a statistical value of an injury of $20,000 to $70,000. De Blaeij (2003) conducted a similar literature research for traffic safety and found an average VOSL of $4.4 million and a median VOSL of $3.2 million. She finds a median value for wounds of €220,000 (p.68).

Table 13.4 Costs and (potential) benefits of enforcing fire safety in horeca establishments after the Volendam disaster

Costs and benefits	In euro's per year
Costs:	
Costs of granting licences	
– catching up backlog	420.000 to 980.000
– new licences	540.000 to 1.260.000
Costs of inspections	1.970.000 to 3.210.000
Costs of further actions (sanctions)	270.000
Overhead costs	
– initial	35.000
– regular	220.000
Costs of horeca sector	
– catching up backlog	1,12 to 2,08 mln
– regular	8,4 to 15,6 mln
Total costs	13,0 to 23,7 mln
Potential benefits:	
(Maximum avoidable) fire damage	30,4 mln
Balance of benefits and costs:	
– for effectiveness of enforcement of 25%	–5,4 à –16,1 mln
– for effectiveness of enforcement of 50%	2,2 à –8,5 mln
– for effectiveness of enforcement of 75%	9,8 à –0,9 mln

There are several arguments to suggest that 40% is quite high. First of all, the interviews showed that inspection revealed a high number of violations. Compliance on some rules (especially regarding decorations and fire extinguishers) has improved. However, for many other rules, non-compliance is still high. Moreover, this is only the result of inspections that were announced or by appointment and which were performed during the day. Secondly, as table 13.5 shows, the figures of the years 2002 to 2005 do not indicate a clear decline in expected fire damage. There is a decrease in the number of fires, but the financial damage has increased, and the number of casualties is constant. Upon further examination, however, the real difference in the number of fires is between the periods 1993-1997 and 1998-2005. Of course, the number of data is very small and such an analysis ignores any background information that might explain the number and severity of fires, but there seems to be no clear evidence of a substantial effect on expected fire damage of the intensified enforcement. Finally, data on the causes of fires in horeca establishments indicate that not all fires can be related to the behavior of proprietors. 11% of the fires are caused by arson; 7% by smoking; 29% by faulty/wrong use of a machine or product; 28% otherwise and 25% unknown.[217] Complying with the use requirements is not likely to reduce all causes of fires, although it may limit the severity of a fire.

Table 13.5 Number of fires in horeca establishments 2002-2005[a]

Year	Number of fires	Of which with damage	Damage (mln euros)	Number of lost lives	Number of wounded
Average 1993-2001 per year	348	297	17,6	2,22	24,78
Average 1993-2001 per fire	-	0,85	50.500	0,00639	0,07122
2002	280	229	34,1	0	16
2003	303	257	38,7	0	20
2004	251	203	24,5	0	19
2005	254	202	28,8	6	8
Average 2002-2006 per year	272	223	31,5	1,50	15,75
Average 2002-2006 per fire	-	0,82	115.900	0,0055	0,0579

a. Source: CBS (according to statements of the municipal fire department). For further explanation see table 13.1.

217 CBS, Webmagazine February 2, 2004. The figures apply to 2002. Horeca includes hotels, snack bars, etc.

13.3.2 *Costs and benefits of alternatives*

Table 13.4 is not the end of the story. There are several ideas for a stricter or more structured approach to this problem. Some municipalities inspect larger horeca establishments twice a year (as prescribed by Prevap for some categories). There are plans in circulation in which use licenses have only a limited period of validity (AVD, 2004*). According to AEF (2002*), every horeca establishment should be inspected at least 2 times a year, and an inspection should take 2.5 hours. A use license should be valid for only 5 years, and judging an application requires 20 hours. Moreover, in essence, AEF judges these norms to be too low. To force the proprietor to comply with the rules, 1 inspection per quarter and continued inspection during special events is necessary. Moreover, horeca establishments which do not need to be in possession of a use license should be inspected by the fire safety department on a regular basis.

The costs of these policies can be calculated. For example, if a use license only has a period of validity of five years, the costs of granting licenses can be estimated at €3.0 to €6.9 million.[218] Compared to the basic scenario of table 13.4 this implies an increase in costs of €2.0 to €4.7 million.[219] To recover these costs alone, the expected fire damage should be reduced with 6.6% to 15%, in addition to the already substantial effect in table 13.4. In these percentages the additional costs to horeca proprietors are not taken into account. These costs should be included, since the additional enforcement efforts intend to cause a behavioral change in horeca proprietors. As the additional costs to horeca proprietors are higher, the required effectiveness of the policy should be correspondingly higher in order to achieve a positive benefit-cost balance. Similarly, if horeca establishments are inspected four times a year (every quarter), the additional costs would be €5.9 to €9.6 million, requiring an additional reduction in expected fire damage of at least 19% to 32%.[220] So if these alternative policies are implemented, the expected fire damage should be reduced by more than 50% as compared to the damage before the Volendam disaster. Under the very strict regime of a five-year license validity period, and inspection every quarter, the expected fire damage may have to be reduced with 125% (under the highest cost estimates) in order to be efficient.

There are also other alternative proposals in circulation (see for example AVD, 2004*). One is that horeca proprietors should obtain a certification for their establishment by some private company, which will replace the use

---

218 If use licenses have a limited period of validity and have to be renewed after five years, the costs are not a once-only catching up plus structured licensing of new establishments, but can be calculated for a cycle of five years. If all 18,060 establishments need a new license once per five years, this requires on average per year 18,060 * 1/5 * (15 to 35 hours) / 1250 = 43.3 to 101.1 fte, or (43.3 to 101.1) * €68,327 = €2,961,570 to €6,910,330.

219 For the lower estimates €2,961,570 – (€420,543 + €541,151) = €1,999,876, and for the higher estimates €6,910,330 – (€981,267 + €1,262,685) = €4,666,378.

220 I.e. the costs of inspections of table 13.4 would be four times higher, and the additional costs would thus be 3 * (€1,974,380 or €3,208,367) = €5,923,140 or €9,625,101.

license. The consequences are unknown. If the price of certification equals that of publicly granting of a license, and the requirements are applied in the same way, it makes no difference. The costs would probably be somewhat higher because the proprietor would then also have to inform the municipality about his certification, and public agencies would have to supervise the certificating companies, leading to higher overall administrative costs. On the other hand, due to market powers, costs might be reduced. Furthermore, there are proposals to not only privatize licenses, but also to let private companies perform the yearly inspections (as in the motor vehicle test). This has of course distributional effects because now it is the proprietor, rather than the tax-payer, who pays for inspection. However, for the total costs and benefits, the effects are similar to those of certification and probably not substantial.

13.3.3 *The potential savings of the Use Decree*

There is a final alternative to consider, which was already suggested by the Alders Committee and will be implemented in 2008. This alternative is the abolishment of the use license and the harmonization of the use requirements into a national Use Decree (by the Ministry of Spatial Planning and Environment). As discussed above in section 11.2.1, all municipalities have adopted more or less the same requirements in their municipal Building Bylaw. Except for the number of visitors, the requirements are not building-specific, so they need not be stipulated in a specific license. If the government would have chosen this alternative from the beginning, it would still have been able to inspect and enforce the use requirements, but without having to catch up on the backlog in granting use licenses. By simply inspecting and enforcing the use requirements directly, the government could have reached the same level of fire damage and compliance costs at lower enforcement costs. According to table 13.4, this would have saved an initial investment of €10.5 to €24.5 million or yearly €960,000 to €2,240,000. This would at least have saved the administrative costs of licenses for horeca proprietors. It is estimated that these costs are €569 per application.[221] Therefore, if policymakers and politicians would have thought better about enforcement policies and would have chosen a system with a national Use Decree without licenses from the beginning, they would have saved over €10 million for horeca establishments alone, and over €100 million for all buildings.[222] This conclusion does not require any information on the effectiveness of inspection and enforcement, so it could have been reached before new policies were initiated. It is unfortunate that the policy is being changed now, at a time that most buildings have been licensed (except the municipal buildings themselves!).

---

221 SIRA Consulting (2004*). The calculation applies to all buildings, not specifically to horeca establishments.

222 Based on the figures of 18,000 horeca establishments and 179,300 buildings.

13.4 Comments and interpretation

The general conclusion seems clear. There is enough doubt about the efficiency of the intensified enforcement policies following the Volendam disaster. Unless the expected fire damage has declined substantially, probably by 50% or more – which is unlikely – the policies do not pass a cost-benefit test. The resources could probably have been spent in a way that saves more lives, for example on traffic safety[223] or on radon in houses.[224] Alternatively, it might have been spent on other enforcement problems. The rules that have to be enforced are numerous and municipalities and other governmental bodies have to carefully select the regulations they want to enforce.[225]

A cost-benefit analysis is helpful in choosing such priorities. The municipal fire safety policies seem especially based on the Volendam disaster. If the government would have conducted a cost-benefit analysis before encouraging and even pressing use license policies, they would probably have concluded that these policies are not efficient. Unlike what happened after the Volendam disaster, the government should have resisted the pressure to take action, at least the substantial actions that were in the end taken.

Of course, this result should be treated with caution. Some possible consequences are unknown and therefore incorporated as PM-item. Moreover, the outcome depends on the reliability of the data, for example the extent to which the statistical value of fire casualties and burns are similar to those of traffic accidents. The differences seem to be limited.[226] Yet several important remarks can be made, especially with respect to the value of life as it is treated in a cost-benefit analysis.

---

223 For example the European Commission (2000) concludes that investing in enforcement of alcohol use in traffic is socially desirable. See also SER (1999*).

224 See RIVM (2003*).

225 See note 173.

226 Savage (1993) compares the willingness to pay for four different hazards. He concludes that people are willing to pay significantly more to lower the risks of stomach cancer than to risks posed by automobile accidents, home fires and aviation (Sunstein, 1997). Stomach cancer appeared to be in a class apart from the other hazards. In terms of willingness to pay, cancer ranks first, followed by automobile accidents, home fires, and aviation. Note that in my analysis the issue is about fires happening elsewhere than at home, which might imply differences in 'control', 'voluntariness', and perhaps dreadfulness. These may influence the willingness to pay. Sunstein (1997) discusses the factors that influence the willingness to pay and concludes that three deserve special attention: those imposing high externalities, those preceded by unusual pain and suffering, and those producing distributional inequity. These three are not important in fire accidents in horeca establishments. All in all, data for traffic accidents seem to be the best available estimate for fire accidents (see also Viscusi, 1993).

13.4.1 *Fires as disaster*

There is one important objection that is often raised. A disaster such as the Volendam disaster falls into a different category than 'common' accidents. Therefore, it is argued, a cost-benefit analysis is either unnecessary or unsuitable. After such tragedies, the public wants the government to do everything in its power to prevent such accidents from happening again, without reference to the costs and benefits. Do disasters like the Volendam, Goteborg or Buenos Aires fires require more stringent regulation and enforcement than 'common' accidents which lead to the same expected number of casualties? Are lives lost in a catastrophe worth more? The calculation above assumes that the data from automobile accidents are fairly accurate estimates for human losses as a result of fire accidents. However, automobile accidents usually lead to one or two casualties at a time, while fire accidents in horeca establishments may result in catastrophes that result in a large amount of casualties.[227] It is argued that these large consequences justify the enforcement policies that were observed after Volendam.

There are two issues here, both relating to the higher media attention that is given to disasters. First, Zeckhauser (1996) argues that to those who die, a catastrophe is no worse than any other accident. However, to those who simply read about it, it is. In all accidents resulting in death, individuals such as family and friends, who care about the victims, suffer pain and grief. However, in the case of a catastrophe, even people who are unconnected or only marginally connected to those who die, grieve and suffer as external observers from empathetic concern with the victims. This empathy arises because the media alert them to these accidents, while common accidents often remain unnoticed. This implies that the (social) costs of these disasters are higher and (more) individuals are willing to spend more on limiting the risk of disasters.

Secondly, people have difficulty estimating the expected harm from accidents, especially catastrophes, biasing them to fear catastrophes more than common accidents.[228] People treat safety as a kind of "all or nothing" matter and are vulnerable to a "zero risk mentality" (Sunstein, 2001a). People suffer from probability neglect (Sunstein, 2001b), implying that they focus on the worst outcome and ignore the (often small) probability of the occurrence of this outcome. Probability neglect occurs especially in the face of emotional, fearful events. Moreover, in estimating the expected harm, the availability heuristic is important. People assess the probability of an event by seeing whether relevant examples are cognitively 'available' (Sunstein, 2001a). Since more (media) attention is given to disasters, they are more 'available' and thus estimated to be more likely than common accidents.

227 Zeckhauser (1996) provides a nice introduction to the economics of catastrophes.
228 See section 1.4.2.

13.4.2 Statistical cost-benefit analysis versus 'irrational' emotions

So people are frightened of disasters and disasters may therefore induce them to be willing to spend more on preventing them. This also explains why the public requires regulation and enforcement after a disaster. The question is whether public policy should be based on the objective, statistical evidence concerning expected (harm from) accidents, or on the public's fear of such accidents. Should the government respond to intense fear that involves statistically remote risks? Or can we maintain that investments in traffic safety or in reducing radon in living houses are more worthwhile than investments in fire safety in horeca establishments?

The general argument

In general, I would argue,[229] public policy should be based on facts, on accurate information, and not on what people think. Since people make large mistakes in estimating accidents and their expected harm, an efficient regulatory and enforcement policy is not one directly, one-to-one, based on the public's perception of priority-setting, but imposes a filter on these perceptions. Moreover, precisely because of probability neglect and other cognitive failures, institutional constraints are required that prevent policies and politicians from being based on public perceptions. An obligatory cost-benefit analysis, or a duty to enforce (section 9.3.1), may prevent us from basing policies on ungrounded fears and from responding too quickly to incidents.

However, ungrounded public fears of accidents should not be completely ignored. People's fear of accidents and deaths creates real social costs, because the fear itself is unpleasant and because it will influence behavior so that people will avoid certain activities or invest in precautions. Of course, if possible, the government should try to inform and educate people, so that they do not persist in their errors. However, if public fear is intractable (in the sense that it is impervious to efforts at reassurance), a failure to respond would impose costs because it would leave people feeling distressed, upset, or fearful. A governmental response in the form of safety regulation and enforcement can be socially beneficial if the benefits in terms of fear reduction justify the costs. Moreover, in order for public policy to be effective, public support and confidence are required. Therefore, regulators and enforcement officials must be aware not only of the actual, but also of the perceived risks.

Fear of fire in horeca establishments

With respect to the fire safety enforcement policies following Volendam, this issue does not seem to play a very important role. Of course, in the period directly following the tragedy, many people questioned the safety of horeca establishments. Moreover, many members of the municipal council asked

229 The following arguments can be found in Zeckhauser (1996) and Sunstein (2001a and 2001b).

questions about the situation in their municipality. However, real feelings of fear and distrust were not very common after the Volendam disaster. In any case, declines in horeca visits as a result of the Volendam disaster are not mentioned anywhere. Moreover, even if people had misperceived the risk of fire in horeca establishments and required public policies, I am convinced that explanation would have been very helpful. The pressure to start enforcement actions came especially from governmental officials themselves (varying from politicians to the Alders Committee, Ministries and Fire Departments) and not from the common citizens.

What does matter in the case of fire safety in horeca establishments are public feelings of empathy with the victims. These feelings can be incorporated into the value of life, because there is a willingness to pay. Unfortunately, there are no data on basis of which to estimate this willingness to pay. Since such feelings are absent or less astute in the case of automobile accidents, they are not fully taken into account in the analysis above. Suppose that, in order to provide some understanding of the order of effects, the expected human losses are twice as high as in case of automobile accidents. This would imply that the total damage from fire rises to €39.2 million. As a result, the minimum required reduction in fire damage varies from 33% to 60%. That is still substantial. I am therefore of the opinion that incorporating the fact that disasters such as Volendam lead to many simultaneous casualties, does not invalidate the conclusion that it is likely that the increase in enforcement resources following the Volendam disaster has been inefficient.

13.5 Conclusion from the cost-benefit analysis

A cost-benefit analysis is useful in 'rationalizing' public policy and preventing it from being founded on unjustified perceptions. In case of disasters and incidents, a cost-benefit analysis is even more, and not less, helpful, precisely because there are cognitive failures and emotions. A cost-benefit analysis provides both the public and the decision-makers with the necessary information about actual risks and alternative policies for reducing this risk. For the public, cost-benefit analysis is important because it may be an instrument in informing and educating individuals about actual risks and help the government defend its priority setting. An obligatory cost-benefit analysis may be a credible constraint for politicians if their voters desire inefficient action.

If the government had conducted a cost-benefit analysis before pressing use license policies, they would probably have concluded that these policies were inefficient. Instead of what happened after the Volendam disaster, the government should have resisted the pressure to take action, at least refraining from the substantial actions they ended up taking. The argument of section 12.5.2 is supported by the fact that fire safety departments seem to have seized the opportunity to increase their claims for budget. Despite the wide use of programmed enforcement, fire safety policies seem primarily based on the Volendam disaster.

From the analysis of the municipal enforcement policies in this and the previous chapter it appears that there is no adequate foundation for the current policy. There is an endless elaboration on tasks, organization, coordination and the required number of personnel, but it remains unclear what the precise objective is, or the desired result. There is not even a trace of a balance of benefits and costs. What seems to be necessary in the field of enforcement is not so much a flood of policy plans, but better analyses of the effectiveness and efficiency of actions and measures. We therefore return to the conclusions of the Netherlands Court of Audit (2005*) as well as Leeuw and Willemsen (2006*) that enforcement has certainly been professionalized at all levels over the past years. However, the effect on actual enforcement and compliance remains unknown. "For example, enforcers do not, as a rule, know to which extent enforcement efforts contribute to compliance of the rules and laws and to the solution of the underlying problems. (...) [I]t is also unknown to which risks citizens are exposed at the chosen level of enforcement. Nor is it clear how and at what price risks can and should be reduced." (Netherlands Court of Audit, 2005*, p.5). This conclusion appears to be directly applicable to the municipal enforcement of fire safety in horeca establishments.

I return to this observation in chapter 16. First, however, I discuss the use of private enforcement.

14 Settlement of liability claims following the Volendam disaster

The previous chapters consider the public enforcement policy following the Volendam disaster and analyze whether it provides an adequate incentive for proprietors to comply with fire safety regulation. Section 2.3 shows that horeca proprietors can also be given an incentive to increase precautions as a result of liability claims from the victims of a fire accident. Under certain conditions, this may induce the proprietor to choose efficient precautions. In this chapter I want to examine the incentives emanating from the Dutch legal system. Is private enforcement by victims effective in inducing proprietors to take precautions against fire damage?

I do not examine the litigation process of liability claims for fire accidents in general. There are no data on these types of court cases, let alone settlements. I focus instead on the settlement of the claims of the victims of the Volendam disaster. For this procedure, a lot of data can be found. Moreover, since it is an important case which received a lot of attention, it is likely to determine the perception for future cases. Therefore it is worthwhile to examine this case-study. Does the settlement of the Volendam disaster provide sufficient incentive to prevent future fire accidents?

Section 14.1 presents the headlines of the litigation process of the damage claims of the victims of the Volendam disaster. Section 14.2 summarizes the payments that the different parties have made. Section 14.3 answers the main question whether the outcome of the litigation process provides an efficient incentive to take precautions. Section 14.4 concludes.

14.1 The facts

In this section I offer a general description of the settlement of the damage of the Volendam disaster. This information is not directly available, but can – with some effort – be gleaned from (annual) reports, newspaper articles, parliamentary documents, and websites. A detailed account of my sources is available upon request.

14.1.1 *The Volendam disaster and the first actions*

Shortly after midnight January 1, 2001, a fire broke out in café "'t Hemeltje" in Volendam. At that moment 356 youngsters were present. As a consequence of the fire, 14 of them died and almost all others were injured, many suffering serious burns (almost 200 youngsters).

In the days following the disaster, an organization was set up, combining fund-raising, medical, financial, tax, legal and insurance knowledge. This foundation, SSNV, initiated by the inhabitants and business community of Volendam, was aimed at providing the casualties with the widest ranging help in the short, but especially the long run. In the course of the following years, the SSNV received donations of approximately € 7.7 million. A few months later, another foundation was established, the BSNV, a special interest group of the victims. Almost all victims (330) were affiliated with this organization.

The proprietor of 't Hemeltje, Veerman, almost immediately donated € 22,700. After two months, his liability insurance company paid out €1,134,450, the full amount for which Veerman's company was insured. Somewhat peculiarly, it was continuously reported that Veerman gave up this money voluntarily and it should be seen as a gift. Apparently, the insurance company had immediately decided to pay out the insurance to get rid off the case. In any case, 25% of the money was donated to the SSNV, the remainder was transferred to the foundation 010101, which paid the money out to the victims.

An information and advice centre was started, later called 'Het Anker'. This centre received several subsidies from the Ministry of Public Health for the years 2002-2004 of a total of € 1,907,641, and from the Ministry of Internal Affairs for the years 2003-2006 of € 5 million. Subsidies were also granted for reintegration courses, but the amount is unknown.

Shortly after the disaster, the Provincial States of North-Holland (to which Edam-Volendam belongs) decided to make € 453,800 available for the victims. This amount was available for specific needs, especially housing, that were not met in another way.

14.1.2 *The investigating reports*

On March 26, 2001, the report of the Committee Polak-Versteden was published which, by order of the governor of the Province North-Holland (and on the request of the municipality of Edam-Volendam), offered advice concerning the governmental lessons to be learned. More important was the final report of the Alders Committee on June 21, 2001, which investigated the café fire by order of the central government. In response to the report the government made a total of €23 million available for the years 2001-2006, divided as follows:[230]

- The Ministry of Education provided a compensation of €6.6 million for six years for additional personnel, tools and facilities for two schools in Volendam.

230 I report here the monetary consequences of the Alders Committee. The results of the investigation into the fire and the enforcement policy are reported in section 11.2.2.

- The Ministry of Public Health paid out €10 million for four years to 'Het Anker' and to a number of health institutes.
- The Ministry of Internal Affairs paid the costs of the Alders Committee of €2.5 million.
- The municipality of Edam-Volendam received a compensation of €2.4 million for the costs of assistance, memorial services, etc. as well as for travel and accomodation costs for family members of the victims.
- The SSNV received €1.1 million.
- The Dutch Burns Foundation received €0.45 million.

On July 4, 2002, the final report of the Committee Financial Settlement New Year's Eve Fire Volendam (CFA Volendam, or CFA II) was published. CFA II advised the government about the support and financial compensation to the victims (only monetary damage was considered). The central government made €30.1 million available, including €4.2 million for implementation costs. Three kinds of support were offered:

1. A centre for reintegration and aftercare, for everyone, €5 million for a period of five years.
2. A (once-only) compensation of functional invalidity, for the victims, in total €17.3 million. This compensation depends on the level of invalidity, with a maximum of €150,000 for 100% invalidity.
3. A compensation of the costs due to injury, for victims and relatives, in total €3.6 million, as far as these costs were not compensated in another way. The maximum compensation was €20,000 for victims, and €7,500 for their relatives.

The committee expected that 336 victims, 1300 relatives and 1700 social/relief workers would make use of these resources. In addition to this arrangement, the SSNV financed complementary measures.

The results of the medical examinations and the corresponding payments were not unambiguously reported. In early 2004, the SSNV reported that 371 decisions had been reached and that €15.5 million of the CFA-II money had been paid out. The SSNV herself had financed an additional €6.6 million (of which €391,018 had been financed by the remainder of the insurance benefit of proprietor Veerman). The SSNV wished to keep €1.3 million available in cash for future arrangements. The distribution of the victims by degree of severity was as follows (based on an interim report of November 2003 of 329 decisions):

1. 50 victims fell in the most severe category with a benefit of €135,000 – 150,000.
2. 30 victims fell in the category with a benefit between €75,000 and €135,000.
3. 85 victims fell in the category with a benefit between €15,000 and €75,000.
4. Approximately 140 youngsters fell in the lightest category with 2-10% permanent invalidity and a benefit of €1,500 – 15,000.

In order to stimulate the youngsters to use the money for the long term (for instance buying a suitable house), the SSNV created a bonus arrangement. If someone put the money on a long-term savings account, he or she received a once-only bonus of 5% of the deposit. This bonus arrangement was financed by the interest on the account on which the CFA-II money had been paid. In total €784,000 were spent on bonuses.

14.1.3 *The settlement of liability claims and other legal procedures*

This subsection describes the settlement of the damage claims against the municipality and the proprietor Veerman.

The damage claims and negotiations

Most victims united themselves in the BSNV. Already as early as August 20, 2001, the BSNV filed a claim to hear witnesses as part of the preliminary investigation into the liability. The intention was to hold both the proprietor Veerman and the municipality liable. These provisional witness testimonies were taken in 2001 and 2002. When questioned, the proprietor Veerman remained silent because of the criminal investigation that was running. In January 2002, the union FNV started an action on the substance against the municipality and Veerman on behalf of 14 victims. One of them claimed compensation in a procedure at the court of Haarlem. Later, the victims represented by the FNV joined those of the BSNV.

The municipality of Edam-Volendam rejected liability from the beginning and was not willing to negotiate about the claims. With respect to the claims against Veerman, there were many skirmishes in the following months and years. The main lines are as follows. In the beginning, Veerman indicated that he was willing to settle. Rumors are that there was talk of a settlement amount of €6 million under the condition that Veerman was able to exploit his properties again. Therefore, the BSNV decided to freeze the current procedures and to invest in an audit into the assets of Veerman. There was much debate concerning the possible exploitation of the cafés by Veerman. Veerman argued that he needed to exploit his cafés to earn money for compensation. However, many victims did not like the idea of reopening the cafés. This led to Veerman suggesting that he could move to neighboring locations. However, the municipality (the council) was not willing to change the zoning plan. At some point, Veerman decided to reopen his old cafés and applied for the necessary licenses. The use license could not be refused, but there were procedures (including court cases) for the liquor and horeca license. The BSNV was so angry about the decision to reopen the cafés that they interrupted the settlement negotiations. Moreover, they argued that Veerman was not willing to cooperate with the audit into his assets. In June 2003, BSNV's lawyers summoned Veerman and the municipality on behalf of 214 victims united in the BSNV. They asked the court to convict them for the consequences of the fire. They reproached

Veerman the following points: (i) the Christmas decorations, installed under supervision of the owner, and hung everywhere, was highly flammable, and hung too low; (ii) the café did not comply with many fire safety requirements, for example with the emergency exits. The reproaches to the municipality were as follows: (i) despite advices they performed no inspections in the period 1994 – April 2000; (ii) they did not act upon the inspection of April 2000 in which many failures were detected; (iii) the municipality knew about the Christmas decorations, but did not act upon this information despite their own statement that it was dangerous. The victims demanded €213 million compensation plus €14 million for non-monetary damages. More specifically, 202 wounded victims demanded €1,054,115 per person for lost income and non-monetary damages varying from €18,750 to €150,000. In addition, 12 parents of dead youngsters demanded €10,000, specifically for the costs of a funeral. In the first instance, however, the procedure was aimed at determining negligence, following which the amount of compensation should have been determined in a separate (and individual) procedure. At the same time, they made a prejudgment attachment on the properties of Veerman for €150 million. Both Veerman and the municipality denied liability.

Criminal prosecution

Meanwhile, the owner Veerman, his daughter and the manager were being criminally prosecuted. On July 18, 2003, the court of Haarlem convicted Veerman on charge of culpable homicide, fire and causing serious physical injury, and condemned him to 240 hours of community service, 12 months of suspended imprisonment and deprivation of the profession of horeca proprietor for two years. The daughter and manager were both acquitted. Both Veerman and the Public Prosecutor appealed against the court decision.

Settlement

On April 22, 2004, the BSNV, Veerman and the municipality presented a draft settlement agreement. The headlines of the agreement were:

1. Veerman c.s. were to sell the properties of Haven 154-156 to the municipality for the following amounts:
   - Haven 156, which belonged to someone else, for € 907,560.43,
   - Haven 154 and the horeca establishment for € 2,885,334.

   After deducting several costs including mortgage expenses, Veerman received an amount of € 1,825,334. Of this amount € 700,000 was to be paid to the foundation Aftercare (see 4). The remainder, approximately € 1.2 million, was to be paid into a frozen account of the SSNV, and paid out by this organization to the victims. The SSNV and BSNV would decide how this money should be spent.
2. The municipality and the BSNV were to set up an idealistic foundation that would exploit the property at Haven 154-156. In this property, there would be no horeca exploitation for at least 50 years.
3. Veerman gave a right to his profit for a maximum of € 1 million until De-

cember 31, 2012 (i.e. for ten years). Supposedly the amount would be € 230,000.
4. The independent Foundation Aftercare was to be set up by the municipality and the BSNV. The municipality was to pay €700,000 as a starting capital and then €40,000 annually for a period of 40 years. Veerman also paid €700,000 (see 1). The Foundation Aftercare was to function specifically as a safety net for future costs that would not be compensated by another way or by benefits of insurance policy.
5. The current civil procedures would be terminated.

All 340 victims had to accept the agreement. The Public Prosecutor and Veerman indicated that they were willing to withdraw the criminal appeal procedure if there was a settlement about the compensation.

On November 1, 2004, the above settlement agreement was signed by all parties. All legal procedures ended. 333 of the 340 members of the BSNV gave a mandate to the BSNV to sign the agreement. Six of the others declared that they did not see themselves as a victim and therefore would not apply for the compensation. All victims who participated in a court procedure (by the BSNV or the FNV) signed. Based on these results, the municipality and Veerman concluded that the agreement could be implemented.

Hereafter, nothing much happened. The appeals in the criminal case were withdrawn by both parties. The administrative procedures concerning the opening of the cafés were no longer relevant, and nothing was heard from the civil procedures. Apparently the one person who did not agree had reconciled himself to the agreement.

14.2 Overview of payments

Based on the facts presented in the previous section, I made an overview of the payments that the different parties made. Not all the information is perfectly clear, but no other information was available. Most of the 'common' costs are not included, such as the costs of fire assistance and casualties, the costs of benefits from health, accident or social insurances, reintegration, etc. It is also not always known whether all the money promised was in fact paid out. Furthermore, it is unknown whether the price the municipality paid for the property of Haven 154-156 is a fair market price. Finally, the amount paid for lawyers and other legal support remains unknown. These amounts were probably paid from the received compensation.

With this in mind, table 14.1 provides the most reliable estimate of the different payments. In total, the different parties paid € 66.8 million. This money was allocated in the following way:

1. The victims received direct financial compensation of € 32.2 million:
   - The SSNV (also on behalf of the BSNV and foundation 010101) distributed €10.1 million. Of this amount, €2.6 million came from the government (after the Alders Committee 1.1, Provincial States 0.5 and

revenues from sale Haven 154-156 1.1), €1.4 million from Veerman (insurance benefit 1.13, gift of 0.023 and profit right of 0.23), and €6.2 million from private donations.
- The Foundation Aftercare had €2.2 million at its disposal (0.7 from Veerman and 0.7 + 0.82 from the municipality).
- The damage fund CFA-II has granted €3.6 + €15.5 + €0.78= € 19.9 million.

2. There was indirect support for the victims of a total of € 27.9 million. The national government provided €19.5 million in response to the Alders Committee for schools, health institutes, 'Het Anker', etc., and another €5 for 'Het Anker' by CFA-II. The municipality established a foundation in Haven 154-156 for fifty years, which cost €3.4 million.
3. The different transaction costs were in total € 6.7 million:
   - The Alders Committee cost €2.5 million.
   - The implementation of CFA-II cost €4.2 million.

Table 14.1 Overview of payments after the Volendam disaster

National government:			**€ 52,5 million**
For the period 2001-2006 (committee Alders)		€ 23,0 million	
(especially aimed at organizations like schools, health institutes etc.)[a]			
Provincial States		€ 0,5 million	
CFA-II money		€ 29,1 million	
Compensation for vicims[b]	€ 24,1 million		
Costs of implementation	€ 4,2 million		
Bonus arrangement[c]	€ 784.000		
Common benefits of medical expenses, social insurance, accident policies etc.		PM	
Municipality Edam-Volendam:			**€ 4,9 million**
Once-only contribution to Foundation Aftercare		€ 0,7 million	
Annual contribution to Foundation Aftercare, for 40 years, € 40.000		€ 0,8 million[d]	
Costs of exploitation Haven 154-156 by idealistic foundation[e]		€ 3,4 million	
Common costs of assistance by police, Fire Brigade etc.		PM	
Private persons:[f]			**€ 6,2 million**
Proprietor Veerman:			**€ 3,15 million**
Insurance benefits donated		€ 1,1 million	
Donation		€ 22.700	
Net profit sale property Haven 154-156		€ 1,8 million	
Right on profit (maximal € 1 million)		€ 230.000	

a. There are also different incidental subsidies for 'Het Anker' of € 6,9 million in total. It is unclear whether these amounts should be counted separately. These amounts are the initial amounts that the government has made available. Possibly the amounts that are made available as a consequence of the "Committee Alders" or "CFA-II" replace the initial subsidies. Therefore I exclude the initial subsidies from the calculations.

b. € 25,9 million has been made available but € 24,1 is actually paid.

c. This bonus is financed by the interest on the CFA-II money. It is therefore open for discussion whether these are really costs of the government or simply interest benefits to the victims. I take the perspective that it takes necessarily some time for medical examinations etc. before implementation of public damage compensation can be realised. Therefore the national government could have decided to make € 25,9 available by about January 1, 2004. Because it decided to donate the money already in July 2002 it loses interest in this period. An argument in favor of this perspective is the fact that to the extent that victims do not make use of the arrangement, the money will (most likely) be returned to the government. Moreover, including the bonus implies that the CFA-II money is calculated in prices of 2004 as holds for most other payments in the table.

d. This is the net present value, calculated with an interest rate of 4% and payment at the beginning of the period. Note that all other values are not discounted

e. The municipality has purchased the property at Haven 154-156 for € 3,8 million. Assume that this is the market value. The transaction itself implies that financial assets of the municipality decrease with € 3,8 million, while the assets of real estate increase with the same amount. There are however costs, because the municipality gives the property a specific destination for 50 years, so that the property does not yield any revenues for the municipality during these years (no rents are paid). If we assume that an investment would yield a return of 4% per year, the costs are about € 150.000 per year. Over fifty years the net present value is then € 3,40 million (€ 3.395.904).

f. In total the SSNV has € 7,7 million at its disposal. This includes € 391.018 of insurance money from Veerman (although this is more than 25%), as well as the donation of the national government of € 1,1 million. So the private donations can be calculated as € 6,2 million.

Victims therefore received financial support for a total of €60.1 million. The largest part of this amount was paid by the national government. The proprietor Veerman only paid €3.15 million, but this amount includes the profit from the sale of his cafés at Haven 154-156. Unfortunately, it is unknown whether the price the municipality paid for the property is higher or lower than the market value. It is therefore unknown what Veerman's real costs were.[231] In the absence of profit from the property sale Veerman only paid €1.35 million, and even this sum includes the insurance benefit of €1.1 million he received.

231 The market value of the property was probably almost zero in the short run because the community would not allow the property to be used freely. Intuitively, a building in which such a disaster has taken place is very difficult to sell. If so, the €1.8 million is not a real contribution of Veerman, but another donation by the municipality.

14.3 Evaluating the Volendam settlement

How must these observations be interpreted? In this section I evaluate the settlement process described in the previous sections.

14.3.1 *The process*

In the end, the damage claim was settled quite smoothly. The year 2003 was important. During that period, BSNV, Veerman and the municipality negotiated with difficulty. Veerman did not want to say much because of the criminal lawsuit. There was also much quarreling about the reopening of Veerman's cafés. On the part of the victims, uncertainty remained concerning the amount of money they had already received and what they should demand from Veerman. Therefore, in June and July 2003, civil, administrative and criminal court cases took place.

At the beginning of the year 2004, the situation changed. The Volendam community wanted to 'move on' and close the case. The criminal lawsuit had ended and Veerman had been convicted. Furthermore, the medical examinations had been performed and the victims had received their money (from the national government and the SSNV). All this opened the way to the settlement of April 2004 that was completed in November 2004.

This immediately leads to the question whether the victims and the Volendam community in general had insufficient incentive to claim damages from Veerman. Apparently, because they had received money from the national government and the criminal conviction had provided the desired statement of Veerman's guilt, they did not have any wish to continue the case. Before addressing this question, I discuss the total social harm from the Volendam disaster.

14.3.2 *Total harm*

The compensation amounts discussed above provide a fair estimate of the monetary damage. The impression is that the mentioned damage compensations were, together with common social insurance and health care benefits, sufficient to cover all the material damage. There were not many unfulfilled desires. In this way, the total monetary damage of victims can be estimated at €60 million. This amount does not include the costs that were covered in some other way (such as the costs of hospital admission, social insurance, reintegration etc.). There were also costs for 'society' in the form of fire brigade and police assistance, the transaction costs of €6.7 million, etc. On the other hand, there are some signals that indicate overcompensation. For example, there were three different arrangements for the compensation of funeral costs. Let us take €60 million to be the most reasonable estimate of the monetary damage.

The non-monetary damage can be estimated at €1.87 million for the dead and €0.19 for the seriously wounded casualties.[232] Therefore, the non-monetary damage from the 14 youngsters who died can be estimated at €26.3 million. If we assume that there were 175 seriously wounded youngsters,[233] the non-monetary damage for injuries can be estimated at €32.9 million. The remaining non-monetary damage of the slightly injured as well as the general social commotion should be considered as PM-item.

All in all, this leads to an estimate of the total social harm of the Volendam disaster of €119.3 million.[234]

14.3.3 Evaluation of Veerman's contribution

The proprietor Veerman only paid €3.15 million, of which €1.1 as insurance benefits and €1.8 in return for the sale of his property. How should we evaluate the fact that Veerman only paid a minor fraction of the total damage, even if we only consider the monetary damage?

If the goal of liability is only compensation or retribution, the results are probably not worrying. Possibly, compensation is most efficiently dealt with by public funds instead of private litigation. The transaction costs of public funds are still substantial (€4.2 million in this case), but probably lower than those of private litigation.[235] Moreover, public compensation can be more effective because it naturally supplements the general public benefits and must satisfy less stringent requirements on causation and the actual occurrence of damage. If the goal is retribution, some 'sufficient' level of punishment of Veerman is effective. It appears that the Volendam community, including the victims, took the view that this punishment has been realized by the criminal sentence and that the additional payments of Veerman were sufficient.

Liability and the incentive to take precautions

Economists, however also, or even mainly, ask attention for the preventive function of liability. Liability provides an incentive to take precautions. As a general rule, explained in section 2.3, this incentive is efficient if the injurer is

232 See table 13.3.

233 I.e. the categories 1-3 named by CFA-II (p.248): (50+30+85)/308 * 329 ≈ 175.

234 If we directly estimate the damage using the figures in table 13.3, the social harm can be estimated at 14 * €2,394,942 + 175 * €232,653 = €74.2 million. This is an underestimation because the 154 (329 – 175) slightly injured are not counted, the medical costs of burns might be higher than those of traffic accidents, the general social costs to others than direct victims and relatives are not counted, etc.

235 It is often suggested that redistributing one euro or dollar through the liability system costs 50 cents or even one euro or dollar. In section 15.2, I calculate that the transaction costs are €3750 per claim. Barendrecht (2002*) claims that liability costs about 50 eurocents per redistributed euro, while redistribution through damage funds and social security costs 20 eurocents.

forced to pay fully for all the damage he inflicted.[236] This is not what happened in the Volendam case. One major problem was that compensation for non-monetary damage was only very limited, but even if we ignore this problem, Veerman did not fully pay for the monetary damage either. There are three potential explanations for this fact.

A first explanation is that the merits of the case were not so large, so that victims expected that the probability of the court convicting Veerman to pay compensation was too low, or that they would have to wait a long time before receiving a compensation. However, the benefits of the case were, in principle, rather high. Since Veerman was held criminally liable, he would also be liable in a civil lawsuit unless he was able to present proof to the contrary.[237] If his behavior was negligent, he would be obliged to compensate the financial damage plus a (less than full) compensation for the non-monetary damage if a victim was physically injured, meaning at the very least the direct financial damage to the victims of €32.2 million. Of course, a court procedure would take more time, but since the threat of a trial was credible, Veerman would probably have been willing to settle the case for a higher amount. It is also not true that the victims could themselves be blamed. This could only be true in the case of the person that lit the packet of sparklers, but he had died. So the merits of the case seem to be quite high.

Secondly, Veerman might have been unable to pay a higher amount of compensation, but this does not seem to have been the case either. Veerman was not completely squeezed. At the time of the Volendam disaster, he did possess several other properties and numerous companies and he did not give up these possessions. He continued to function in the Volendam community and to own several companies, as well as horeca establishments. He was also unwilling to cooperate in the audit into his assets. His assets were estimated at several tens of millions of euros. Even the company to which ''t Hemeltje' belonged had assets of at least €10 million. Therefore, although Veerman may have been unable to pay €32.2 million, let alone the full damage of €119.3 million, he could have paid substantially more than he did.

The final and most likely explanation is that the Volendam victims and community did not have an incentive to proceed with the case because the

236 This is ignoring the fact that under negligence, the injurer might choose efficient precautions even if the damage compensation is less than full because he then completely escapes liability by choosing due care. However, the gap between the compensation of €3.15 million and the total damage of more than €100 million seems too great for this point to apply. Moreover, the general rule that the compensation should equal total damage is a sufficient (although not necessary) condition for efficient precautions under negligence too.

237 Article 161 of the Civil Procedure Code [Wetboek van Burgerlijke Rechtsvordering] stipulates that a fact that a criminal court has declared to be proven produces imperative evidence of that fact. According to Article 151, the civil court is obliged to hold imperative evidence as being true unless proof to the contrary is presented. Therefore, in this case, the fact that the criminal court convicted Veerman also provides proof of his negligence in a civil procedure. See also the verdict of the Dutch Supreme Court of December 12, 2003 (no. C02/139HR), *NJ 2004/102*.

national government and other parties had provided them with sufficient compensation. Several factors may play a role here. A court may be unwilling to award any further compensation if the victims have already received full compensation in some other way. Therefore, the case may have clear merits of several millions of euros on paper, but these merits disappear if courts are afraid to award 'double' compensation. In addition, because of Veerman's position in the Volendam community, the victims did not like to deal further or more firmly with Veerman. Veerman's companies provided employment, his horeca establishments provided opportunities to go out, and he personally sponsored the Football Club Volendam, the restoration of a church, etc. There was pressure on the part of non-victims, but also some of the victims, to refrain from further seizing Veerman's assets and to close the case. This might have been stimulated by the fact that the victims had received all the monetary compensation required to resume their lives. Put differently, by providing extensive support to the victims, the national government took away the incentive to claim damages from Veerman. The bill of the Volendam disaster was largely paid by the tax-payer. The government did not exert any effort to recover its expenses from Veerman. This is a bad signal to other horeca proprietors who must decide about taking precautions. As a consequence, the incentive to take precautions is inefficient.

Thus, the instrument of liability, which looks so effective on paper (see section 2.3), becomes a weak tool once the government provides the victims with the compensation they desire. As a consequence, they do not feel the need to make a negligent proprietor responsible for the damages he inflicts.

14.3.4 *The cocktail of incentives*

It is, however, not possible to judge the incentive from liability to be insufficient without considering the other incentives at work. Even if one instrument like liability is insufficient, the joint use of different instruments might be an effective stimulus to take precautions.

First, the insurance payment complicates matters. The insurance contract might provide the proprietor Veerman with an incentive to take precautions, by requiring that some precautions are taken. However, it is not likely that this will reverse the conclusions above. The insurance amount of €1,1 million was included in full in the contribution of Veerman of €3,15 million. This assumes that he has a perfect incentive to prevent the damage for which he is insured, because his insurance company fully controls moral hazard. Whether this is true, remains to be seen. It is remarkable that the insurance company paid out so soon, without much investigation. Apparently, it was clear that Veerman had fulfilled his obligations, suggesting that there might have been no such obligations; in which case, the contract did not provide an incentive to take precautions either. Alternatively, there might have been several obligations, but these obligations did not match those of (criminal) courts (see main text to note 237) and at least failed to provide an incentive to prevent potential disasters (for example because the insurance contract was aimed more at preventing monetary losses to the proprietor than damage to victims, or because it had a maximum payment). By including the insurance benefits in Veerman's contribution, I assume that these benefits do include an incentive to take precautions. If this is incorrect (because of the arguments given), the conclusions are even worse with respect to Veerman's incentive to take precautions.

Other incentives to take precautions might come from public enforcement. Administrative enforcement, however, was absent. Apart from a letter that warned against dangerous decoration, the municipality of Edam-Volendam did not undertake any action towards Veerman or his establishments. Criminal enforcement leads to a suspended imprisonment term of two years. This only creates incentives to prevent future accidents. Veerman was also convicted to 240 hours of community service and is not allowed to practice the profession of horeca proprietor for two years. These are non-negligible sanctions, but it is uncertain whether they are sufficient to deter. In relationship to a total harm of more than €100 million it seems too small, unless his wage rate is really high. That public enforcement was insufficient appears from the fact that Veerman was judged to be in gross non-compliance by both the Alders Committee and the criminal court.

14.3.5 *The government payments*

There is also an alternative explanation, one in which the outcome can be judged to be efficient.

The government is responsible for supervising the fire safety in horeca establishments. The municipal government is supposed to inspect the establishments, and the national government is supposed to supervise whether municipalities really have an active enforcement policy. It is possible to argue that the government caused the disaster and is responsible for its consequences because it failed to adequately enforce fire safety (see for example the investigating report of the Alders Committee, 2001*). From this perspective, it would be efficient if the government had to pay for the damage, because this would provide it with an incentive to enforce fire safety, which will in turn induce proprietors to take efficient precautions.

The national government has recognized its failures and therefore launched a great action program to stimulate municipalities to adopt enforcement policies, especially in the field of fire safety. As a result of its failures, the government also felt responsible for compensating the victims of the Volendam disaster, and paid heavily for it, both directly and indirectly. For the municipality of Edam-Volendam, a similar argument applies. The investigating committees indicated that the municipality was to blame for not actively enforcing fire safety. By paying for the damage, the municipality's incentive to adopt an active enforcement policy was improved.[238]

Again, the incentive for the government to enforce compliance is only efficient if the government is made to pay fully for all harm. Although the national government in particular ended up paying a lot of compensation, both directly and indirectly, the non-monetary damage remains largely uncompensated. This is particularly noticeable when looking at the victims who passed away. For deadly victims, the compensation is generally at the most €26,700,[239] while only the non-monetary damage is estimated at a minimum of €1.5 million. Neither the government nor the proprietor is given an incentive to prevent non-monetary damage.

Another problem is that the government suffers from internal organization problems as a result of which it does not maximize social welfare even if it has to fully compensate all damage (chapter 10). The monetary bill of the compensation is primarily paid by tax-payers. Responsible bureaucrats and politicians can be induced to take precautions in two ways. First, the fact that high compensation must be paid might hamper their career (and hence wage rate). For example, the mayor of Edam-Volendam resigned as a result of the disaster. Secondly, they might be unable to spend money on other desired projects. None of the different departments will be willing to pay, and they will try to pass on the bill to another department. However, the incentive for bureaucrats and politicians to take care of efficient enforcement will probably be imperfect, even if they are forced to fully pay for all harm because they are only in service for a limited number of years, and therefore primarily interested in the short run. If they fail to adequately enforce regulations, it is their successors, and not they themselves, who will have to pay for it. Moreover, policy is formulated by several individuals and departments together, so it is not always possible to point to a single individual responsible. Note that the financial burden of the Volendam disaster was spread over several Ministries and their departments, so that no one bore the full responsibility

---

238 Note that this does not imply that the government will actually be convicted and made to pay damages by a court. The compensation payments of the government can be seen as a settlement amount that the government offers because it feels responsible for the damage and feels that it (rather than the victims themselves) has to pay. It is questionable, if not unlikely, that a court would convict the government and force it to pay compensation. For example, in the settlement of the fireworks explosion in the municipality Enschede, similar problems played a role, but the court did not convict the municipality nor did the central government have to pay damages. See *LJN: AZ4247* (Court The Hague, December 13, 2006, case number 170232) published on www.rechtspraak.nl. This decision is disputed by legal scholars. See for instance Van Maanen (2007*).

239 I.e., a parent couple of a passed away youngster receives €9,076 from the SSNV, €4,500 from Foundation 010101 and €4,084 from the municipality (for funeral expenses). If relevant, a compensation for lost income of €9,076 is available. This amounts to a total of €26,736. Parents and other relatives of fatalities hardly profit from the indirect compensation to schools, hospitals, etc. Maybe the organization 'Het Anker' can help them to come to terms with their sorrow.

for inadequate enforcement. Finally, departments which fail in their duties are not always punished, but – on the contrary – are sometimes rewarded with higher budgets. The responsible inspection agencies, for instance, which could have prevented the Volendam disaster had they applied a more active, on-the-streets, strict enforcement policy, were granted higher budgets and increased importance following the Volendam disaster.

In short, tax-payers are not always in possession of the right instruments for inducing politicians and bureaucrats to prevent the compensation claims they will have to pay. Many questions remain concerning the role of the government. Why did both the central government and the municipality of Edam-Volendam decide to provide almost all the required compensation to the victims? Why did the municipality participate in the settlement agreement and bear the brunt of most costs? Why did the central government fail to undertake any attempt to recover its expenses from Veerman? Was that because it is easier to let tax-payers pay rather than invest in such legal procedures? If the reason was society-wide empathy with the victims, why could compensation not be left to private donations? What was the value of the property at Haven 154-156 for which the municipality paid €3.8 million? And what was the role of lawyers in the litigation process? What advice did they offer? Why was there no information concerning their (financial) interests?

14.4 Conclusion

Section 2.3 discusses the fact that holding an injurer liable provides an efficient incentive to take precautions if the injurer is actually forced to pay for all the harm he causes. In the case of fire safety in horeca establishments, the proprietor and the government can both be viewed as injurers. However, the analysis of the settlement of the Volendam case shows that neither of the two had to pay fully for the harm done. In fact, the payment of Veerman is far below the harm done. From the point of view of prevention as the objective of liability, the result is therefore rather unsatisfactory.

First of all, non-monetary damage was not compensated but placed on the account of the victims themselves. Therefore, neither the proprietor nor the government had an incentive to prevent non-monetary damage.

Secondly, the proprietor Veerman did not have to pay any significant damages at all. He paid €3.15 million of the total damage of €119.3 million and of the total monetary damage of €60 million. Of this €3.15 million, €1.1 million was financed by his insurance company, and another €1.8 million by the municipality Edam-Volendam for a property of questionable market value. There is no good explanation available for why the victims did not try to obtain higher damages from the proprietor, except that they no longer expected higher damages since these had already been provided by the government and/or that the pressure from the community to close the case was high. As a consequence, the signal to horeca proprietors is that they do not have to care much about paying damages for fire accidents in their establishments.

Thirdly, the government paid heavily for the monetary damage. We have seen large increases in enforcement of fire safety in the years following Volendam. This might be a response to the high compensations the government is expected to have to pay for disasters such as Volendam. However, the increase in enforcement effort might also result from private interests by the national and local governments who are able to increase their importance by enforcing fire safety. In this context, the previous chapter demonstrates that the increase in enforcement effort was not preceded by a careful analysis of the costs and benefits, or of the expected accidents and compensation. Moreover, it is unknown whether bureaucrats and politicians are ever really affected by compensation claims.

The latter point demonstrates a remarkable point about the enforcement of fire safety. On the one hand, the enforcement of fire safety is excessive. The previous chapter shows that the increase in enforcement effort is not efficient because the benefits are (probably) low relative to the costs. On the other hand, the incentive for the government to enforce fire safety is inefficient, because it does not have to fully pay for all damage. This suggests that enforcement is insufficient. There are several explanations for this result. First, this chapter does not consider the costs of enforcement by the government. It only argues that the incentive to enforce is inadequate because the government does not have to fully pay for all harm done. This does not suggest that if the government had an adequate incentive, it would increase enforcement effort. The government might conclude that the costs of enforcement are not worth the reduction in expected compensation, i.e. in social harm. Secondly, as discussed in the previous chapter, the increase in enforcement is inefficient because it focuses too much on paperwork, instead of actual contact with proprietors, too much on granting licenses instead of actual inspections. A third explanation is that the enforcement effort is inefficiently allocated. The enforcement of relatively small accidents and safe establishments is excessive, while the incentive to prevent severe disasters as in case of Volendam is too low.

All in all, the proprietor is given an incentive to take precautions through a combination of administrative enforcement, civil and criminal liability, and the insurance contract. We should be careful in judging the value of all these instruments individually, since the combination might be an effective stimulus to take precautions, even if each of the instruments is insufficient by itself. At least for proprietor Veerman, this mix of factors was unlikely to provide adequate incentives. Administrative enforcement was almost absent. Civil damages were limited and insurance benefits were easily paid out. It is hard to determine to what extent the criminal sentence of 240 hours community service plus the deprivation of the profession of horeca proprietors for two years created deterrence.

I studied the case of the Volendam settlement because there are sufficient data available for this case, but not for other fire accidents. And the Volendam settlement is likely to affect perceptions about expected damages that proprietors have to pay relatively much, because it is such an important case that received a lot of attention. Therefore, if in the case studied, the proprietor does not have to pay sufficient damages, this will not induce many proprietors to take sufficient precautions. Moreover, many arguments for an insufficient incentive in this specific case, for instance the one on the insufficient award of non-monetary damage, will apply to private enforcement in general. Therefore, the analysis of the case-study provides conclusions on the effectiveness of liability in general. It is not expected that the situation has improved a lot since the Volendam disaster. Only the administrative enforcement is likely to have been improved over the past years. Therefore, although there certainly is an incentive for proprietors to take some, possibly substantial precautions, there seem to be too many barriers for making such an incentive fully efficient.

15 Efficient enforcement in a representative municipality

In the previous chapters, I examine whether the current enforcement policy of fire safety regulation, both public and private, is efficient. Definitive conclusions are impossible to draw due to data problems. It is unknown how enforcement precisely affects proprietor's behavior, and to which extent compliance affects expected damage.

Since this information is crucial in determining the efficient enforcement policy, I take a different approach in this chapter. I make two simplifications. First of all, I focus on a hypothetical, self-made municipality. I create a case of a municipality and its horeca establishments, based on as much information as available, so that it is as representative as possible. Secondly, I assume that the horeca proprietor acts according to the rational choice approach of economics. Moreover, since the influence of moral and social concerns is unknown, I assume that the proprietor is only interested in financial concerns (which may include time). With these two steps, I can overcome the data problems. I can discuss how proprietors respond to enforcement and how this affects the benefits and costs of compliance. Of course, this is done at the expense of reality. The analysis is more a simulation of the effects of enforcement policies than a real empirical analysis. However, it allows me to establish whether a given enforcement alternative is promising, and what is the direction of the effects.[240]

I first present the case of the representative municipality, including an explanation of the efficient level of compliance and of the behavior of a proprietor (section 15.1). Thereafter I examine different types of enforcement, respectively private, administrative and criminal law enforcement (sections 15.2 through 15.4). In section 15.5, I compare the alternatives and discuss the simultaneous use of enforcement alternatives.

15.1 Description of the case

In this section, I present the characteristics of the representative municipality. This case intends to be as representative as possible. I therefore use as much as possible the data available from different sources.[241] With the help of these

240 This chapter is an update of Suurmond (2005*).

241 In many cases, I have had to combine data from different sources and/or to calculate estimates. Here, I only briefly refer to the sources and mention the final results. An extensive description of the data, their sources and processing can be found in the appendix to Suurmond (2004*).

data, I provide an estimate of the expected damage, the costs of compliance, and the costs of enforcement.

15.1.1 *Horeca establishments and the municipality*

The case concerns a representative, medium-sized municipality of 75,000 inhabitants with a standard composition of the population. There are 100 horeca establishments in this municipality, classified into 15 groups. Table 15.1 below provides a general overview of these establishments. The establishments are opened 250 days a year.

Table 15.1 Overview horeca establishments case[a]

Type	# of esta-blishments	Description	sfa[b]	Annual turnover
1	10	Local bar	65	70.000
2	8	Local bar	100	125.000
3	8	Pub	125	200.000
4	4	Disco/music café	300	800.000
5	8	Café (other)	150	225.000
6	10	Pub serving food	175	300.000
7	10	Pub serving food	175	300.000
8	8	Restaurant (Dutch/French)	140	240.000
9	2	Fast food restaurant	185	700.000
10	5	Restaurant (Asian)	200	350.000
11	5	Restaurant (Asian)	100	110.000
12	8	Restaurant (Mediterranean)	175	400.000
13	4	Party / conference centre	500	400.000
14	5	Restaurant (luxury, special)	250	500.000
15	5	Restaurant (other)	100	225.000

a. Sources: Several reports by the Association of the Catering Industry, to name Horeca in numbers 2003; Drinking in the Dutch Catering Industry (2001); Dining in the Dutch Catering Industry (2002); Trendreport Catering Consumer 1998-2000. With these publications, combined with data of the CBS (Statline), calculations can be made on the number and type of horeca establishments in a municipality, the average sales floor area, the average annual turnover, the number of visits per type of establishment and the average spending per visitor.

b. sfa = sales floor area

15.1.2 *Compliance costs*

In order to estimate the compliance costs and expected damage in each category of establishments, I divide the use requirements into two groups:

1. *Technical requirements*. These include all requirements relating to the servicing, maintenance and correct functioning of technical devices (especially requirement number 2, 3 and 4 on page 187). These must be carried out approximately once a year. The costs of compliance with these requirements are several hundreds of euros per year, in money and in time that has to be spent on maintenance, check-up, etc. I assume that half of these costs is related to problems that arise during the year, for example an escape-route sign light that fails and has to be replaced. The other half relates to costs that are made at the beginning of the year, and which guarantees compliance during the whole year, for example the inspection of fire extinguishers.
2. *Number of visitors*. These requirements require continuous attention on the part of the proprietor during the use of the building (especially requirements 1, 5 and 6 on page 187). I estimate these costs by considering the number of visitors. In the course of a year, a proprietor has for several days a year the possibility to admit more visitors than allowed. This includes periods such as Carnival, New Year's Eve, the weekends before Christmas, rental for private parties, soccer matches, etc. If the proprietor complies with the maximum number of visitors, he will miss the additional turnover which could be gained during these days by allowing more visitors. I assume that for all horeca establishments, there are three days a year on which it is generally known that they can admit more visitors than allowed, divided over two days of Carnival and a one-day local event. The other days fall on weekend evenings which are not known in advance (100 potential days).

The estimated costs of compliance per type of establishment are reported in table 15.2. Based on this distinction the proprietor is able to choose between four strategies with respect to compliance:

Stv: Comply with both the technical prescriptions and the number of visitors.
St: Only comply with the technical prescriptions.
Sv: Only comply with the number of visitors.
Sn: Not comply with any of the requirements.

Table 15.2 Overview of costs of compliance per establishment[a]

Type	sfa[b]	max. # visitors[c]	# days visitors[d]	Costs per day visitors[e]	Costs visitors (per year)	Technical costs (per year)[f]
1	65	110	2	60	120	250
2	100	160	5	105	525	300
3	125	200	5	140	700	500
4	300	425	20	375	7500	700
5	150	200	5	150	750	400
6	175	300	5	125	625	400
7	175	300	5	125	625	500
8	140	225	4	150	600	600
9	185	250	2	75	150	500
10	200	250	3	150	450	800
11	100	175	3	100	300	400
12	175	300	2	300	600	500
13	500	800	20	110	2200	1000
14	250	400	4	150	600	800
15	100	150	5	125	625	300

a. Costs are in euros.
b. sfa = sales floor area.
c. This is an estimate based on the sfa. The maximum number of visitors can never exceed twice sfa.
d. This is the number of days the proprietor has the possibility to admit more visitors than allowed.
e. This is the additional profit (per day) the proprietor is able to receive if he admits more visitors than allowed. This is estimated on the basis of data on the expenditures per visitor per visit (Suurmond, 2004*), adjusted with taxes and profit margin. I assume an excess of 10-15%.
f. These are the costs of compliance with the technical requirements. Based on the estimate of RHN of the structural costs of the (merely technical) use requirements, these costs are on average €410 to €760 per year per establishment. Also based on the time needed for inspection and servicing several hundreds of euros is realistic.

15.1.3 Expected damage

The expected damage can be divided into three components. In section 13.2.3, I provided data from Statistics Netherlands (CBS) about the number of fires, financial damage and number of victims (table 13.1).[242] These can be used to provide some average information on the expected damage. The data show that in the period 1993-2001, there were on average 348 fires a year, which

242 See also table 13.5.

together caused €17.6 million of financial damage, 2.22 fatalities and 24.78 seriously wounded victims. A very rough estimate of the probability of a fire per day is 61.5929 $*10^{-6}$.[243]

1. *Monetary damage to the building.*[244] This damage is likely to be related to the sales floor area. The financial damage equals on average €720 per year per establishment, or €4 per m^2 sales floor area per year.
2. *The damage resulting from fatalities.* This damage consists of medical costs, output loss and non-monetary losses. Per fatality, these costs are estimated at €2,395,000.[245] The number of (fatal and non-fatal) victims is likely to be related to the number of visitors. In a horeca fire, an average of 0.006 people die. Per establishment this is on average 98.3284$*10^{-6}$ per year. 1 in 250 million visits to an horeca establishment results in someone's death.
3. *The damage to (seriously) wounded victims.* Likewise, this damage consists of medical costs, output loss and non-monetary losses, in total €232,600 per victim. In a horeca fire, an average of 0.071 people is wounded. Per establishment, this implies an average of 0.0011 per year. 1 in 22 million visits to an horeca establishment results in someone getting wounded.

The magnitude of the expected damage depends on the extent to which the proprietor complies with fire safety regulations. For all four strategies of the proprietor, I have made an estimate of the probability of a fire and the corresponding damage in terms of financial damage and the number of casualties. This enables me to determine the total expected damage in euros per strategy, as shown in table 15.3.[246]

243 I.e., 348 fires / (22,600 establishments * 250 days).

244 In addition, there is also monetary damage due to the loss of profit following a fire. Because of a lack of data, this factor is ignored here. Moreover, since this loss mostly concerns the proprietor, it is not fundamental to the enforcement problem (section 2.2).

245 See table 13.3.

246 Amounts for individual establishments are rounded off to the nearest ten euros, for the total to the nearest hundred of euros. In the remainder of this chapter – if not mentioned otherwise – money sums refer to the number of euros per year.

Table 15.3 Overview of costs and damage per establishment per year in euros[a]

Type	Nr	Costs technical	Costs visitors	Stv dam.	St dam.	Sv dam.	Sn dam.	Stv total	St total	Sv total	Sn total
1	10	250	120	160	170	460	470	530	420	580	470
2	8	300	525	570	670	7980	8160	1390	970	8510	8160
3	8	500	700	1210	1400	13350	14370	2410	1900	14050	14370
4	4	700	7500	660	1860	3970	6320	8860	2560	11470	6320
5	8	400	750	850	1100	9910	10040	2000	1500	10660	10040
6	10	400	625	1630	2390	12320	13370	2660	2790	12950	13370
7	10	500	625	630	700	2200	3000	1750	1200	2820	3000
8	8	600	600	1660	1780	3230	3390	2860	2380	3830	3390
9	2	500	150	710	730	1000	1030	1360	1230	1150	1030
10	5	800	450	1880	2000	3140	3310	3130	2800	3590	3310
11	5	400	300	1130	1550	2430	2980	1830	1950	2730	2980
12	8	500	600	1250	1300	6780	6820	2350	1800	7380	6820
13	4	1000	2200	1070	1160	4770	5040	4270	2160	6970	5040
14	5	800	600	1820	2620	14180	15090	3220	3420	14780	15090
15	5	300	625	800	840	1040	1080	1720	1140	1660	1080
Total		**49.200**	**88.100**	**105.000**	**131.200**	**620.800**	**663.300**	**242.300**	**180.400**	**708.800**	**663.300**

a. Explanation:
dam. = damage.
Total. = compliance costs (technical plus visitors) plus damage.
The grey highlighted cells indicate for each type of establishment which strategy leads to the lowest sum of damage plus compliance costs.

15.1.4 *The efficient level of compliance*

Table 15.3 also provides the costs of compliance of table 15.2. Table 15.3 shows that, even if the enforcement costs are ignored, it is not efficient for all proprietors to comply with all requirements. For a large proportion of the proprietors, it suffices if they comply with only the technical requirements (St). For two types of proprietors, it is optimal if they do not comply with any requirement (Sn).

Suppose we construct an indicator for the level of compliance. Every proprietor (100 in total) who complies with a group of requirements (technical requirements and number of visitors), is counted as one unit, so that the level of compliance varies from 0 (no one complies with any requirement) to 200 (every proprietor complies with all requirements). The first-best level of compliance then equals 113 and consists of 73 proprietors who comply with the technical requirements (compliance level 1), 20 proprietors (type 6, 11 and 14) who comply with both groups of requirements (compliance level 2), and 7 proprietors (type 9 and 15) who do not comply with any requirement (compliance level 0). See table 15.4. In this optimum, the costs of compliance are €57,500, with an expected damage of €119,400, in total €176,800.

Tabel 15.4 Expected damage plus costs of compliance

	Optimum	**Without enforcement**
Expected damage	119.400	663.300
Compliance costs	57.500	0
Total	176.800	663.300
Level of compliance	113	0

I assume that in this municipality each proprietor acts according to the rational choice approach of economics. A proprietor decides about his level of compliance by maximizing his utility. I assume that proprietors are risk-neutral and only care about their own level of wealth. Under these assumptions, proprietors are expected to minimize their costs of compliance and the costs of possible sanctions. These assumptions allow me to focus on the importance of the effect of enforcement on compliance. Abstracting from other factors does not mean that they do not exist. However, by assuming risk neutrality, I am able to focus on the monetary costs instead of specifying a utility function. We know that if proprietors are risk-averse their choices will be drawn away from the uncertain to the more certain alternatives. In this case, it is likely to increase their level of compliance because a fire and a sanction are risky alternatives, the probability of which can be reduced by choosing a higher level of compliance. Assuming that moral and social motivations to comply are not important has a similar function. It is uncertain whether they are important to proprietors if there is public enforcement. As discussed in section 2.2, they are unlikely to survive if there is no public enforcement. If someone is never called to account for norm-violating behavior, the norms are likely to be destroyed. In a market setting where violating behavior results in higher profits, this tendency is strengthened. Abstracting from moral and social motivations relieves me from the problem of monetizing these factors. If it appears that moral or social motivations to comply are relevant, they are likely to increase the level of compliance, because they make non-compliance a less attractive alternative.

The analysis of this case is a kind of benchmark. Here enforcement has 'maximal' effect on compliance, because the proprietor only considers his costs of compliance and the costs of sanctions. Different risk attitudes and other factors than costs of compliance and sanctions are likely to complicate matters so that the effect of enforcement lies somewhere between no effect and maximal effect. If the influence of these complicating matters is so great as to reduce the effect of enforcement to zero, enforcement will no longer be an interesting alternative as it only leads to costs and no effects. In the other cases, enforcement does have some effect and the analysis in this chapter throws light on the effect an enforcement policy has, the magnitude of this effect and whether an enforcement policy is efficient.

According to this set-up, if there is no enforcement, all proprietors will choose a level of compliance of zero. As discussed in section 2.2, if there is no public enforcement, it is likely that (in the end) no one will comply. In that case, there will be no costs of enforcement or of compliance, but the expected damage will be rather high, i.e. €663,300. Comparing this figure with the optimum of €176,800, we see that there is much scope for improvement of social welfare by increasing the level of compliance. However, this requires enforcement costs to be made, so that we need to consider whether the increase in the level of compliance is worth the costs of enforcing this level of compliance. This is the subject of the next sections. In this analysis, we can distinguish three major enforcement methods: by victims (private – or civil – enforcement), by the municipal Executive (Mayor and Aldermen, administrative enforcement), or by a Public Prosecutor (criminal enforcement).

15.2 Private law enforcement

One of the enforcement possibilities is to leave it to the victims. If they suffered damage, victims can file a claim for damages against the proprietor. If the proprietor is forced to pay damages, he will consider this when reaching a decision about his level of precautions for preventing fire damage. He will balance the costs of compliance and the damage he is expected to compensate.

As explained in section 2.3, we expect that the incentive from liability is only efficient if (1) courts enforce efficient levels of compliance in a predictable way, (2) insurance companies can control moral hazard problems, (3) the magnitude of liability involves full compensation for both monetary and non-monetary losses, (4) the proprietor is able to pay damages, (5) litigation costs are not a barrier for efficient litigation and settlement. These issues are discussed below, sometimes in a combined way.

15.2.1 *Enforcing efficient levels of compliance through liability*

The level of due care

The proprietor is obliged to pay damages if he fails to take sufficient precautions. What is meant by sufficient precautions is determined by the judge. A proprietor acts unlawfully if he violates a statutory duty or a rule of unwritten law pertaining to proper social conduct, or if he otherwise violates a visitor's rights. In determining whether (insufficient) precautions result in danger, the judge is expected to balance the probability and severity of damage and the costs of taking precautions.[247] However, the extent to which the judge is able to determine efficient precautions depends on the information he has concerning any given individual case. If there is no information available (ex-post) about the individual efficient compliance level of a proprietor, the judge can only determine a general rule for all proprietors. Moreover, it is possible that the judge sets levels of due care which differ from the most efficient levels. This would, of course, lead to inefficient results. I discuss the following three levels of due care which can be set by the judge:[248]

NLT: The standard is to comply with the technical requirements.

NLTV: The standard is that the proprietor has to comply with both the technical requirements and the number of visitors. It implies that the judge follows the statutory standard.

NLES: The standard is that the proprietor has to comply if the costs of compliance are smaller than the reduction in expected damage. The proprietor has to choose the efficient level of precautions where the sum of expected damage and compliance costs is the lowest. Table 15.3 defines this level for each proprietor.[249]

Besides these forms of negligence liability (NL) I provide the results for the alternative strict in which strict liability (SL) is chosen, so that the proprietor is always obliged to pay damages, irrespective of his level of precautions. Under strict liability, visitors can always claim their damages. Under negligence, no claims are filed, and no enforcement costs made, if the proprietor has satisfied the standard. If there is a fire, the behavior of the proprietor (the precautions he has chosen) is generally quite easily apparent. Which precautions are acceptable to the judge is a matter that becomes clear after a few trials.

The compensation claim

Under the existing law, a negligent proprietor is primarily obliged to compensate property damage. For losses other than property losses, the injured person is entitled to a fair compensation if he suffered physical injury. This implies that, in principle, only the medical costs and funeral costs are compensated to the relatives of fatalities. Other casualties receive a limited amount of non-monetary compensation. More specifically, this leads to victims being entitled to the following damage compensation:

- Medical costs and funeral costs are fully compensated, €6,400 for each fatality and €8,800 for each wounded.

---

247 HR 5 November 1965, NJ 1966, 136 (Kelderluik-arrest).

248 Strictly speaking, there is also the alternative of only prescribing the standard of the number of visitors. However, as can be deduced from table 15.3, this standard is a suboptimal standard for all proprietors together and it is therefore ignored.

249 This standard can be extended to include the enforcement costs. This might be relevant if the proprietor does not pay all damages so that he is likely to choose inefficient precautions despite efficient standard-setting (see tables 15.5 and 15.7). Therefore, lowering the standard might not alter the behavior of the proprietor but save enforcement costs.

- Loss of income (output loss) to relatives is only compensated if they were financially dependent on the dead person, and is therefore estimated at 1/3 of the real loss of €510,200.[250] For other victims, the entire loss of €36,000 is recoverable.
- Non-monetary losses are not recoverable in case of fatalities. Wounded victims receive a compensation of €20,000.[251]
- In total, relatives are entitled to €176,500 of damages if someone dies, while victims that are wounded are entitled to €64,800 of damages.

Furthermore, I use the following data:

- The enforcement costs equal €3750 per claim.[252]
- If the claim is awarded, the enforcement costs have to be paid in full by the proprietor.
- On average, 95% of the claim is awarded. The claim is settled for 95% of the claim for damages plus enforcement costs (more in section 15.2.4) This percentage is lower than 100% because there might be discussion about the victim's own fault, about the magnitude of damage and the extent to which this damage was caused by the fire.[253]
- The proprietor himself must pay the financial damage he suffers. In section 15.2.3, I discuss this assumption.

Analysis under current law

All in all, under strict liability, the proprietor expects to pay a compensation of 95% * (€176,500 + €3750) = €171,200 per fatality, and 95% * (€64,800 + €3750) = €65,100 per wounded. Consider for example a proprietor of type 10. Under Sn, he would expect 0.000012 fatality per day and under St 0.0000082. Making costs of compliance for technical requirements of €800 per year yields a reduction in the expected compensation of (0.000012 – 0.0000082) * 250 days * €171,200 = €161. Similarly, there is a reduction in wounded casualties of €704 and in monetary damage to the building of €340. All together, making costs for technical precautions is more than profitable for the proprietor.

Under the negligence rule, a proprietor of type 10 may escape compensation claims completely by making costs of technical compliance of €800. He then saves on fatalities 0.000012 * 250 days * €171,200 = €514. In total, he would avoid costs of €3100 by complying with the technical requirements.

In a similar way, each proprietor's behavior can be determined under

250 This is due to the fact that most visitors are youngsters who still live at home and are not permanently engaged to a partner.

251 Based on ANWB, *Smartengeldgids 2003*. This report contains compensation amounts which judges have awarded to burn victims. In addition on p. 24/25, there is a general table for compensation for different classes of injuries.

252 This estimate is based on Weterings (1999*). He estimates the transaction costs for the settlement of injury claims, i.e. the costs of (1) the pre-claim stage, (2) the determination of liability, (3) the medical examination, and (4) the determination of the damage.

253 In any case, the magnitude of this percentage has little impact, as long as it remains above 60%.

the different liability systems. Table 15.5 provides an overview. It shows that under strict liability the enforcement costs are higher than under the negligence rule. The reason is – as mentioned before – that under strict liability, victims can always claim damages. Under the negligence rule, most proprietors end up complying with the standard, and hence no litigation occurs. Victims have to bear the damage themselves. However, under both strict and negligence liability, the enforcement costs are very low compared to the expected damage and compliance costs. The enforcement costs are only made if there is a fire and this fire results in casualties. The probability of such an event is very small.

Table 15.5 Liability under current compensation schemes

	SL	NLT	NLTV	NLES
Expected damage	171.300	142.500	155.300	142.500
Compliance costs	30.400	40.200	35.400	40.200
Subtotal	*201.700*	*182.700*	*190.700*	*182.700*
Enforcement costs	990	150	550	230
Total	**202.700**	**182.800**	**191.300**	**182.900**
Level of compliance	60	78	70	78

The different liability systems force a substantial number of proprietors to take precautions, although the level of compliance is significantly less than 113. The reason is that proprietors only have to compensate a very small amount of the damage to victims. They take into account only a part of the social harm they cause. For example, under the current compensation scheme, no proprietor will comply with the maximum number of visitors, neither under strict liability nor under negligence with full compliance as the standard. Moreover, under NLTV, the level of compliance is lower than under NLT. Under NLT, it is beneficial for proprietors of type 8 to comply with the technical requirements to escape liability (costs of compliance of €600, no compensation of €610), but this incentive is insufficient under NLTV where they have to comply with both standards. In these cases, choosing Sn (paying compensation of €610) is more profitable than choosing St (costs of compliance of €600 plus paying compensation of €24, in total €624) or Stv (costs of compliance of €1200, no compensation).

The incentive to comply is higher under negligence than under strict liability, because under a negligence rule, proprietors can escape liability completely by complying with the standard. This is reflected in the fact that in table 15.5 the level of compliance under negligence is higher than under strict liability. Since there is underdeterrence, a stronger incentive to comply is beneficial for social welfare. If there is too little information about individual proprietors, applying technical requirements as the standard level of care is the best alternative. The existing enforcement costs arise from proprietors of type

1, 9, 10 and 15 who do not comply with this standard, so that there are litigation costs in case of casualties as a result of fire.

If there is sufficient information to determine an individual standard for every type of proprietor, social costs are higher when the efficient standard is applied to each proprietor than when the technical requirements are the standard for everyone. The reason is that under NLES, proprietors of type 6, 11 and 14 do not comply with the maximum number of visitors. Therefore, the enforcement costs are higher, while the level of compliance remains the same. The enforcement costs can be avoided by lowering the standard for type 6, 11 and 14 to the technical requirements. In this way, the maximal feasible level of compliance of 78 can be obtained without enforcement costs, so that a social welfare of €182,700 results.

Predictability and efficient standards

The analysis above assumes that judges apply optimal standards given their information and that victims are fully aware of the standards that are applied by the judge. If judges set their standards in a predictable way, it is likely that victims will be able to estimate their chances of obtaining compensation correctly. However, in general, standard-setting is surrounded by uncertainty, as was discussed in section 2.3.1. In these cases, proprietors may choose higher levels of compliance in order to be sure that they escape liability. Since there is currently under-compliance, this would improve social welfare. However, if uncertainty is high, proprietors may also decide to decrease their level of compliance because making compliance costs would not lead to a sufficiently large reduction in compensation.

The problem of uncertainty is unlikely to be important in the present case. As discussed in sections 2.3.1 and 2.3.5, there are three sources of uncertainty. First, there might be uncertainty about the level of due care. It might not be easy for judges to determine the efficient level of precautions. However, table 15.5 shows that in the case of fire safety, all proprietors applying the standard of technical requirements would lead to an almost identical result, and at the very least not a worse one. The second source of uncertainty is the actual level of precautions taken by a proprietor. If judges misperceive this level, it might affect compliance and litigation. However, in general after a fire (that resulted in damage), it is quite easy to check which level of compliance the proprietor had chosen. There are witnesses and a technical investigation by firemen and technical police departments. The final source of uncertainty is that proprietors are unsure which precautions to take. As explained in section 12.4.1, proprietors generally know the rules very well – or they ought to. Under the negligence rule, they also know how to escape compensation. The only problem might be that under strict liability, proprietors do not know how much compliance will pay off in terms of reduction in compensation.

15.2.2 *The magnitude of liability*

The previous subsection showed that if victims cannot obtain full compensation for damage, proprietors have insufficient incentive to prevent damage. If the compensation award increases, proprietors are more inclined to make compliance costs. Especially the non-monetary damage is to a large extent excluded from compensation.

However, there is a bill under consideration by the parliament to award compensation for emotional loss (loss of affection) to the relatives of fatalities and of victims with severe and permanent damage.[254] The idea is that relatives will receive a fixed amount of hedonic damages as set by an Implementing Order. The figure under consideration is €10,000 per entitled person. The amount of compensation paid under this regime is shown in column 3 of table 15.6. As can be seen in table 15.7, the effect of compensating hedonic damages in this way is very limited. Under SL and under NLTV, there is one additional type of proprietor who would increase compliance. For the remaining, the proposal for compensating hedonic damages in this way is insufficient from the perspective of prevention and unlikely to yield significant effects. In order to do so, the compensation for non-monetary damage would have to be based on real human losses, as estimated by the willingness to pay.

Table 15.8 shows that, if compensation is equal to the full damage to victims, an efficient outcome might result. Under such a compensation system, full output loss and full non-monetary damage is compensated (see column 4, table 15.6). If there is no information about individual proprietors, the best result is obtained under strict liability. As follows from section 2.3.3, this will induce proprietors to choose optimal levels of compliance. The resulting enforcement costs are limited. Under the negligence rule, the same standard would have to be applied to each proprietor (either NLT or NLTV), resulting in higher total costs. If there is information about each proprietor's individual optimal standard, a first-best outcome of €176,800 will result if judges apply this standard (NLES). Since proprietors have to compensate the full damage, every proprietor will comply with this standard. Therefore, no victim will be able to claim damages and no enforcement costs will be made.

254 *Kamerstukken II* 2002/03, 28 781, nrs. 1-3. This bill has been approved by the Second Chamber, but it is held up in the First Chamber. See for a discussion Suurmond and Van Velthoven (2005*).

Table 15.6 Overview of compensation awards (in euro's)

	Current[a]	**Hedonic**[b]	**Full**[c]
per fatality:			
Medical costs[d]	6.400	6.400	6.400
Output loss[e]	170.100	170.100	510.200
Human losses	0	15.000	1.878.300
Total	176.500	191.500	2.394.900
per casualty:			
Medical costs	8.800	8.800	8.800
Output loss[f]	36.000	36.000	36.000
Human losses[g]	20.000	27.500	187.800
Total	64.800	72.300	232.700

a. Current means: A compensation to which victims and their relatives are entitled according to current law. To relatives of fatalities only the medial and funeral costs are paid, and to casualties only a limited amount of non-monetary compensation.

b. Hedonic means: A compensation as it would be if the current proposal for compensating emotional loss is passed. See *Kamerstukken II 2002/2003, 28781, nrs. 1, 2 and 3.* In this bill it is proposed to compensate relatives of fatalities or of severe and permanent wounded victims a fixed amount of €10.000 per entitled person. The circle of entitled persons is limited to spouses/partners, parents of children underaged or living at home, or these children themselves. Therefore I assume that per fatality of casualty there are on average 1,5 entitled relatives that may claim hedonic damages. For wounded victims there may be hedonic damages if there is severe and permanent injury, which anyhow corresponds to a functional loss of 70% or higher. I assume that half of the victims are victims that fall into the category of the bill. All together this means that the hedonic damages equal €15.000 per fatality and €7500 per casualty.

c. Full means: A compensation in which relatives and victims can claim the full damage from the proprietor. The amounts in this column correspond to those of table 13.3.

d. I assume that medical costs (that include funeral costs) can be completely claimed from the injurer. This is likely to be an over-estimation.

e. The loss of income, i.e. the gross output loss, is under current law only awarded to relatives if they were economically dependent of the victim. I assume that under *current* and *hedonic* 1/3 of the output loss can be claimed.

f. I assume that for wounded casualties the whole income loss is recoverable, for example because this whole loss is on account of the victim.

g. Also under current law wounded casualties are entitled to a compensation for pain and suffering that increases with the severity of the injury. I use an estimate of €20.000, based on *ANWB, Smartengeldgids 2003.*

Table 15.7 Liability with hedonic damages

	SL	NLT	NLTV	NLES
Expected damage	155.300	142.500	142.500	142.500
Compliance costs	35.400	40.200	40.200	40.200
Subtotal	*190.700*	*182.700*	*182.700*	*182.700*
Enforcement costs	870	150	390	230
Total	**191.600**	**182.800**	**183.100**	**182.900**
Level of compliance	70	78	78	78

Table 15.8 Liability with full damage compensation

	SL	NLT	NLTV	NLES
Expected damage	119.400	131.200	117.600	119.400
Compliance costs	57.500	49.200	60.000	57.500
Subtotal	*176.800*	*180.400*	*177.500*	*176.800*
Enforcement costs	460	0	120	0
Total	**177.300**	**180.400**	**177.600**	**176.800**
Level of compliance	113	100	120	113

15.2.3 *Ability to pay damages and insurance*

The analysis so far assumes that the proprietor is able to pay for all the damage he causes. This is generally not the case. This section discusses the importance of wealth constraints and of insurance.

Wealth constraints

The amounts that must be compensated to victims or their relatives can be rather high, especially if full compensation is chosen. Proprietors will probably be unable to pay for these damages, and the effect of liability will hence be weakened.[255] The extent to which this problem is of actual relevance is questionable. There are some local proprietors who exploit and possess several horeca establishments (and often other property too), as for instance the proprietor Veerman in Volendam (section 14.3.3). Since they possess these buildings, they are likely to have sufficient assets too. Other proprietors possess only one establishment. Whether that building is sufficient to guarantee

255 In fact, an important factor is left behind in this case, because I only focus on expected damage. However, in reality, there are probably several, possibly dozens, of casualties together.

that damages will be paid is unclear. It all depends on the type of building and the type of accident. Moreover, these proprietors often have a high mortgage on the building. A final category of proprietors concerns those who rent the establishment from a brewery, investment holding or some other concern. These proprietors are not likely to have large assets at their disposal.

In all instances, the problem is whether the proprietor himself or the proprietor's corporation is held liable. In the latter case, the possession of buildings and other assets in a different corporation does not guarantee sufficient wealth to be able to pay for the damages. See section 2.3.6.

An example of limited wealth

Suppose, as an indication of what might happen, that half of the proprietors have insufficient assets to pay for damages. More specifically, suppose that each uneven type of proprietor has a level of wealth that is so problematic that the proprietor reduces his level of compliance with 1 point (unless it is already zero). Table 15.9 shows how this would affect the level of compliance and the total costs.

If the asset constraint is binding for half of the proprietor types, the level of compliance will also be approximately halved: 35 to 43. Many proprietors will choose Sn. The total costs to society increase substantially because the expected damage is much higher.

Insurance

A solution to the possible wealth constraints is obligatory liability insurance. Whether this will induce proprietors to take efficient precautions depends on whether insurance companies are able to handle the moral hazard problem (section 2.3.2). They must be able to ensure that proprietors take some precautions, which is problematic if the insurance company pays for the damage resulting from insufficient precautions. If proprietors suffer from wealth constraints, supervision ex-post will not do. However, insurance companies do require for instance that certificates be sent in for fire extinguishers, fire detection alarms, etc. Moreover, the additional costs might be limited. In the current situation, insurance companies also have to act as supervisor, because proprietors are generally insured for their financial damage.

In this case, I assume that proprietors are risk-neutral and do pay for their own financial damage. However, the outcome would be the same if proprietors were risk-averse and insured for financial damage, and if insurance companies could control the moral hazard problem. Controlling the moral hazard problem seems to be possible at a reasonable cost as far as the technical requirements are concerned, as these are quite easily observable and do not vary from day to day. On the other hand, the requirement for the number of visitors depends on the daily behavior of the proprietor and is therefore less easy to control (only ex-post). However, compliance with this requirement is generally not efficient and not required by judges, and should therefore also not be required by insurance companies. I therefore assume that insurance companies are able to control the moral hazard problem. The costs might not be negligible (compare public inspections, section 13.2 and 15.3), but also not unreasonably high.

Table 15.9 Liability under limited wealth (current law)[a]

Type	Number	Limited assets?	Choice under SL	Choice under NLT	Choice under NLES
1	10	Yes	Sn	Sn	Sn
2	8	No	St	St	St
3	8	Yes	Sn	Sn	Sn
4	4	No	St	St	St
5	8	Yes	Sn	Sn	Sn
6	10	No	St	St	St
7	10	Yes	Sn	Sn	Sn
8	8	No	Sn	St	St
9	2	Yes	Sn	Sn	Sn
10	5	No	Sn	Sn	Sn
11	5	Yes	Sn	Sn	Sn
12	8	No	St	St	St
13	4	Yes	Sn	Sn	Sn
14	5	No	St	St	St
15	5	Yes	Sn	Sn	Sn
Level of compliance			35	43	43
Expected damage			369.200	356.400	356.400
Compliance costs			17.200	22.000	22.000
Subtotal			*386.400*	*378.400*	*378.400*
Enforcement costs[b]			2.800	2.200	2.300
Total			**389.200**	**380.600**	**380.700**

a. See the main text for more explanation. If the proprietor has limited assets, he chooses a level of compliance that is one point below the level that he would choose absent this constraint.

b. It is assumed that litigation costs are always realised, because victims are able to claim some damages and it is worthwhile to try to obtain this amount (although less than absent wealth constraints). The problem is only that this amount is insufficient to take precautions. For example, suppose that a victim is able to obtain €10.000 to €20.000.

Finally, it can be demonstrated that if proprietors do not pay for the financial damage (for example because they are insured and insurance companies do not control moral hazard), the level of compliance under current law is 10 to 20 points lower. This does lower the incentive to take precautions, but to a limited extent. Only the decisions of proprietors of type 11 and 13 and sometimes of proprietors of type 7 and 8 are affected by their own financial damage. Under full compensation, it only matters for proprietors of type 1 under strict liability who chooses Sn instead of St.

15.2.4 *Litigation costs and the incentive to file a claim*

The analysis in the previous subsections assumes that a victim is able to claim the compensation he is entitled to. I do not analyze the incentives for victims to claim damages. I implicitly assume that each legitimate claim is settled against 95% of the claim for litigation costs of €3750, and an illegitimate claim is not filed (or settled for €0, against costs of €0). There is no financial barrier, nor any other constraints on claiming damages.

Are victims able to claim damages?
As discussed in section 2.3.7, a settlement can only be obtained if the victims are able to go to court should the injurer fail to pay. If a proprietor knows that – in the end – a victim is unwilling or unable to go to court, he will not agree to a settlement, and hence will also not feel the need to take precautions to prevent damage. Victims will not go to court if the costs of a trial fail to outweigh the expected compensation, or if they are unable to advance the money.

Table 15.10 shows the costs of a trial faced by a seriously wounded victim. The different costs are:

- Court fees.
- Costs of a lawyer: the contribution if the victim uses the system of legal aid, and otherwise the (full) commercial price.[256]
- The remaining costs that the lawyer makes for the victim, including costs of conducting official proceedings, calling in witnesses, experts, etc.
- The remaining (opportunity) costs of the victim, such as costs of travelling, time, etc.

If there is a strict order for litigation costs, the victim bears costs of €7,290 to €10,430 if he loses. On the other hand, there is a compensation award of €64,800 if he wins. The victim expects to benefit from a trial as long as the probability that he receives his claim is at least 11-16%. Therefore access to the court seems to be sufficiently guaranteed. This holds as long as victims are not too risk-averse, do not face credit constraints if they are not entitled to legal aid or do use legal aid if they are entitled to it, and are willing to claim compensation from the proprietor.

---

256 An alternative not explicitly analyzed here is that involving a victim who is insured against legal expenses. On the one hand, his situation is comparable to that of a victim who is entitled to legal aid. On the other hand, the position of the insurance company that decides about the proceedings is that of a 'No Legal Aid' victim who bears full costs. In addition, several personal injury firms operate on a 'no cure no pay' basis. However, this is only possible as long as a claim is settled. Since the focus here lies on the incentive to go to trial, I ignore 'no cure no pay'.

Table 15.10 Litigation costs for a casualty under current law (euro's, 2004)[a]

	Lowest Legal aid[b]	Highest Legal Aid[c]	No Legal Aid[d]
Victim:			
– Own contribution[e] / lawyer[f]	89	761	1911
– Court fees[g]	109	219	1426
– Remaining costs lawyer	PM	PM	PM
– Remaining costs victim	PM	PM	PM
Total victim	200	980	3340
Total victim plus injurer[h]	3540 + PM	4320 + PM	6680 + PM
Total victim plus injurer if PM = €3750[i]	7.290	8.070	10.430
– Costs of legal aid[j]	756	84	-
– Costs of civil law suit[k]	469	469	469
Total social costs[l]	6.980	6.980	8.040

a. The table refers to a claim of €64.800 to which a casualty is entitled under current law
b. I.e. those who are entitled to legal aid and have to pay the lowest nonzero contribution.
c. I.e. those who are entitled to legal aid and have to pay the highest contribution.
d. I.e. those who are not entitled to legal aid, but have to pay their own trial costs.
e. See Legal Aid Act.
f. The number of hours that a lawyer spends on a case can be estimated at 9 hours, based on the paid legal aid cases. The commercial lawyers tariff is €212.
g. Court fees on January 1, 2004 (see Civil Cases Fees Act). If someone is entitled to legal aid, he is also entitled to a deduction in court fees.
h. I assume that the injurer is not entitled to legal aid and that his lawyer spends as much time and money on the case as the victim's lawyer. So the injurer faces the same costs as a victim who is not entitled to legal aid.
i. An estimate about the remaining trial costs is missing. These costs largely consist of the costs of calling in experts, so correspond to the costs of settling the case. Therefore I use the estimate of Weterings (1999) of €3750 that was used in section ??? as the estimate for the remaining costs for both parties. These are at least the costs that have to be made for a settlement as well as for a trial.
j. On January 1, 2004 the remuneration equals €93,84 per hour. The costs for the government (society) equal 9*€93,84 minus the contribution.
k. In Suurmond (2004*, table B4.2) I calculate the costs for an average civil law case at €469.
l. These are the total social costs that are spent on courts and legal assistance by victims, injurers and the government. Court fees do not count, because they are expenses for victims and injurers, but income for the government. That these costs are smaller than the total costs for private parties stems from the fact that the court fees for 'No legal Aid' are higher than the estimated costs of an average law suit. The problem is not important because the aim is to give an impression of the magnitude of the trial costs and to discuss whether or not the incentive to file a claim is efficient.

Trial or settlement and the social costs of litigation

In a court case, the social costs are at least €6,980. These costs can be decreased – and social welfare therefore increased – if the victim and the proprietor agree to settle the dispute outside the court-room. The victim and the proprietor save at least the court fees, and at least part of the cost of lawyers, so that settlement can also be in their best interest. If the cost of a settlement is €3,750 (as assumed in section 15.2.1), they can jointly save approximately €3,540 (Lowest Legal Aid) to €6,680 (No Legal Aid). If both agree that the probability of compensation is 95%, the minimal settlement range lies between €61,215 and €68,505.[257] The settlement amount in section 15.2.1 (€65,100) lies within this range. Therefore, the assumption that the claim is settled for a compensation of €65,100 against settlement costs of €3,750 is reasonable given the threat of a trial. The crucial assumption here is that courts apply efficient standards. In that case, the probability of 95% that compensation must be paid by a non-compliant proprietor (and 0% by a compliant one) is realistic and agreed on by both parties.

Some comments

Of course this calculation is a simplification. First, I assume that the victim does or does not receive compensation. In general, the dispute is not only about whether or not the victim receives compensation, but also about the level of compensation. The importance of the analysis above is that it shows that the trial costs are manageable as compared to the magnitude of the compensation claim, so that the victim can credibly threaten to go to court if there is a serious dispute about the compensation claim. This also applies to disputes about the magnitude of the claim.

A related point is that I assume that the level of litigation costs does not depend on the magnitude of the claim. This is unlikely and a simplification of the true nature of enforcement costs. In general, the higher the claim, the higher the expenses individuals will be prepared to make obtaining or escaping the claim. For example, if there is compensation of hedonic losses, a discussion might arise concerning who is entitled to this compensation. Alternatively, full compensation of losses might lead to more compensation claims or to more trials so that the enforcement costs will rise and the outcome will possibly be inefficient.[258] Moreover, the higher the victim's litigation expenses, the higher the expected compensation. The probability of 95% might depend on the level of enforcement costs.

It may also be that the litigation costs per claim depend on the number of victims. For example, in the Volendam disaster, approximately 300 people

257 The victim with lowest trial costs will want at least 95%*€64,821 – 5%*€7,290 = €61,215. The proprietor will not want to pay more than 95%*(€64,821 + €7,290) = €68.505.

258 However, it can be shown that in the current case an increase in enforcement costs under full compensation of damage will have little effect. Given the high compensation and probability of award, enforcement costs can rise tremendously before the victim refrains from a claim. For the proprietors too, the enforcement costs have little impact on the level of compliance. Only under strict liability might some proprietors choose more than first-best precautions in order to save on litigation costs, if the enforcement costs become larger than €147,000 per claim under full compensation. Below this amount, the only effect of a multiplication of litigation costs will be the same multiplication of enforcement costs in table 15.8.

jointly settled their case with the proprietor and the municipality (section 14.1). If there are multiple victims, questions of negligence have to be decided only once, and a lawyer may realize economies of scale by representing several victims. This is likely to reduce the litigation costs per victim.

All in all, the crucial assumption is that courts do apply efficient standards so that for legitimate claims the probability that a victim receives compensation is high (95%) relative to the litigation costs, but that illegitimate claims do not stand a chance. Only if this relationship fundamentally changes will the conclusions be different. In general, there is no reason to assume that standard-setting by Dutch courts is inefficient. Moreover, although specifying the damage does require the costs of (medical) experts, the compensation to which a victim is entitled is clearly defined. Therefore, a probability of 95% is not necessarily unrealistic. Besides, the effect is limited as long as the probability is higher than 30%.[259]

15.2.5 *Concluding on private enforcement*

The analysis of liability claims by victims shows that a reasonable level of social welfare is obtained. The level of enforcement costs is low because enforcement occurs only once a fire has taken place. Some potential problems have been dismissed. Definitive conclusions concerning some other problems could not be drawn. Whether private enforcement really leads to an efficient result depends on (1) an adequate compensation award that reflects full losses, (2) sufficient assets available on the side of the proprietor, (3) efficient standard-setting by courts.

15.3 Administrative enforcement

Let us now turn to administrative enforcement. Under administrative enforcement, enforcement takes place at a time when damage can still be prevented (act-based).

15.3.1 *Administrative enforcement in general*

Under administrative enforcement, the municipality is able to induce the proprietor to comply by imposing a duty under penal sum or administrative coercion, or by threatening to close the establishment for a short or longer

259 In case of strict liability, under current law, the level of compliance remains 60 as long as the probability is 80%. Between 64% and 79% it drops to 47 (proprietors of type 4, 11 and 13), a level at which it remains until a probability of 28% is reached. Under NLES the levels of compliance are the same. The difference is only that it drops from a level of 78 to 47.

period. For example, a proprietor of type 1 can be forced to make the technical compliance costs of €250 if he is charged a penal sum of €250 and the municipality realizes a probability of apprehension of 100% with respect to the forfeiture of the penal sum. The proprietor will be better off making compliance costs (costs €250) than violating the technical requirements (costs of more than €250). I assume that the municipality is able to impose sufficiently high penal sums (or otherwise close the establishment).

A drawback of this type of administrative enforcement is that the proprietor is only confronted with a sanction once a violation has been discovered. The proprietor has an incentive to be in non-compliance as long as he is not inspected. Once inspected, the proprietor has a sufficient incentive to comply, for instance because there is a 100% probability of re-inspection and the proprietor forfeits a sufficiently high penal sum if he is still (or again) in non-compliance.

15.3.2 Costs of enforcement and level of compliance

Let us start by analyzing the policy currently implemented in most municipalities. Each establishment is inspected once a year on technical requirements. As a consequence, all proprietors comply with the technical requirements after a first inspection and realize the corresponding costs. AVD (2004*) provides the necessary information about the enforcement costs (see section 13.2.1):

- On average an inspection requires 3.25 hours, including re-inspection and paperwork. The hourly tariff is €51.30.
- In 1% of the cases, it is necessary to actually impose penal sums and other sanctions. This procedure requires on average 20 hours of work, with an hourly tariff of €70.[260]

I assume that these costs apply to the inspection of technical requirements as well as the number of visitors. Both requirements can be inspected simultaneously in the course of one inspection.

With respect to compliance with the requirements, the following applies:

1. The technical requirements:
   - Half the time, this concerns expenses on technical devices that have to be fixed every year, but are then certainly in compliance for a whole year. Examples are the impregnation of decorations or the yearly certification of fire extinguishers. If there is at least one annual inspection, the proprietor is forced to realise these costs every year and the level of compliance is 1.

260 For convenience and due to lack of data, I refrain from including the costs of these procedures for the proprietor, such as the costs of objection and appeal, a lawyer, etc. In practice, this does not seem to happen. The proprietor uses these procedures if the costs are lower than the resulting expected reduction in penalties.

- The other half concerns expenses on devices that may break down during the year. If that happens, the proprietor is obliged to repair the device, but he may decide to wait and only repair once he is forced to do so. For example, the proprietor can choose to annually replace the lamps of the escape-route signs, but he can also choose to wait until they fail and then possibly decide to replace them. In the first case, or if the proprietor immediately replaces the lamps, there is full compliance (level 1). However, under administrative enforcement, the proprietor is better off waiting and replacing only once an enforcement official has ordered him to do so (under threat of penalties). The lamps may fail the day after the inspector has come, so that compliance during the next year is (almost) zero. Only after the next inspector's visit a year later will the proprietor comply again. The lamps may also fail a day before the inspector comes, so that compliance is (almost) 1. On average, then, compliance by the proprietor is 0.5.
- All in all, in case of an annual inspection, the expected level of compliance with the technical requirements equals 0.5*1 + 0.5*0.5 = 0.75. For multiple inspections per year, it can be calculated in a similar way, assuming that the lamps will not break down more than once a year.

2. The number of visitors. This can be enforced by clearing the establishment on site whenever a violation of the number of visitors is discovered. Therefore, on days when there is an inspection, the level of compliance is 1. On days when there is no inspection, the proprietor will be in violation (the level of compliance is zero), so that an average compliance level can be calculated.[261]

From table 15.3, we can deduce that if the municipality is unable to distinguish between different types of proprietors (because it lacks the information to do so), it is inefficient in inspecting the number of visitors, even if we ignore the costs of enforcement. Moreover, it appears that inspection of any type of proprietor is inefficient if it is only aimed at enforcing the number of visitors. The costs of inspection (€181[262]) are higher than the potential benefits of an increase in compliance with the number of visitors requirement. Only inspecting the number of visitors decreases social welfare. However, given that there is an inspection of the technical requirements, it may be efficient to use this inspection to also enforce the number of visitors for proprietors of type 6, 11 and 14. In that case, the inspection of the technical requirements should take place during a day (evening) when there is a possible excess of the number of visitors. The marginal costs of inspecting the number of visitors would then be zero.

261 The level of compliance equals the number of days the number of visitors is complied with divided by the potential number of days this requirement might be violated. For example, if a proprietor of type 6 complies 2 days with the maximum number of visitors, the level of compliance equals 2 / 5 = 0.4.

262 The costs of an inspection are 3.25*€51.30 + 1%*20*€70 = €181

15.3.3 *The result under administrative enforcement*

If all proprietors are inspected once a year on the technical requirements, this leads to a level of compliance of 75 and a corresponding level of social welfare of €331,500, divided into €18,100 worth of enforcement costs, €49,200 worth of costs of compliance, and €262,200 worth of expected damage. See table 15.11.

If an establishment is inspected more frequently than once a year, proprietors will sooner replace the escape-route signs and other devices, so that the expected damage will be reduced. The question is whether this reduction in expected damage compensates for the increase in enforcement costs. Table 15.11 shows the level of social welfare for different numbers of annual inspections. It appears that it is optimal to inspect establishments three times a year.

Table 15.11 Administrative enforcement – technical inspections

Number of inspections per establishment	**0**	**1**	**2**	**3**	**4**
Expected damage	663.317	262.217	197.700	175.528	164.442
Costs of compliance	0	49.200	49.200	49.200	49.200
Subtotal	*663.317*	*313.417*	*246.900*	*224.728*	*213.642*
Costs of inspections	0	16.673	33.345	50.018	66.690
Costs of imposing sanctions	0	1.400	2.800	44.200	5.600
Enforcement costs	*0*	*18.073*	*36.145*	*54.218*	*72.290*
Total (social welfare)	**663.317**	**331.489**	**283.045**	**278.945**	**285.932**
Level of compliance	**0**	**75**	**88**	**92**	**94**

If the municipality is informed about the expected damage and costs of compliance of each type of proprietor and knows a proprietor's type, proprietors will not have to be inspected with the same frequency. For each type of proprietor, an optimal number of inspections can be determined. Table 15.12 contains the results under this assumption. Column 2 contains the results when only the technical requirements are inspected. For types 9 and 15, Sn is optimal, so that no inspection will always be best. For types 1 and 10, no inspection is also optimal because the costs of inspection do not balance out the reduction in expected damage minus the costs of compliance. For other types, the optimal frequency varies from 1 to 4. A total number of 233 inspections is optimal.

For proprietors of type 6, 11 and 14, it is optimal to perform inspections on days when they might violate the number of visitors. This will not affect the enforcement costs, but will lead to a reduction of expected damage plus costs of compliance of about €3400.[263]

Table 15.12 Administrative enforcement – Specific info

	Optimal per establishment only technical	Optimal per establishment including visitors
Expected damage	181.696	173.835
Costs of compliance	40.200	44.700[a]
Subtotal	*221.896*	*218.535*
Costs of inspections	38.847	38.847
Costs of imposing sanctions	3.262	3.262
Enforcement costs	*42.109*	*42.109*
Total (social welfare)	**264.005**	**260.644**
Level of compliance	**70**	**79**
Number of inspections	233	233

a. This consists of € 40.200 of technical compliance costs and € 4500 of costs for complying with the number of visitors (proprietors of type 6, 11 and 14).

15.3.4 *Concluding on administrative enforcement*

The analysis above shows that it is possible to reduce the expected damage from fire in horeca establishment by means of administrative law enforcement. However, the disadvantage of this approach is that proprietors will wait until there is a real threat of sanctions. Therefore, the proprietor will only comply with the requirements if there are sufficient inspections, which involves substantial enforcement costs.

If the enforcement costs rise, the efficient level of compliance will be reduced. This is apparent for example from the fact that it is not efficient to inspect only for the number of visitors. In addition, it is not efficient to inspect on the technical requirements for proprietors of type 1 and 10, while St is efficient for them in the absence of enforcement costs.

The disadvantage of administrative enforcement is the expectant compli-

263 The three commonly known days on which excessive numbers of visitors can be expected are the two days of Carnival and one day for a local event six months later. Since only inspecting the number of visitors is inefficient, inspection only takes place on two of these days. Inspecting on both Carnival days is unlikely to improve technical compliance, because the probability that the lamps fail that day is negligible. For proprietors of type 11, it is optimal to have just 1 inspection, which should take place at a time when there is a possible excess of the number of visitors.
The level of compliance for the number of visitors for proprietors of type 6, 11 and 14 respectively is 0,40, 0,33 en 0,50.

ant behavior of the proprietors. The extent to which this behavior shows up depends on the extent to which compliance with the technical requirements can be postponed or requires continued attention.[264]

15.4 Criminal enforcement

The expectant behavior of proprietors under administrative law enforcement can be prevented by immediately punishing their behavior with a criminal fine.[265] If the fine is high enough, the proprietor will choose to immediately replace the failing lamps of escape-route signs, thus fully complying with the technical requirements.

15.4.1 *Conceivable policy: a fine of €250*

Such a criminal law enforcement policy is not currently implemented, but it is actually possible.[266] If we consider similar offences, a fine of €250 is a realistic assumption for the magnitude of the fine.[267] This fine could be levied per detected violation of the technical requirements or the number of visitors.

The enforcement costs of criminal law enforcement
The corresponding enforcement costs consist of the costs of apprehension and detection, the costs of conviction and the costs of execution.

264 I assumed a ratio of 50%-50% for annual and continued attention. To compare, if 1 inspection of the technical requirements would suffice to guarantee compliance with the technical requirements for a whole year, a level of compliance of 100 could be achieved with enforcement costs of €18.100. Social welfare would be €198.456 (expected damage of €131.200 and compliance costs of €49.200).

265 Another alternative is the use of an administrative fine. I did not discuss the use of an administrative fine in the previous section, because its effect is similar to that of a criminal fine. For a proprietor, it probably makes little difference whether he has to pay a fine of one kind or another. Analyzing the different enforcement costs would not much improve our understanding. If an administrative fine is not immediately imposed, but is preceded by warnings, its effect is similar to that of a penal sum and the result would be similar to that discussed in the previous section.

266 According to article 12.1 of the model Building Bylaw (VNG, 2002*), violation of the use requirements is a criminal act and is punishable by imprisonment for a term not exceeding four months or by a fine of the third category.

267 For example, according to the fines database of the Public Prosecutors Office, violations of closing times by proprietors are punished with a fine of €125, noise pollution by horeca establishments with a fine of €75, and violations of the Licensing and Catering Act (for instance selling alcoholic beverage to minors) with fines of €400 to €500.

The costs of apprehension are those of police resources.[268] These can be estimated at €65 per hour including administrative support, training, accommodation, etc.[269] An inspection requires half an hour of police-effort.[270] I assume (again) that the probability of detection is 100%. In case of an inspection, possible violations can be established with certainty.

The costs of conviction are more difficult to determine. There are two alternatives. Based on average figures for criminal disposals, the picture is as follows:[271] The costs for the police of making a report of the offence are 0.25 hour (15 minutes) at €65. Of these reports, 94% is dealt with by the Public Prosecutors Office (PPO) and 6% through a trial. If the case is dealt with by the PPO (costs €745), it will involve a transaction in 82% of the cases and a dismissal in the remaining 18%. If the case is dealt with in a court (costs of the government: €1360[272]) the probability of conviction is 95%. The proprietor then spends €848 on a lawyer (i.e. four hours at an hourly rate of €212). The total conviction costs thus equal 0.25*€65 + 94%*€745+6%*(€1360 + €848) = €850. The proprietor has a probability of 80% of paying a fine of €250 and a 6% probability of paying a lawyer, on average 80%*€250 + 6%*€848 = €250.

There are several problems with this estimate. First, the conviction costs are rather high relative to the fine of €250. Moreover it only takes the average disposal as given, without providing an analysis of these probabilities. For example, why is not every case settled by transaction, especially given that the probability of detection is 100% and that there is little room for discussion concerning offenses? Why is it beneficial for the proprietor to refuse a transaction? Why does the PPO dismiss some cases? I therefore consider another

---

268 I use average police figures. Enforcement could be delegated to special officers (so-called *boa*: special investigating officer) instructed with the enforcement of Economic Offenses. On the one hand, these officials may be cheaper because they can be lower skilled. On the other hand, they may be more expensive, because they have to be more specialized.

269 According to table 4.8 from *Kerncijfers van de Nederlands Politie 2002* (Ministry of Internal Affairs, see www.politie.nl), the average costs per fte equal €51,996. This document also states that 79.3% of the personnel are primary personnel. The total costs of the police force consist of 77.7% personnel costs and 13.3% accommodation, transport, communication, ICT, etc. Therefore,, the total annual costs per fte equal €51,996 * 1/0.793 * 1/0.777 = €84,387. With approximately 1300 productive hours in a year (see note 202), the hourly tariff is €65.

270 This estimate is based on the average time that the fire department now spends on inspections (section 12.2) which largely corresponds to the 3.25 hours of AVD (2004*). Recall that, in this case, administrative support etc. is incorporated in the tariff and not in the number of hours for inspection and that there is no need for re-inspections under criminal law enforcement. Moreover, there is usually less time for explanation. On the other hand, if they are not carried out by special officers (note 268), police inspections may require the presence of fire safety officials.

271 CBS, *Rechtspraak in Nederland 2002*. The data refer as much as possible to the Economic Offenses Act. The time spent on writing a report and the time spent by lawyers is my own estimate.

272 I.e., costs of PPO of €745 plus the costs of a criminal court case, computed as in table 15.10 for civil law cases.

alternative, which is particularly relevant for small fines. In this alternative, the conviction (plus execution) costs equal €100, without any costs for the proprietor. There is always a transaction of €250.

The costs of the collection of fines are €7 for a transaction and €37 for a fine imposed by a court, on average €9.[273]

The relevant information

The proprietor will comply with the regulations if his costs of compliance are lower than the expected sanction. The magnitude of the sanction is €250. Therefore, the relevant question concerns the socially optimal number of inspections (probability of apprehension), so that the efficient sanctioning risk can be realized. The answer depends on the available information. There are three possible situations with corresponding policy alternatives (in addition to the no-inspection option)[274]:

1. *Perfect information*: for every type of proprietor, the enforcement officials have the correct information about the expected damage, the costs of compliance for the technical requirements and the number of visitors. Moreover, they know a proprietor's type. As a result, they are able to determine for every type of proprietor which requirements are enforced and how many inspections are carried out.
2. *Compliance costs information*: for every type of proprietor, the enforcement officials know the costs of compliance of (at least) the technical requirements and they know a proprietor's type. They therefore have to enforce the same level of compliance at each proprietor, and therefore decide to enforce only the technical requirements.[275] They are, however, able to vary the number of inspections for proprietors.
3. *No information*: the enforcement officials have no information available that might help to vary enforcement. They therefore have to enforce the same level of compliance and have to inspect each proprietor with the same frequency.

In situations 1 and 2, a sanctioning risk is realized for every type of proprietor. For example, a proprietor of type 3 will comply with the technical requirements if there are at least 2 inspections per year: costs of compliance of €500

273 Source: *Justitiebegroting 2004*, part 03 Centraal Justitieel Incassobureau. Based on the figures discussed before, the average costs equal (94%*82%)/83%*€7.06 + (6%*95%)/83%*€37.23 = €9.11.

274 It is possible to think of one other alternative, namely that the enforcement officials do know the expected damage, but are not informed about the costs of compliance. In this situation, they also have to enforce the same level of compliance and will choose to enforce the technical requirements for all proprietors. They are not able to realize a sanctioning risk that differs between proprietors so the outcome is as under "no information".

275 As discussed above, it is not efficient – in the absence of specific information – to enforce only the number of visitors. This is always second-best to only enforcing the technical requirements. Moreover, enforcing both requirements can only be efficient if the enforcement costs for 'everyone St' are much higher than for 'everyone Stv'. This is unlikely to be the case.

versus a sanctioning risk of at least 2 * €250 ≥ €500. The costs of a proprietor of type 4 are €700, so that he will comply if there are at least 2.8 inspections per year. In situations 1 and 2 it is optimal either to make the sanctioning risk so large that a proprietor complies, or to refrain from inspections altogether. In both cases, no fines are imposed and no conviction and execution costs realized. The only enforcement costs are the costs of inspection. This is different in situation 3. If, for instance, there are 2 inspections per year, a proprietor of type 3 will comply with the technical requirements, but a proprietor of type 4 will not. If a proprietor of type 4 is inspected (2 times a year is the expectation) there is an offence detected and he will have to pay a fine with corresponding enforcement costs as a consequence. In which case the distinction between the high (I) and low (II) conviction costs is relevant.

Which situation is most realistic may vary. Of course, over the years, enforcement officials can find out about the relevant costs and benefits and a proprietor's type. However, the situation within an establishment might not be stable and there is turnover in the enforcement personnel. The enforcement officials I interviewed (section 12.2), in any case, indicated that they only generally know the costs of compliance.

The results of a fine of €250

The results are reported in tables 15.13, 15.14 and 15.15. Table 15.13 describes the result for the different situations. When the enforcement officials have *perfect information*, a level of compliance is enforced for every proprietor, which minimizes the sum of the expected damage and the costs of compliance. This implies that proprietors of type 9 and 15 are not inspected at all. Proprietors of type 6, 11 and 14 are also inspected on the number of visitors, but only on the three generally known days. It is not efficient to also inspect on the other 100 weekend days that these proprietors may be in violation of the maximum number of visitors. This would require so much funding in terms of enforcement costs that social costs would rise on balance. Therefore, the level of compliance of 108 is somewhat smaller than the first-best optimum of 113. For none of the proprietors are the enforcement costs so high that it becomes efficient to enforce a lower lever of compliance with the technical requirements than the social optimum in table 15.3.

If the enforcement officials only have *information about compliance costs*, they will enforce the technical requirements for everyone. There is no inspection of the number of visitors for proprietors of type 6, 11 and 14, and proprietors of type 9 and 15 are also inspected for compliance with the technical requirements. Table 15.14 shows the expected number of inspections on the basis of which a proprietor will choose to comply with the technical requirements. In total, the enforcement officials choose to perform 197 inspections. This should induce all proprietors to comply with the technical requirements.

Table 15.13 Criminal enforcement with fine € 250

	No enforcement	Perfect info	Info compliance costs	No info
Expected damage	663.300	123.400	131.200	131.200
Costs of compliance	0	54.200	49.200	49.200
Subtotal	*663.300*	*177.600*	*180.400*	*180.400*
Costs of inspection	0	6.100	6.400	13.000
Costs of conviction[a]	0	0	0	0
Enforcement costs	*0*	*6.100*	*6.400*	*13.000*
Total costs	**663.300**	**183.700**	**186.800**	**193.400**
Level of compliance	**0**	**108**[b]	**100**[c]	**100**[d]
Number of inspections (of which visitors)	0 (0)	187[b] (4,5)	197[c] (0)	400[d] (0)

a. Including costs of execution of sanctions.
b. For proprietors of type 9 and 15 the regulations are not enforced. Table 15.14 shows how many inspections have to be performed to induce the (other) proprietors to comply with the technical requirements. For proprietors of type 6, 11 and 14 the number of visitors are enforced on the three generally known days. This requires a probability on such a day of respectively 0.5, 0.4 and 0.6 (costs per day divided by €250). The level of compliance is equal to the number of the days they comply divided by the total number of potential days. For all other type of proprietors only the technical requirements are enforced.
c. See table 15.14. The number of inspections is the same as under ***perfect information*** plus the inspections for type 9 and 15. For type 6, 11 and 14 the required number of inspections for the technical requirements is larger than for enforcing the number of visitors on the three known days. The number of inspections for these types can therefore not be lowered compared to ***perfect info***.
d. Table 15.15 shows which number of inspections is optimal in these situations.

If the enforcement officials have *no information*, it is efficient to inspect so many times that every proprietor will comply with the technical requirements. Therefore, the magnitude of the conviction costs is irrelevant. Table 15.15 explains this result. If the number of inspections per establishment increases, the costs of inspections rise. The effect on the costs of conviction is twofold. On the one hand, there are more proprietors who choose to comply, so that fewer fines have to be imposed. On the other hand, the probability of apprehension increases, so that non-complying proprietors are more frequently fined. On balance, the conviction costs decrease almost over the entire range. We might expect that this increase in costs of inspection and decrease in costs of conviction would lead to an optimal level of compliance that is smaller than 100. This is not the case, however, because the decision to comply is determined by the compliance costs compared to the sanctioning risk. The higher the costs of compliance, the more inspections are required to enforce compliance. In this case, for proprietors with high costs of compli-

Table 15.14 Number of required inspections for technical requirements under fine € 250

Type	#	Technical compliance costs	Minimum # of required inspections per year	Enforcement costs per year	Total social costs per year
1	10	250	1,0	33	450
2	8	300	1,2	39	1010
3	8	500	2,0	65	1960
4	4	700	2,8	91	2650
5	8	400	1,6	52	1560
6	10	400	1,6	52	2840
7	10	500	2,0	65	1270
8	8	600	2,4	78	2460
9	2	500	2,0	65	1300
10	5	800	3,2	104	2910
11	5	400	1,6	52	2000
12	8	500	2,0	65	1860
13	4	1000	4,0	130	2290
14	5	800	3,2	104	3530
15	5	300	1,2	39	1180
Total	**100**	**49.200**	**197**	**6.400**	**186.800**

ance, expected damage can be reduced so much that it is socially desirable for them to comply with the requirements, even if it implies high costs of enforcement and proprietors of type 9 and 15 being also forced to comply. To illustrate: a proprietor of type 15 has compliance costs of only €300. He will therefore choose to comply (inefficiently) for a relative low probability of inspection. From a social point of view, it is optimal to increase the number of inspections so much that all proprietors, also those with high costs of compliance, comply with the technical requirements.

In all three situations, it is possible to induce proprietors to comply by criminal law enforcement.

Table 15.15a Result for 3-I: no information, high conviction costs

Number of inspections From	To	Level of compliance	Expected damage	Costs of compliance	Sub total	Costs of inspection	Costs of conviction[a]	Enforcement costs	Total social costs
0,0	1,0	0	663.300	0	*663.300*	0	0	*0*	**663.300**
1,0	1,2	10	660.300	2.500	*662.800*	3.300	77.200	*80.500*	**743.300**
1,2	1,6	23	599.200	6.400	*605.600*	3.900	79.300	*83.200*	**688.800**
1,6	2,0	46	410.700	15.600	*426.300*	5.200	74.100	*79.300*	**505.600**
2,0	2,4	74	246.200	29.600	*275.800*	6.500	44.600	*51.100*	**326.900**
2,4	2,8	82	233.400	34.400	*267.800*	7.800	37.100	*44.900*	**312.600**
2,8	3,2	86	215.600	37.200	*252.800*	9.100	33.600	*42.700*	**295.500**
3,2	4,0	96	146.700	45.200	*191.900*	10.400	11.000	*21.400*	**213.300**
4,0	Infinite	100	131.200	49.200	*180.400*	13.000	0	*13.000*	**193.400**

Table 15.15b Result for 3-II: no information, low conviction costs

Number of inspections From	To	Level of compliance	Expected damage	Costs of compliance	Subtotal	Costs of inspection	Costs of conviction[a]	Enforcement costs	Total social costs
0,0	1,0	0	663.300	0	*663.300*	0	0	*0*	**663.300**
1,0	1,2	10	660.300	2.500	*662.800*	3.300	9.000	*12.300*	**675.100**
1,2	1,6	23	599.200	6.400	*605.600*	3.900	9.200	*13.100*	**618.800**
1,6	2,0	46	410.700	15.600	*426.300*	5.200	8.600	*13.800*	**440.200**
2,0	2,4	74	246.200	29.600	*275.800*	6.500	5.200	*11.700*	**287.500**
2,4	2,8	82	233.400	34.400	*267.800*	7.800	4.300	*12.100*	**279.900**
2,8	3,2	86	215.600	37.200	*252.800*	9.100	3.900	*13.000*	**265.800**
3,2	4,0	96	146.700	45.200	*191.900*	10.400	1.300	*11.700*	**203.600**
4,0	Infinite	100	131.200	49.200	*180.400*	13.000	0	*13.000*	**193.400**

15.4.2 *Trading off the probability of inspection and the magnitude of the fine*

The previous subsection demonstrates that proprietors can be forced to comply with the regulation by means of criminal enforcement. Therefore, a sanctioning risk is required that exceeds the costs of compliance. For the technical requirements, this can be realized with a fine of €250 and – in case of *no information* – four inspections per establishment per year. However, the same sanctioning risk can also be realized with a fine of €500 and two inspections per year. In both cases, the sanctioning risk is €1000 so that all proprietors are better off complying with the technical requirements. In the second alternative, however, the costs of inspection are halved as the number of inspections is halved. The costs of conviction remain unchanged as there is still no proprietor who violated the requirements and therefore no sanctions are imposed.

On balance, doubling the fine will lead to halving the costs of inspection. Since the level of compliance and therefore the expected damage and the costs of compliance remain unchanged, this implies that social welfare is improved. Table 15.16a-d shows the result for fines of €75, €250, €500 and €4500.[276] [277]

Table 15.16a Criminal enforcement: perfect info

Fine	**€ 75**	**€ 250**	**€ 500**	**€ 4500**
Expected damage	133.900	123.400	123.400	123.400
Costs of compliance	45.700	54.200	54.200	54.200
Subtotal	*179.600*	*177.600*	*177.600*	*177.600*
Costs of inspection	13.100	6.100	3.400	400
Costs of conviction	0	0	0	0
Enforcement costs	*13.100*	*6.100*	*3.400*	*400*
Total social costs	**192.700**	**183.700**	**181.000**	**178.000**
Level of compliance	**88**	**108**	**108**	**108**
Number of inspections (of which visitors)	402 (2,7)	187 (4,5)	104 (2,5)	13 (0,3)

Table 15.16b Criminal enforcement: info compliance costs

Fine	**€ 75**	**€ 250**	**€ 500**	**€ 4500**
Expected damage	131.200	131.200	131.200	131.200
Costs of compliance	49.200	49.200	49.200	49.200
Subtotal	*180.400*	*180.400*	*180.400*	*180.400*
Costs of inspection	14.500	6.400	3.600	400
Costs of conviction	0	0	0	0
Enforcement costs	*14.500*	*6.400*	*3.600*	*400*
Total social costs	**194.900**	**186.800**	**183.900**	**180.800**
Level of compliance	**100**	**100**	**100**	**100**
Number of inspections	447	197	109	13

Results

Table 15.16a shows that, if there is perfect information, raising the fine will result in the same level of compliance, at lower enforcement costs. With a fine of €4500, it is still unprofitable to inspect the number of visitors outside the

276 On January 1, 2004, €4500 is the maximum for fines of the third category (article 23 Criminal Code).

277 The expected sanctioning risk equals for fines of €75, €500 and €4500 respectively 80%*€75 + €50 (i.e. 6%*848) = €110; 80%*€500 + €50 = €450; 80%*€4500 + €50 = €3650. Under low conviction costs of €100, there is always a transaction for, respectively, €75, €500 and €4500.

three known days. Only if the fine is raised further, to for example €15,000, is it is efficient to create a probability of inspection that induces proprietors of type 6 and 14 to comply with the number of visitors on all days.[278] Then a level of compliance of 113 is realized at minimal enforcement costs. However, this would require a substantial change in the legal environment.

Table 15.16c Criminal enforcement: no info I (high conviction costs)

Fine	**€ 75**	**€ 250**	**€ 500**	**€ 4500**
Expected damage	131.200	131.200	131.200	131.200
Costs of compliance	49.200	49.200	49.200	49.200
Subtotal	*180.400*	*180.400*	*180.400*	*180.400*
Costs of inspection	29.500	13.000	7.200	900
Costs of conviction	0	0	0	0
Enforcement costs	*29.500*	*13.000*	*7.200*	*900*
Total social costs	**209.900**	**193.400**	**187.600**	**181.300**
Level of compliance	**100**	**100**	**100**	**100**
Number of inspections	909	400	222	27

Table 15.16d Criminal enforcement: no info II (low conviction costs)

Fine	**€ 75**	**€ 250**	**€ 500**	**€ 4500**
Expected damage	131.200	131.200	131.200	131.200
Costs of compliance	49.200	49.200	49.200	49.200
Subtotal	*180.400*	*180.400*	*180.400*	*180.400*
Costs of inspection	43.300	13.000	6.500	700
Costs of conviction	0	0	0	0
Enforcement costs	*43.300*	*13.000*	*6.500*	*700*
Total social costs	**223.700**	**193.400**	**186.900**	**181.100**
Level of compliance	**100**	**100**[d]	**100**	**100**
Number of inspections	1.333	400	200	22

It is also possible to consider a lower fine. A fine of €75 might be more realistic in the eyes of the parties concerned. The estimate of €250 was a not-so-conservative estimate of comparable fines. Especially in the beginning, a mild start might be favored, since fines are currently not in use. In addition, a potential administrative fine (a question under debate) would probably be more in the order of €75 than €250.

278 For proprietors of type 6 it is beneficial to always enforce the number of visitors beginning with a fine of €9925, and for those of type 14 beginning with €11.485. For proprietors of type 11, there are only 3 days on which there is a potential excess of the number of visitors.

If the fine is €75, it will be too little to force proprietors of type 6 and 14 to comply with the maximum number of visitors. Their costs of compliance of €125 respectively €150 are so high that they are always better off non-complying, even if the probability of inspection is 100%.[279] Moreover, proprietors of type 1 have to be inspected so frequently that it is inefficient to enforce the technical requirements for these proprietors. Therefore the level of compliance drops to 88.

In situations involving other than *perfect information* the effects are comparable. If the fine is increased the enforcement costs decrease and social welfare is improved (tables 15.16b-d). In case of *no information*, the efficient level of compliance remains 100.

Comments

An increase in the fine and a decrease in the number of inspections such that the sanctioning risk is unchanged will improve social welfare. This is along the lines discussed in the theoretical part (especially section 3.2, following Becker, 1968). There are some important comments on the recommendation to impose as high a fine as possible and correspondingly as low an inspection rate as possible.

A fine that exceeds the assets of a proprietor is ineffective. The proprietor will only take into account the part of the fine he is able to pay. As explained in section 15.2.3, the constraint on the assets is not likely to be so high that proprietors will be unable to afford €4500. For much higher fines, this may become problematic. One solution would be to replace fines by imprisonment. If we are willing to impose imprisonment sentences severe enough to deter proprietors from non-compliance, the costs of executing the imprisonment sentences in question will not be realized. It is doubtful, however, whether there is enough willingness to do so at present. If we are willing to impose only small imprisonment sentences, which do not sufficiently deter, the outcome is likely to be worse: high costs of enforcement without any effect on the level of compliance.

Several factors were abstracted from in this case, which might in practice be relevant to some extent. A maximal fine and a minimal probability might have consequences in terms of wrongly punishing proprietors who are not in violation, as well as on the effects of informal sanctions, or the valuation of the risk of sanctions by proprietors.

It might be useful not only to punish offences with fines, but also to instruct proprietors to restore compliance. If criminal enforcement is adequately carried out, this will be irrelevant because proprietors will choose the optimal level of compliance. However, if, for some reason, proprietors choose inefficient compliance, it might be useful to supplement the sanction with an instruction to restore compliance.

279 For proprietors of type 11, it remains optimal to inspect on the three generally known days. The costs of compliance are €100, so that the probability of inspection should equal €100/€110 = 91%. This requires costs of €30. The total costs of €100 + €30 = €130 are lower than the reduction in expected damage of €138.

Harm-based criminal enforcement
So far, I only discussed act-based criminal enforcement. However, it is also possible to sanction proprietors after some damage has occurred. This requires much lower enforcement costs. Suppose for example that after a fire has occurred and it appears with 100% certainty that the proprietor has not complied with the technical requirements, a fine of €50,000 can be imposed. For Sn, the sanctioning risk is equal to the yearly expected probability of a fire * €50,000. For example, for a proprietor of type 1, this is equal to 0.00825*€50,000 = €410. This is higher than the costs of compliance of €250, so that a proprietor of type 1 will comply with the technical requirements. With a fine of €50,000, all proprietors will comply with the technical requirements except those of type 4 and 13. Raising the fine further will also deter them from violating the technical requirements. If the assets are insufficient, imprisonment might be helpful, as described above. The other problems are the same as under liability: the proprietor must be able to estimate the probability that he will have to pay such a high fine.

15.4.3 *Concluding on criminal law enforcement*

Choosing to enforce fire safety regulations by means of criminal law enforcement is a good alternative. With relatively low enforcement costs, good results can be obtained. It is best to choose rather high fines as this will allow for more limited costs of inspection. If the information of enforcement officials is limited, they must enforce the same level of compliance for each proprietor. This might be valued as 'unfair' or 'disproportionate' by some of the proprietors.

A word of warning in order: Criminal enforcement is only a good alternative if it is implemented in the right way. Otherwise, it is better to completely refrain from enforcement. In the short term, it is not expected that fines will exceed €250. Moreover, the police and the Public Prosecutors Office show little interest in the enforcement of fire safety in horeca establishments and clearly choose other priorities. Nevertheless, if they are forced and provided with the financial resources to inspect all establishments once per year and to impose fines of €250 if necessary, this will only induce proprietors of type 1 to comply. For all other proprietors, the sanctioning risk is too small to enforce compliance with the technical requirements. As a consequence, such a policy will lead to a level of compliance of only 10 with €660,300 of expected damage and €2500 of costs of compliance (see table 15.15). The corresponding enforcement costs are, depending on the high or low costs of conviction, €80,500, respectively €12,300, so that the total costs are €743,300, respectively €675.100. Irrespective of the precise costs of conviction,, society is better off if the police and Public Prosecutors refrain from enforcement, so that the social costs are only €663,300.

It turns out (especially from table 15.16a) that proprietors cannot be induced to comply with low fines. The problem is similar if the budget for inspections is insufficient to realize an adequate sanctioning risk (given the magnitude of the fine). In such circumstances, a higher level of compliance might be obtained by not immediately punishing violations (the so-called compliance strategies discussed before). For example, in such cases, warning the proprietor first might lead to a better outcome as demonstrated for administrative law enforcement in section 15.3. Other alternatives were discussed in sections 8.2 and 12.4. However, these alternatives only improve social welfare given the magnitude of the fine. Social welfare could be improved much more by simply increasing the fine.

Table 15.17 Overview of alternative enforcement methods

Table 15.17a Overview of damage and compliance costs

Level of compliance	0% (0)	100% (200)	First-best (113)
Expected damage	663.300	105.000	119.400
Costs of compliance	0	137.300	57.500
Total	663.300	242.300	176.800

Table 15.17b Results under civil law enforcement (total social costs)

	Current damages[a]	Full damages
Strict liability	191.600	177.300
Negligence rule	± 183.000	176.800 to 180.400

Table 15.17c Results under administrative enforcement (total social costs)

Annual inspection	331.500
Three inspections per year	278.900
Optimal per establishment	260.600 to 264.000

Table 15.17d Results under criminal law enforcement (total social costs)

Fine € 75	192.700 to 223.700
Fine € 250	183.700 to 193.400
Fine € 500	181.000 to 187.600
Fine € 4500	178.000 to 181.300
Infinite high fine	176.800

a. results under hedonic damages are comparable.

15.5 Enforcement alternatives compared

In the previous sections I analyzed the consequences of different methods of enforcement. It turns out that in all three alternatives, a better result can be achieved than in the absence of enforcement, at least if an enforcement policy is properly implemented. Without enforcement, the total social costs are €663,300. Under the private law enforcement approach, the social costs will vary from €176,800 to €191,600. With administrative enforcement, results of €260,000 to €331,500 can be achieved. Finally, criminal law enforcement involves social costs of €176,800 to €223,700. Therefore, in all three cases, substantially lower total social costs can be achieved than under a no enforcement policy (see table 15.17). These results are obtained when, given the magnitude of the sanctions, the efficient enforcement policy is carried out. If an inefficient policy is chosen, the total costs will of course be higher.

15.5.1 *The optimal enforcement policy*

Table 15.17 shows a preference for harm-based enforcement, as under liability. Private enforcement is only effective if there are no barriers for victims to claim damages from the proprietor, if the proprietor is able to pay damages and if the proprietor has sufficient information about the possibilities for preventing damage. A harm-based criminal enforcement approach is the best alternative if the assets of the proprietor are insufficient and additional imprisonment is necessary.

If a harm-based enforcement policy is ineffective because a high sanction is not possible or one of the other problems is persistent, an act-based enforcement policy must be chosen. Under act-based enforcement, a lower sanction will suffice to induce compliance. Moreover, detailed knowledge on the part of proprietors concerning the relationship between ex-ante precautions and ex-post sanctions is not required. A criminal law enforcement policy leads to an efficient result, provided that there is sufficient (political) willingness to make the sanction and the budget for inspections. Only if the willingness to immediately impose these sufficiently high sanctions is absent is an administrative enforcement policy a good alternative.

It is also possible to choose a combination of different enforcement methods. An obvious example is the joint use of private and criminal harm-based enforcement. Criminal enforcement may be used as a supplement to private enforcement if imprisonment is needed to create an adequate sanctioning risk, or if public investigation and prosecution is required to create an adequate probability of sanctions. Criminal enforcement can also be used to punish more severely deliberate offences and/or gross negligence. Such a joint use of civil and criminal enforcement will induce proprietors to comply by at least one of the two methods. Once proprietors comply, the enforcement costs will be relatively low and so will the social cost. The only condition is that the proprietor has sufficient information ex-ante about ways of prevent (ex-post) damage.

The benefits of supplemental administrative enforcement are that a municipality is able to close an establishment if the proprietor perseveres in the violation. An injunction to use the establishment might also be imposed by a criminal (and even civil) court but this is likely to require much more time and effort. However, as demonstrated, if a criminal enforcement policy is adequately carried out, proprietors will not choose non-compliance, at least not persistently, and such closure will never be necessary.

Let me discuss two additional reasons for administrative enforcement as a supplement to harm-based enforcement.

15.5.2 *The problem of information and act-based enforcement as supplement*

If a proprietor does not have sufficient information about ways of preventing damage, a harm-based approach (a combination of private and criminal enforcement) will probably lead to an inefficient level of compliance because the proprietor will be unable to properly balance the costs and benefits of compliance. It might then be optimal to choose for act-based enforcement. If administrative law enforcement is chosen, it will lead to relatively high enforcement costs because many inspections will be needed to reduce the social costs. Under administrative enforcement, proprietors can postpone compliance. This disadvantage can only be reduced by inspecting more often. It might therefore be optimal to choose a combination of private and administrative enforcement. Administrative enforcement can be used to provide proprietors with the relevant information, so that then private enforcement can be used (if necessary supplemented with criminal enforcement). Suppose that one annual inspection, for example before Christmas or during Carnival, suffices to inform the proprietor about the requirements and possible compensation claims. These inspections might be announced and preceded by letters and brochures. This will induce proprietors to take the correct level of precautions under liability. Under strict liability and current damages it would lead to social costs of €220,800, of which €171,300 of expected damage, €30,400 of costs of compliance, and €19,100 of enforcement costs (€18,100 of administrative inspections, table 15.11, and €1000 of private litigation, table 15.5). Similarly, under the negligence rule with efficient standards, the total costs are €201,000 (table 15.18). This is a better result than under administrative enforcement alone: €260,000 at the best (table 15.12). Since the objective of enforcement is specifically to advise and inform, an administrative enforcement policy might be more appropriate than a criminal enforcement policy. Besides, in case of serious violations the municipality can close the establishment.

This is the usual defense of the use of administrative enforcement. However, the fact that a joint use of ex-ante administrative enforcement and ex-post private enforcement can solve the problem of insufficient information is only valid under the following two conditions. First of all, it must be the case that information cannot be provided in any other way. As discussed in section 12.4.1, it might very well be possible to provide the necessary information by means of general information campaigns. This is likely to be much cheaper than an annual inspection on site. Private organizations such as insurance companies, advice and certification firms, etc. can also do at least part of the work in this respect. Moreover, enforcement officials and municipalities often stress that administrative enforcement is a leading factor of vital importance, rather than supplemental to private enforcement. Secondly, this result is only valid if there is no willingness to impose criminal act-based sanctions. Sufficiently high criminal fines would achieve a higher level of social welfare.

Table 15.18 Joint private and administrative enforcement under information problems (current law)

	Strict liability (SL)	Negligence, efficient standards (NLES)
Expected damage	171.300	142.500
Costs of compliance	30.400	40.200
Subtotal	*201.700*	*182.700*
Private enforcement costs	990	230
Public costs of inspection	16.700	16.700
Public costs of conviction	1.400	1400
Enforcement costs	*19.100*	*18.300*
Total social costs	**220.800**	**201.000**
Level of compliance	**60**	**78**

15.5.3 *The problem of wealth and act-based enforcement as supplement*

The joint use of private and administrative enforcement is also useful if, for some reason, a criminal enforcement policy fails and some but not all proprietors have insufficient assets to pay damages (section 10.2.2).[280] Insufficient assets make private enforcement ineffective, but administrative enforcement alone might be expensive. Administrative enforcement with one annual inspection may force all proprietors to comply at least with the technical requirements which have to be restored once a year and for which an annual inspection suffices. Under administrative enforcement, proprietors with insufficient assets will take at least some precautions. Additional private enforcement induces those proprietors that have sufficient assets to take higher precautions. They will – depending on the specific liability rule – comply with all technical requirements and possibly also with the number of visitors.

As demonstrated in section 15.2.3, the total social costs will be €380,600 (level of compliance of 43) if there is private enforcement under a negligence rule with the technical requirements as the standard (NLT) and if all uneven types of proprietors have insufficient assets. Under administrative enforcement (where asset problems are absent) the total costs are at least €264,000 (level of compliance of 70) or €278,900 (level of compliance of 92). The joint use of an annual inspection under administrative enforcement and private

280 Or, similarly, if not all victims always claim damages.

enforcement under a negligence rule with the technical requirements (NLT) will lead to lower total costs. Under this approach, the level of compliance will be 86 with total social costs of €255,900. These consist of €187,500 of expected damage, €49,200 of compliance costs and €19,200 enforcement costs (€18,100 for one annual public inspection and €1,100 for litigation costs against proprietors with limited assets who are sometimes found to be in non-compliance) (see table 15.19). This result can be further improved by varying the public inspections per establishment or by choosing different liability rules.

15.5.4 *Conclusion*

This chapter analyzes the costs and benefits of several enforcement alternatives based on a necessary abstraction from reality due to lack of more accurate data. An important simplification is that the requirements and therefore the level of compliance, the use of sanctions, etc. are reduced to two groups. Furthermore, I simply assume that proprietors are rational, risk-neutral wealth maximizers. By using all available data sources, I have made as realistic an estimate as possible of the costs and benefits of enforcement. These provide relevant and actual understanding of the effects of enforcement.

The analysis shows that it is indeed worthwhile to invest in enforcement. Harm-based enforcement, both private and criminal, should be used as much as possible. In general, a more strict enforcement policy of higher compensation awards or higher public sanctions will improve social welfare. Currently, enforcement takes place primarily through administrative enforcement. Municipalities usually choose an annual inspection which is announced beforehand. This can be defended if administrative enforcement is especially a supplement to private enforcement. However, it is not clear why private enforcement fails. Information problems do not seem sufficiently severe to justify the intensive use of administrative enforcement. Potential wealth problems are more easily solved by criminal enforcement. Moreover, the supplemental function of administrative enforcement is not disseminated by the government. The role of the government (municipalities) is stressed. It grants licenses and bears responsibility for fire safety. The efforts following the 'Volendam' disaster also create the impression and expectation that the government takes responsibility for the enforcement of fire safety, while according to the analysis here, the role of public enforcement is primarily informative and supplemental. The analysis shows that social welfare will be higher if harm-based enforcement is more important and if a more strict enforcement policy with higher compensation or fines is used.

Table 15.19 Joint private and administrative enforcement under limited wealth (current law)

Type	**Number**	**Limited assets?**	**Choice under NLT[a]**	**Choice under optimal admin. alone[b]**	**Choice under joint use[c]**
1	10	Yes	Sn	Sn	St (0,75)
2	8	No	St	St (0,92)	St (1,00)
3	8	Yes	Sn	St (0,94)	St (0,75)
4	4	No	St	St (0,92)	St (1,00)
5	8	Yes	Sn	St (0,94)	St (0,75)
6	10	No	St	St (0,94)	St (1,00)
7	10	Yes	Sn	St (0,88)	St (0,75)
8	8	No	St	St (0,88)	St (1,00)
9	2	Yes	Sn	Sn	St (0,75)
10	5	No	Sn	Sn	St (1,00)
11	5	Yes	Sn	St (0,75)	St (0,75)
12	8	No	St	St (0,92)	St (1,00)
13	4	Yes	Sn	St (0,88)	St (0,75)
14	5	No	St	St (0,94)	St (1,00)
15	5	Yes	Sn	Sn	St (0,75)
Level of compliance			*43*	*70*	*86*
Expected damage			356.400	181.700	187.500
Compliance costs			22.000	40.200	49.200
Subtotal			*378.400*	*221.900*	*236.700*
Enforcement costs[d]			2.200	42.100	19.200
Total			**380.600**	**264.000**	**255.900**

a. Liability under the negligence rule with the technical requirements as the standard, with damages as under current law. See table 15.9.

b. Administrative enforcement with the optimal number of inspections on technical requirements (table 15.12, second column). This is the best result under administrative enforcement. If this outcome is worse than a joint use, this will be true for all results under administrative enforcement. To compare, under enforcing St for each proprietor, total social costs are €331.489 for one inspection per year (St, 0,75 for each proprietor) and €278.942 for three inspections per year (St, 0,88 for each proprietor) (table 15.11). Here choosing St means that they choose St but only in as far as they do not postpone compliance until they are inspected. In parentheses the specific level of compliance.

c. A combination of administrative enforcement with inspection once per year and negligence liability with technical requirements as the standard (under current law). If the wealth of a proprietor is larger than his technical compliance costs, he can be forced to choose a level of compliance of 0,75 (table 15.11). Proprietors who have sufficient assets anyway will choose the same as under NLT, i.e. St completely (compliance level 1,00).

d. It is assumed that litigation costs are realised if a proprietor non-complies even if he had insufficient assets.

16 Conclusion

On 1 January 2001, a fire broke out in café "'t Hemeltje", in the municipality Volendam, killing 14 and wounding approximately 300 youngsters. This accident drew widespread attention to the Dutch municipal enforcement policies in general, and to fire safety regulation in particular. This thesis attempts to assess the effects of the resulting increase in enforcement actions. The central question is: What is an effective and efficient enforcement policy for fire safety regulation in the catering industry, i.e. in horeca establishments, in the Netherlands?

The typical situation to which this question applies is one in which the proprietor of a bar, restaurant, discotheque or similar establishment is both the establishment's manager and its owner. At any one time, there may be from 50 to 250 people present in the establishment, but the number of visitors varies. On several days of the year, the number of people whishing to visit the establishment might be higher than allowed. The proprietor is expected to comply with the use requirements concerning fire safety, including the following: (1) escape-routes and emergency exits must be kept free at all times; (2) furnishings and decorations must be correctly hung and impregnated; (3) escape-route signs must be clearly visible and burning; (4) fire extinguishers must be clearly indicated, ready to hand and annually certified; (5) candle lights, trash, ash-trays, etc. must be used safely; (6) the maximum number of persons allowed may not be exceeded; (7) the remaining use requirements, for example those concerning technical devices or evacuation plans, must be adhered to.

The optimal enforcement policy for fire safety regulation in horeca establishments is analyzed in two parts. In part I, this policy is described by analyzing the economic literature on the enforcement of safety standards at firms. In part II, I collect and analyze the data concerning the actual enforcement policies. I summarize the conclusions of part I and part II in, respectively, sections 16.1 and 16.2. I conclude the analysis by describing the efficient enforcement policy in section 16.3. Finally, in section 16.4, I reflect on the description of the efficient enforcement policy in section 16.3 by considering the questions requiring further answers.

16.1 Summary part I: theory

The enforcement of fire safety regulation can basically be organized in three ways. Under private enforcement, enforcement is left to the victims of the harm. Liability rules enable visitors to require compensation for their damage. Under administrative enforcement, regulation is enforced by municipalities through a system of licenses, inspections, administrative sanctions, etc. Such administrative sanctions are based on the expected harm. Finally, public prosecutors can prosecute an offense, so that a criminal law sanction can be imposed on a proprietor failing to meet the use requirements. In part I, I analyze, on the basis of the economic literature, the conditions under which the different enforcement methods are effective in inducing compliance and are efficient in reducing the total social costs of compliance and enforcement.

More specifically, I extend the overview of the economic literature on administrative enforcement and/or the use of informal enforcement actions in general. The standard economic model stresses the importance of punishing firms in order to deter them from committing a violation. In practice, however, many enforcement authorities, especially when operating under administrative enforcement, seem to apply more cooperative, informal, advisory enforcement policies, the so-called *compliance strategies*, to be distinguished from *deterrence strategies*. In part I, I analyze the economic literature with a focus on the need for and the effects of policies of advice, persuasion and warnings. Under what conditions is a compliance strategy more effective than a deterrence strategy? This analysis is valuable because an economic (literature) review of these policies has not, so far, been available. It is important to confront economic models with the actual enforcement policies. Moreover, other disciplines do extensively debate optimal enforcement strategies. Adding the economic perspective to this discussion is valuable for several reasons. The economic perspective more explicitly focuses on the interactions between the characteristics of offenders and the enforcement actions. Economists are good at calling attention to the general effects of enforcement, the conditions for effective enforcement and the constant balancing of the benefits and costs of enforcement.

16.1.1 *The need for public enforcement*

Chapter 2 describes the enforcement problem. There are two relevant problems for the enforcement of fire safety regulation in horeca establishments. First, are visitors and proprietors provided with the optimal level of compensation for losses (*risk-spreading*)? If they are risk-averse, they will have a preference for certainty concerning their level of wealth or income, thus requiring compensation for monetary losses. However, since wealth is more valuable when someone is in good health, the optimal level of compensation is lower than the monetary losses. Secondly, do proprietors take the socially optimal level of *precautions* to prevent fire damage? The socially optimal level of precautions balances the (marginal) costs of taking precautions and the (marginal) benefits of a reduction in expected harm.

Since generally individuals are able to obtain compensation by insuring themselves, the analysis focuses on the incentive to take precautions. Proprietors might choose some precautions voluntarily, but for a number of reasons, they are unlikely to take sufficient precautions. First of all, visitors are not informed about the expected damage and therefore unable to force the proprietors to take precautions. Secondly, there are hardly any other effects of (non-)compliance for the proprietor's reputation. Moreover, in general, such effects would only appear if there is a public detection of non-compliance. Thirdly, the intrinsic motivation to comply with the rules and to prevent damage is not likely to survive if there is no enforcement of the rules. A lack of enforcement stimulates a culture in which compliance is seen as unimportant. Fourthly, insurance companies are helpful in stimulating compliance with the requirements concerning technical devices, but they are unable to control the daily behavior of proprietors, including issues such as keeping emergency exits free, being attentive, etc. Moreover, they will only try to prevent the losses to the proprietor that do generally not include the losses to visitors. Finally, proprietors have difficulty estimating expected harm from fire because this information is rather complex and technical in nature, and/or because they are cognitively constrained and therefore insufficiently aware of the importance of fire safety regulation and of compliance with the regulation. They are therefore likely to underestimate the probability of fire.

Due to all of these reasons, the level of precautions is very low if there is no enforcement of the regulations. Given the magnitude of the damage, it is therefore worthwhile to consider making enforcement costs in order to increase compliance. The enforcement costs of obtaining a higher level of precautions should be weighed against the balance of compliance costs and expected damage.

16.1.2 *Private enforcement by liability rules*

Victims are able to claim compensation for their damages from the proprietor based on the liability rules of the Civil Code. Section 3 of chapter 2 discusses the fact that private enforcement by victims is only effective if injurers are really forced to internalize all the harm they inflict. Therefore private enforcement by visitors is only effective under the following conditions:

1. Most of the harm done must consist of monetary damages, or, if there are non-monetary damages, courts must be able and willing to award adequate compensation for these damages. A complication lies in the fact that awarding non-monetary damages provides the victim with inefficient compensation. Buying 'negative insurance' is generally not possible.
2. The injurer must have sufficient wealth to pay for the damages or should be fully insured for liability. Higher damages than the available assets do not further stimulate precautions. If the injurer is insured, insurance companies should be able to easily monitor and enforce precautions. Otherwise, the outcome will be 'reasonable', but not first-best. Monitoring costs

directly decrease social welfare. Indirectly, the injurer will choose partial insurance and less than first-best precautions.

3. Courts and injurers must have sufficient information about possible accidents so that courts are able to enforce socially optimal levels of precautions and injurers are able to anticipate on the courts' decisions. This might be problematic especially when courts deviate from the regulatory standards in determining negligence. Uncertainty about the law, or about the facts, has two opposing effects. The injurer increases the level of precautions in order to reduce the probability that he will be found negligent, but decreases it because the benefits of precautions are lower. Hence, such uncertainty might create both under- and overdeterrence.
   Being held liable gives the injurer an optimal incentive to acquire information about taking precautions. However, problems might arise if the acquisition of information has the characteristics of a public good because it concerns technically complicated issues.
4. Victims must have sufficient access to the legal system, so that the probability of sanctions is sufficiently high. Therefore, victims must be able to identify the injurer and the costs of claiming damages must be sufficiently low to find it worthwhile to file a claim. Otherwise, the injurer will not be inclined to take precautions.
   If these conditions are not satisfied, the injurer will choose less than efficient precautions. Under the negligence rule, some of the conditions discussed are less severe, especially when it functions perfectly. If the injurer chooses to be non-negligent, he will not have to pay damages.

16.1.3 *Public law enforcement*

In chapters 3 to 9, I examine public law enforcement policies. Building on Becker (1968), an economic analysis of law enforcement leads to the following results: (1) a higher expected sanction decreases the level of non-compliance, (2) costly non-monetary sanctions should only be used when costless monetary sanctions are used to the maximum, (3) decreasing the probability of detection and increasing the magnitude of the sanction might keep deterrence constant but save on enforcement costs, hence it is socially desirable.

The empirical challenge is that these results are not commonly observed in practice, especially not for regulatory crime. This is referred to as the 'Harrington Paradox': (for environmental regulation) (1) the frequency of inspection is low; (2) even if a violation is detected, penalties are hardly ever imposed and are rather low (sometimes even lower than the costs of compliance); yet (3) the level of compliance seems to be pretty high. The review of public law enforcement analyzes the reasons the economic literature provides for the absence of sanctions, at least immediate sanctions. These reasons can be used to defend compliance strategies.

In chapter 3, I analyze the conventional economic literature which focuses on the costs of enforcement, the detection technology and on the available information. If prosecution and punishment are costly, extreme sanctions are inefficient if the marginal social costs of sanctions increase or the marginal private costs of sanctions decrease. If full information is available concerning all relevant benefits and costs, it is possible to sanction only inefficient non-compliance. Otherwise, a sanction must be made dependent on the available information, for example on the level of harm. An expected sanction that is equal to the harm caused results in the first-best level of precautions.

Sometimes, the enforcement authority can observe who caused the damage without any (additional) effort, for example in the case of a fire accident. In other cases, investments in detection are required. Because of the costs of detection, some underdeterrence is desirable. If enforcement is carried out by means of specific enforcement (investigation), the enforcement authority will be able to observe a violation and its harm, but it will have to invest in detecting the injurer. By lowering the probability of investigation and increasing the sanction, enforcement costs can be saved and deterrence can be held constant (as long as the sanction does not exceed wealth or other constraints). On the other hand, if enforcement is carried out by means of general enforcement (monitoring), detection of non-compliance will only occur once a firm is inspected. Under monitoring, it is not optimal to impose maximum sanctions because the expected sanctions should increase with harm (marginal deterrence). If the maximum (monetary) sanction is not known to the enforcement authorities (before detection), the probability of detection must be high enough to deter firms with low wealth. Therefore, the sanction on firms with higher wealth should not be maximal.

In chapter 4, I focus on offenses that can be repeated or can last over several periods of time. If the penalty is optimal for the first offense, there is no reason to increase the sanction in the case of a second (identical) offense. Only if there is underdeterrence might it be optimal to impose lower sanctions for first offenses in later periods. This might be done for example by being tough on young companies that have just been established as well as on firms that have a record of violation, while punishing existing companies that have good compliance records only moderately. Another reason for increasing sanctions is that an offense reveals valuable information to the enforcement authority about the firm's willingness or ability to comply.

In many situations, especially with respect to offenses by firms, (non-) compliance is not a discrete event, but rather a continuing state of affairs (Veljanovski, 1984) lasting for several periods of time until it is ended deliberately or naturally. If the enforcement authority detects a violation, it can take action to restore compliance by imposing so called 'repair sanctions' such as a duty by penal sum or administrative coercion.

Chapter 5 extends the analysis by considering whether it is optimal to increase detection for repeated offenses. Immediate and equal punishment of all offenses is not a good idea if the level of enforcement resources and the magnitude of the sanction are sufficiently restricted and therefore fail to implement optimal deterrence. If the expected sanction is too low, it is optimal to target enforcement resources. The optimal targeting strategy is one in which enforcement depends on past compliance (state-dependent enforcement policies; Harrington, 1988), for example in the form of issuing warnings. By tolerating a given level of non-compliance, or by sometimes tolerating non-compliance, the enforcement authority 'buys' or 'negotiates' at least partial compliance.

Chapter 6 considers the uncertainty concerning standards and expected harm. If the law is complete and perfectly predictable, a strict enforcement policy will be efficient – provided that the expected sanction is large enough (Xu and Pistor, 2002). If, however, standards are uncertain, more flexible enforcement might be required. Public inspections might be needed to inform firms about the standards and the optimal compliance techniques, so that enforcement is aimed at advice, in addition to or instead of deterrence. Cooperation might be needed to buy or negotiate precautions not included in the regulation in exchange for tolerating violations of other standards. Moreover, enforcement is hampered by the incentive of firms to try to escape sanctions. Informal enforcement might on balance achieve higher levels of compliance than the strict punishment of every offense, especially when legal procedures are costly for enforcement authorities.

Chapter 7 analyzes the effects of informal sanctions and intrinsic motivation to comply on the optimal enforcement policy. Informal sanctions and intrinsic motivation make compliance more attractive. However, they also impose limits on the deterrent effect of (formal) enforcement. First, informal control may only work if there is some public enforcement, limiting the possibilities of reducing detection. Secondly, informal control strengthens the effectiveness of public enforcement, when informal control depends on the proportion of firms complying. Therefore, public enforcement must be strong and strict enough to sustain a compliance culture. Thirdly, informal control weakens the effectiveness of public enforcement, while formal control diminishes the intrinsic motivation to comply. Especially for low sanctions, an increase in formal control might therefore decrease compliance. If informal sanctions and intrinsic motivation depend on the compliance rate, the deterioration and recovery of compliant behavior are not asymmetric processes.

Chapter 8 summarizes the literature so far by discussing the use of compliance strategies. A compliance strategy can be beneficial when regulation is incomplete or when sanctioning is costly. If regulation is incomplete, cooperation and advice may be more effective than strict, legalistic enforcement. If the enforcement authority does not have sufficient powers to enforce full compliance, imposing sanctions on every violation may be an expensive process. Under such circumstances informal standards, warnings, self-reporting, advice and negotiation might be helpful.

The benefits of alternative enforcement actions also depend on the incentives for enforcement officials. Chapter 9 considers the role of the enforcement authority as a strategic player with its own objectives and incentives. Since law-enforcers are human beings, they are not necessarily seeking efficient enforcement, but may have different, even counterproductive, objectives. Of course, this may harm social welfare. However, the one-sided objectives of law-enforcers – including a strict enforcement of all violations – may be beneficial for several reasons. First, a social welfare-maximizing law-enforcer may be unable to commit itself to an efficient enforcement policy. Social welfare may be improved if law enforcers are instructed to enforce violations regardless of the individual circumstances. Secondly, social welfare may be better served by a biased, political regulator than an independent judge, because the first is more inclined to investigate and acquire the necessary information of a case. Furthermore, if the enforcement resources are insufficient to ensure optimal deterrence, the enforcement officials can confine their efforts to maximizing compliance or minimizing expected harm. The optimal level of discretion that is granted to the enforcement authority balances the problem of commitment and of private interests with the flexibility required to evaluate the desirability of enforcement (due to incomplete law and insufficient resources).

Chapter 10 concludes part I. It concludes on several distinctions that can be made between the different enforcement methods, such as the stage of intervention (harm-based, i.e. ex-post, or act-based, i.e. ex-ante), the type of sanction (administrative or criminal) and the initiating party (private or public). Moreover, it argues why there should be a joint use of enforcement methods. Criminal law enforcement might be needed as an *ultimum remedium*. Regulation and liability might be combined to reduce uncertainty or wealth problems. Chapter 10 also concludes by providing a scheme for deciding whether a deterrence or a compliance strategy should be used. However, without filling in the analysis with actual data on fire safety in horeca establishments, the optimal enforcement policy can not be described.

16.1.4 *A compliance or deterrence strategy*

It is important to limit the distinction between deterrence and compliance strategies to a distinction between the enforcement *styles*. The style describes what an enforcement authority will do when it detects a violation. A deterrence style is a style in which every violation is immediately punished and the enforcement authority bases its decisions solely on its own information. A compliance style is a style in which the enforcement authority might forego some violations (for some time) when this induces the firm to reveal important information necessary for enforcement.

Whether a deterrence or a compliance style is optimal, does not (directly) follow from the rationality of firms or individuals, the voluntary willingness to comply or the level of ignorance about the regulations. Moreover, the enforcement style should not be confused with what might be called the objective of enforcement. In regulatory enforcement of firms maximizing social welfare may include several intermediate (not mutually exclusive) enforcement objectives: deterrence (signal that non-compliance does not pay), remediation (restore compliance), advice (inform about compliance), education (create voluntary compliance), and completion (supplement private enforcement). A different question is which enforcement style does achieve an objective the best. My assertion is that there is no relationship whatsoever between objectives and strategies. There is no necessary relationship between the objective of deterrence and a deterrence style, or between other objectives and a compliance style. Each of the objectives can be served by both a deterrence and a compliance style. It is important to start with identifying the objective of enforcement and to subsequently examine which enforcement style can best achieve this objective. Which style is optimal, is not related to the objective, depends on the specific conditions and can not be stated in general. The current literature does not sufficiently acknowledge the difference between firms' characteristics, enforcement objectives and enforcement styles.

Deterrence

Under deterrence (the focus of most economic literature) the objective is to induce compliance by making non-compliance unattractive. The deterrent effect of enforcement relies critically on the expectation by potential offenders that a detected offense will consistently be prosecuted (Fenn and Veljanovski, 1988). Under a compliance style, the firm is able to delay compliance or to negotiate a lower level of compliance.

However, a deterrence style assumes that enforcement authorities are granted sufficient resources to enforce compliance. In practice, they are usually not, so that there is serious underdeterrence and the maximum achievable compliance rate is a partial compliance rate. It might then not be optimal to directly impose severe sanctions because such sanctions might – on balance – increase underdeterrence. In such cases, it is optimal to target enforcement on some subgroup of the population. The best targeting scheme is one that depends on past compliance, for example by issuing warnings. Another possibility is to adjust the standard downwards and to enforce some informal standard.

Because law is incomplete, a deterrence style that legalistically enforces the legal standard to the letter will not induce the first-best level of precautions if the enforcement authority is imperfectly informed and/or is granted insufficient discretionary powers. If there is uncertainty about the standards and the compliance methods, a cooperative, flexible enforcement style can be efficient. The firm will be required to reduce the most important safety problems in exchange for reduced sanctions for minor violations.

A final reason why a deterrence style may fail is that enforcement officials are insufficiently interested in pursuing deterrence and in consistently sanctioning every violation.

Remediation
Remediation implies that enforcement is aimed at restoring compliance. For example, repairing a technical device that causes non-compliance or cleaning-up the environmental harm resulting from some emission. If offenses are purely accidental and if it is impossible to make enforcement dependent on the level of remediation, a strategy of warnings is efficient. If, however, it is possible to sanction insufficient remediation and/or insufficient prior prevention of failures, a deterrence style that imposes sanctions dependent on the level of remediation or prevention is efficient. It is hard to think of any example in which offences are truly accidental and in which sanctions cannot be made dependent on the level of remediation.

Advice
Advice implies that the objective of enforcement is to ensure that the firm is sufficiently informed about all the benefits and costs of (non)compliance, the standards that apply and the least costly method to create compliance. If the firm is able to acquire the relevant information at a reasonable effort, sanctions for non-compliance provide the incentive to inform oneself. Especially under technically complicated regulation, the costs might be unreasonably high. If all firms lack the same kind of information, general information campaigns will be sufficient to inform them. If the law is (very) incomplete, so that an individual firm cannot learn from others, it might be efficient for public inspection to have advice as its objective, characterized by speech, explanation and cooperation.

Education
Enforcement may also have as objective the stimulation of norms and values that enhance compliance. Public inspection may be needed to maintain, strengthen or even create norms and to realize the costs and benefits of norms.

Public inspection might be necessary to detect offences and to realize the costs of norm-violation (especially for informal social sanctions). If the costs of norm-violation depend on the level of compliance in the relevant population, public enforcement through sufficient deterrence is needed to prevent a breakdown in compliance and norm-subscription. Norm formation can occur by observing that non-compliance is punished, or by observing that compliance is valued positively. Imposing sanctions can crowd out the intrinsic motivation to comply.

Completion
Another objective of public enforcement is to supplement private enforcement. First, public enforcement might be needed as advice to reduce the uncertainty concerning the levels of due care or to prevent cognitive failures. Public enforcement might also be needed when private enforcement fails to deter efficiently because of wealth constraints or litigation barriers. In gener-

al, a strict enforcement policy is needed to induce the privately underdeterred firms to take precautions.

16.2 Summary part II: empirical analysis

In part II, I conclude on the enforcement of fire safety regulation in horeca establishments by collecting and analyzing the data concerning actual enforcement policies. Let me summarize the main results.

16.2.1 Empirical analysis: data and methods

Ultimately, balancing the costs and benefits of enforcement is an empirical matter. Therefore, in part II, I focus on analyzing the enforcement policy for fire safety regulation in horeca establishments making use of actual data. The problem with analyzing actual enforcement policies is finding or collecting the right data on compliance and the benefits of compliance. I try to do so in a number of ways. First, and quite uncommonly for an economist, I have conducted interviews to acquire insight into actual enforcement policies. A disadvantage of such qualitative research is that it is more difficult to verify the results. Conclusions concerning the effects of enforcement do not directly follow from a statistical analysis but follow from the verbal interpretation of the researcher. Another problem with such interviews is that they are time-consuming. However, given the lack of other tangible data, interviews with enforcement officials are necessary to collect data on enforcement policies, their effects and the interaction between enforcement officials and proprietors. Unfortunately, these data could not be complemented with the impressions and opinions of proprietors. In chapter 12, I use these interviews, together with the available reports and other written material, to determine whether the municipal enforcement policies in the years following the Volendam disaster are effective in inducing compliance. In particular I investigate whether the observed use of a compliance strategy is justified given the arguments identified in part I.

Secondly, I combine bits and pieces of available evidence into a cost-benefit analysis of the increase in enforcement efforts following the Volendam disaster. A cost-benefit analysis attempts to list all the effects of a policy intervention. Monetizing these effects allows them to be compared and to evaluate whether the intervention leads to a positive or negative contribution to social welfare. Due to lack of more reliable data, the analysis does not lead to a clear figure of thé balance of costs and benefits (and its alternatives), but only to a more general rule on the required effectiveness of enforcement.

Thirdly, in order to gain insight into the settlement of claims for compensation, I carry out a case-study of the settlement of the Volendam disaster. The problem here is that there are simply not enough compensation claims and not enough knowledge concerning these claims to carry out a more general analysis. We must of course be careful in extending any conclusion from a single case-study to the more general problem. However, this case-study is of particular interest since it received relatively much attention and is therefore likely to have a relatively large influence on proprietors' (and victims') future expectations. Sources of information are newspapers, parliamentary documents, (committee) reports, websites, etc.

Finally, I construct a representative, hypothetical municipality based on all available data. Assuming that the proprietor of a horeca establishment is a risk-neutral wealth-maximizer, I analyze the effects of different enforcement policies. Of course, the disadvantage of such an analysis is that it abstracts from reality. However, without such an abstraction, it becomes impossible to further analyze the effects of enforcement on compliance. The simulation of enforcement policies provides insight into the effectiveness and efficiency of possible enforcement policies.

Such analyses do not answer all questions. Further research is required to provide unambiguous answers to the question of which level of compliance, and hence which level of enforcement resources, is desirable. However, I am convinced that it is more worthwhile to invest in (improving) these analyses than in extending our theoretical models. These are essential steps in further analyzing enforcement policies.

16.2.2 Public enforcement between 2001 and 2007

In chapters 11 to 13, I analyze the public enforcement policy of the period 2001-2007. This period begins with the Volendam disaster and ends with the introduction of the national Use Decree and the abolishment of the use licenses mid 2008.

Enforcement deficits and action programs

During the 1990s and early 2000s, large 'enforcement deficits' were detected in many fields of administrative enforcement, thanks to academic pressure, government policy reports and several important disasters. From the national level municipalities were stimulated to professionalize their enforcement policies by adopting 'programmed enforcement'. This was to guarantee that enforcement would proceed according to plan and not to incidents. Different committees (for example the Alders Committee that investigated the Volendam disaster) formulated major action plans to improve enforcement policies and regulations. These initiatives also affected the enforcement of fire safety regulation by municipalities (chapter 11).

The municipal enforcement policy for fire safety regulation

In 2001, there was a backlog in granting use licenses for more than 70% of the buildings. This backlog was accompanied by a personnel shortage for the granting as well as inspection of licenses. In almost no municipality was there a systematic and adequate inspection of the licenses. In most municipalities, a good foundation of the enforcement policy was lacking. After 2001, there was a major increase in municipal fire safety enforcement effort, especially with respect to horeca establishments. Municipalities started to eliminate the backlogs in use-licenses and to regularly inspect the licensed buildings.

Once a license is granted, there are three different types of inspections: periodical inspections, special actions (during events like Carnival), or inspections following a complaint or report. The sanctions that a municipality can impose are a penal sum, administrative coercion or a (partial or temporary) closing of the establishment. As described in chapter 12, bars and restaurants are generally inspected once a year (barring re-inspections). Bars and some restaurants in the centre of the city often face an additional inspection during special events or feast periods. Municipalities usually begin with informal notifications and re-inspections. Formal sanctions are hardly ever imposed.

A Harrington Paradox?

These observations do not lead to a Harrington Paradox. The level of compliance is not unexpectedly high relative to the expected sanction. Enforcement officials do find many minor violations during first inspections. Inspections in which the report contains no failures at all make up less than 10% of all cases. In more than 50% of the cases, the enforcement officials announce re-inspection. This is a logical consequence of the fact that proprietors are not punished, but only warned that sanctions might follow in the future. The enforcement policy is in line with the state-dependent targeting scheme (Harrington, 1988). In the first instance, violators are not punished, but moved to a target group in which they are more closely monitored by means of re-inspections until they comply. In response, proprietors choose to delay compliance until they are inspected. In the end, the municipality manages to induce compliance. The only important point for proprietors is that they should not willingly and deliberately create a fire-hazardous situation. For such behavior, there is a credible threat of sufficiently high sanctions so that proprietors are deterred from these major violations. As a consequence, the actual imposition of sanctions is hardly observed.

A compliance strategy?

Most scholars would argue that the field of fire safety in horeca establishments is a classic example of a field in which enforcement authorities should primarily apply a compliance strategy and not a deterrence strategy. Moreover, the enforcement officials themselves argue that they do use and should use a policy of advice, persuasion and warnings. They claim (1) that inspection is needed to explain to the proprietor his failures and the danger of non-compliance (as a result of ignorance); (2) that inspection is needed to restore compliance and fire safety (rather than punishing violations); (3) that inspection is needed to persuade the proprietor that compliance is socially and morally desirable (because most proprietors are willing to comply); (4) that explanation and cooperation are more effective than sanctions (because of the severe legal requirements); (5) that they choose to be cooperative as long as the proprietor is seen to be cooperative too; and (6) that inspection is only a random indication while fire safety is the personal responsibility of the proprietor.

Upon further examination (chapter 12), these different claims have been found to be usually unfounded. On the contrary, most arguments imply that in the current situation, the balance should be shifted towards a more deterrent strategy. The arguments do not sufficiently distinguish between the 'objective' of enforcement and the 'enforcement style'. For example, the fact that proprietors are willing to comply or that they fail to comply because of mistake or misinformation does in itself not provide a sufficient argument in favor of a compliance strategy.

The use of a compliance strategy is so widespread because it is difficult to realize a more deterrent strategy for two reasons. First, the fire safety department is not allowed to impose direct fines on non-compliance. A compliance strategy is applied because capacity constraints and the necessity to reserve enforcement resources for the most severe offenses do not allow for higher enforcement budgets and more severe sanctions. The point here, however, is that within the legal framework, the fire safety department is able to be stricter by sooner imposing sanctions than in the current situation. The number of re-inspections can be reduced. The fire safety department can save time by reducing advice, explanation and persuasion during a (re)inspection. Secondly, a compliance strategy is used when enforcement officials are not necessarily interested in higher deterrence, but might derive more pleasure from advising and persuading non-compliant proprietors. This could be mitigated by limiting their discretion but that is hardly an option in enforcing fire safety regulation.

A cost-benefit analysis

In chapter 13, I analyze the (overall) benefits and costs of intensified enforcement policies of fire safety regulation for horeca establishments following the Volendam disaster. The annual costs of the additional enforcement actions and additional compliance are estimated at €13.0 to €23.7 million. The potential savings are a total damage of €30.4 million per year. Hence, the required reduction in expected fire damage for a positive balance of benefits and costs is at least 43%. This figure of 43% is a lower bound that applies if the costs of compliance and enforcement are limited to the lower estimates. If the real costs are higher, the minimum required effectiveness increases correspondingly, up to 78% under the highest estimates. It is unlikely that the expected fire damage has decreased with more than 40%. This leads to the conclusion that the enforcement policies for fire safety in horeca establishments following the Volendam disaster have been and/or are inefficient.

Conclusion on the action programs

The analysis of the municipal enforcement policies shows that there is no adequate foundation for the current policy. There is an endless elaboration on tasks, organization, coordination and the required number of personnel, but it remains unclear what is the precise objective and the desired result. There is not even the beginning of a balance of benefits and costs (see also Netherlands Court of Audit, 2005*, and Leeuw and Willemsen, 2006*). Despite the wide use of programmed enforcement, the fire safety policies seem primarily based on the Volendam disaster. For example, the introduction of the national Use Decree without use licenses immediately after the Volendam disaster would have saved over €10 million for horeca establishments alone, and over €100 million for all buildings. Now, use licenses have been abolished just after we have exerted a major effort to grant one to most buildings. It is striking that specifically the action program of the Alders Committee leads to many policy documents that are produced behind the desk. Attention for actual enforcement actions is drawn away to policy development, municipal (re)organization, and granting licenses. Compliance is induced not by desk-work, but by actual enforcement actions.

16.2.3 *Evaluating the settlement of the Volendam disaster*

In chapter 14, I examine the outcome of the damage claims of the victims of the Volendam disaster. It was estimated that victims received financial support for a total of approximately €60 million, largely paid by the national government. The proprietor Veerman paid €3.15 million at most. The total damage to the victims, including non-monetary damage, was estimated at almost €120 million. There is no valid explanation for why the victims did not try to obtain higher damages from the proprietor. The only explanation that cannot be refuted is that victims no longer expected higher damages because the government had already provided these, stimulated by the pressure from the community to close the case. As a consequence, the signal to horeca proprietors is that they do not have to care much about paying damages for fire accidents in their establishments. The financial bill of the Volendam disaster was largely paid by the tax payer. Moreover, since the non-monetary losses remain largely uncompensated, neither the proprietors nor the government have any incentive to prevent these damages.

16.2.4 *Enforcement in a representative municipality*

The analysis of the actual enforcement actions shows that important shortcomings exist. The question then is whether alternative policies will be more effective and efficient. In chapter 15, I combine and develop the available data to examine the annual costs of enforcement in a representative municipality.

According to the set-up, all proprietors choose to non-comply if there is no enforcement, in which case, there are no costs of enforcement or compliance, but the expected damage is rather large: €663,300. In the absence of enforcement costs, the best result would be a level of compliance of 113 (on a scale of 0 to 200) with total costs of €176,800 (costs of compliance of €57,500 and an expected damage of €119,400). It turns out that all three possible enforcement alternatives can lead to a better result than refraining from enforcement, if, that is, it is properly executed.

The analysis of liability claims by victims shows that a reasonable level of social welfare is obtained (total costs varying from €176,800 to €191,600). The level of enforcement costs is low because enforcement is only carried out once a fire has taken place. Victims have sufficient access to the legal system. Whether private enforcement really leads to an efficient result depends on (1) an adequate compensation award which reflects full losses, (2) sufficient assets available on the side of the proprietor, (3) efficient standard-setting by courts.

It is also possible to reduce the expected damage by means of administrative law enforcement. However, the disadvantage of this approach is that proprietors will postpone compliance until there is a real threat of sanctions. Therefore administrative enforcement requires substantial enforcement costs and results in total costs of €260,000 to €331,500.

Choosing to enforce the fire safety regulation by means of criminal law enforcement is a good alternative. Good results can be obtained with relatively low enforcement costs (total costs of €176,800 to €223,700). Higher fines allow a reduction in the costs of inspection. If the information available to enforcement officials is limited, they must enforce the same level of compliance for each proprietor. This might be seen as being 'unfair' or 'disproportionate' by some of the proprietors. Moreover, fines should be high enough to induce proprietors to comply (not lower than €250). Otherwise, it is better to completely refrain from enforcement or to try to obtain a higher level of compliance by not immediately punishing violations (the so-called compliance strategies).

The analysis shows a preference for harm-based enforcement, both privately and criminally. The benefits of harm-based enforcement are the low enforcement costs. A harm-based criminal enforcement approach is the best alternative if the assets of the proprietor are insufficient and additional imprisonment is necessary. The other problems with harm-based enforcement are not of major importance or can be mitigated by adding (a low level of) ex-ante administrative enforcement. The role of ex-ante (administrative) enforcement is primarily informative and supplemental. The major incentive to comply can be created by ex-post enforcement.

The analysis confirms that full compliance is not efficient. Enforcing compliance with the technical devices that require periodical attention is more worthwhile than enforcing compliance with the number of visitors and other regulations that require continued attention.

16.3 Enforcing fire safety in horeca establishments

The central question of this thesis is: What is an effective and efficient enforcement policy for fire safety regulation in horeca establishments? Now it is time to answer this question.

16.3.1 *The efficient enforcement policy*

The conclusions on the optimal enforcement policy for fire safety regulation in horeca establishments are the following.

First of all, some level of enforcement is required to stimulate proprietors to take precautions to prevent fires. Without enforcement, there would be insufficient voluntary compliance (chapters 2, 12, 15). Visitors are not much concerned with fire safety nor do other reputational effects play a role. Moreover, refraining from enforcement sends a signal that compliance with the regulation is not important. Therefore, the motivation of the proprietor to comply will disappear. In addition, note that insurance companies cannot induce sufficient compliance. They are especially helpful in creating compliance with the standards concerning technical devices, but they cannot really control the daily behavior of the proprietor which is also part of the use requirements. Moreover, currently, insurance does not cover the non-monetary damage to visitors. Proprietors themselves are interested in preventing their own (financial) losses, but are not sufficiently interested in preventing damage to visitors and therefore not sufficiently interested in for instance keeping emergency exits free.

Secondly, to stimulate proprietors to take precautions we need some public money, but not too much. We should spend some public money because private enforcement alone is not likely to induce efficient compliance. Because of cognitive failures and because of the technical nature of fire safety, some information is needed to stress the importance of fire safety and explain the standards (chapters 12 and 15). But this does not necessarily mean that money should be spent on inspections. It should be spent on information campaigns to make proprietors aware of fire safety (regulation). Moreover, we should not spend too much money on fire safety regulation in horeca establishments, because there are other social problems that have a higher need of public resources (chapter 13). Full compliance with all standards and especially *enforcing* full compliance with the standards is too costly.

Thirdly, the enforcement policy should rest on a smart combination of all possible enforcement methods. No single enforcement method is likely to create sufficient deterrence, but all methods together may induce efficient compliance (chapters 14 and 15). Ex-ante action is needed to make proprietors aware of fire safety (regulations). This is partly done by insurance companies, but additional action is required by fire brigades, who are the most likely candidates for the information campaigns. Ex-ante administrative enforcement is also needed to create deterrence for those proprietors who display very high and broad levels of non-compliance. The establishments of these proprietors can be closed by a municipality (chapter 12). This does not imply that a municipality needs to perform regular and time-consuming inspections in all establishments, but rather that it can benefit from the activities of other enforcement officials. If enforcement officials in the domain of environmental regulation, public order, hygiene, etc. detect serious non-compliance or if the municipality notices other signs of non-compliance, the municipality can start enforcement procedures. Moreover, the municipality can perform joint actions, involving different enforcement officials, in which fire safety is part of the (quick) inspection. Preferably, these actions should be carried out during periods of full use.

Deterrence can especially be achieved by means of ex-post enforcement, a combination of private and criminal enforcement (chapter 15). Private enforcement provides an incentive to take precautions because the proprietor will have to compensate the victims for their monetary losses (and a small amount for non-monetary losses) if he behaves negligently by failing to comply with the standards. In principle, this is likely to work because victims know the proprietor who has to pay damages and they expect sufficiently large damages to spend on litigation, and because the proprietor is usually able to pay for a significant proportion of the damage. The problem that the standards are unknown or uncertain is solved by means of the ex-ante actions described above. Supplementary criminal enforcement is needed to improve deterrence through the threat of imprisonment. This is especially helpful if the proprietor is unable to pay damages and/or if there are significant non-monetary damages.

Note that the enforcement style for public enforcement (either ex-ante or ex-post) should be primarily deterrent (chapter 12). This is important because a strict enforcement policy provides an incentive to restore technical failures as soon as possible and to acquire information concerning the standards, because it signals that compliance with the regulation is important, and because it creates (additional) deterrence by imposing sanctions on offences. A compliance style will only allow the proprietor to delay compliance and await real enforcement actions. There are only two important comments on the use of a deterrence style. Given the limited resources and sanctions the municipality needs to target its enforcement actions (1) on the most serious offences, and (2) on repeated offences. It might do so by enforcing some informal standards and (especially under ex-ante enforcement) by starting with a warning.

16.3.2 *Comparing the efficient policy with actual enforcement policies*

The description of the optimal enforcement policy has some important implications for the current policy.

First, in the current policy, ex-ante administrative enforcement plays a central role (chapters 11 and 12). It is promoted as the most important method for creating compliance. All municipalities inspect or intend to inspect all establishments regularly, that is at least once, often twice a year. They take their time for these inspections. There are however two problems. First, these inspections are relatively costly (chapters 13 and 15). It is more efficient to rely on ex-post enforcement and/or accept that some fire damage may result by cutting back the intensive ex-ante inspections. As described, ex-ante administrative enforcement should be used (only) for important signals of non-compliance and in close collaboration with other enforcement actions. A second problem with ex-ante administrative inspections is that these inspections and their subsequent sanctions do not easily create deterrence because of their informal, repairing character (chapter 12). After detecting non-compliance, the municipality should be stricter in starting enforcement actions.

Secondly, the analysis shows that it is necessary for the invested resources to become available for actual enforcement actions. It seems that many initiatives and action programs of the past years primarily led to more deskwork and paper-safety instead of actual field-work and real safety improvements (chapters 12 and 13).

A third remark concerning actual enforcement policies concerns the use of private enforcement through litigation. Relying more on private enforcement is only beneficial if victims really do claim their damages from a negligent proprietor. Otherwise, it will fail to create sufficient deterrence. The incentive for victims to claim damages is removed by a government that does largely compensate the victims. The (national) government should only do so, if it tries to recoup the compensation it pays from the proprietor (chapter 14).

With respect to criminal enforcement, it should be noted that there is currently little attention for enforcing fire safety in horeca establishments by the Public Prosecutor and the police officials. Only in case of a severe accident such as the Volendam disaster are resources spent on this problem (chapters 12 and 14). Moreover, for criminal sanctions there is a heavier burden of proof, because it is important to have a more direct causal relationship between the acts (precautions) and the damage. Therefore, it is likely that there will be underdeterrence of accidents that result in relatively small harms. This is not necessarily a problem since we mostly want to prevent the major disasters, and the acts that clearly result in danger.

16.3.3 *Further remarks on enforcing fire safety*

There are three relevant remarks on the recommendations for the most effective and efficient policy.

First, the description of the efficient enforcement policy of fire safety regulation applies to horeca establishments and cannot necessarily be extended to other sectors. The question however is whether it is possible to have different policies for different sectors. For example, in more complex organizations than horeca establishments, ex-post enforcement, especially criminal enforcement, might not work well because of the diffused responsibilities. In such cases, intensive ex-ante administrative enforcement might be needed. That means that each sector has its own enforcement policy. It must be clear to the sector which policy applies to them. That horeca proprietors are not intensively exposed to ex-ante administrative enforcement should not lead them to believe that they can get away with non-compliance, as they will be subjected to ex-post enforcement. The question is whether such different enforcement policies for different sectors are feasible and credible.

Secondly, the current enforcement policy can be further improved by enlarging the legal possibilities (chapters 12 and 15). If public enforcement was to apply higher fines, or administrative fines, a higher level of compliance is feasible. However, it remains to be seen whether higher fines are desirable. First of all, such solutions improve ex-ante enforcement, but I argue that intensive ex-ante inspections are unnecessary. Another problem is that enforcing fire safety regulation necessarily requires a lot of discretion (chapter 12). As the analysis reveals, fire safety enforcement officials also use this discretion to promote their own interests and status. This incentive should be restrained by limiting the possibilities they have in enforcing fire safety. Proprietors who feel they have been treated 'unfairly' or 'disproportionately' are likely to use court proceedings to claim their right. Hence, higher sanctions might be very costly without really leading to higher levels of precautions. Moreover, the analysis revealed (chapters 12 and 15) that imposing fines is only efficient if these fines are high enough. Fines of less than €100 will only be counterproductive. A more promising way to improve compliance by changing the legal environment, is to increase the award for non-monetary losses, especially for fatalities. This will improve the effectiveness of private enforcement.

Finally, it is important to note that the analysis has of course some important limitations concerning the analysis of the accident problem. For example, the analysis of obligatory insurance for proprietors, the use certification or private periodical tests, the optimal method of compensating victims, the complications of principal-agency problems in more complex horeca organizations, limited liability etc. is insufficient to provide founded conclusions concerning these issues. However, I do consider these issues of minor importance. For example, obligatory insurance can not be expected to really improve compliance as long as the proprietor is only obliged to pay a small part of the damages (especially when it concerns non-monetary damages).

16.4 Further research to the enforcement of regulatory crime

In short, this section states that further empirical analysis to the enforcement of regulatory crime is needed.

16.4.1 The economic perspective

In practice, theories from the socio-legal sciences are very popular with enforcement authorities, witness examples like the enforcement pyramid, responsive regulation, best-practices, etc. (Van de Bunt et al., 2007*). This is often very useful. This thesis provides an overview of the economic analysis of enforcement, which uses the assumption of rationality of potential offenders to examine the effects of enforcement. In many respects, this leads to similar recommendations for the optimal enforcement policy. Economists do not necessarily favor harsh, deterrent, legalistic enforcement, but have broadly investigated the conditions that determine the optimal policy, i.e. the most welfare-enhancing policy. Advice, persuasion and warnings, compliance strategies, have their merits for economists too. The conclusions about optimal enforcement policies do not always deviate much from those of other disciplines.[281] A difference is that the economic analysis more explicitly focuses on the interactions between characteristics of offenders and enforcement actions. Moreover, economists continuously require attention to the balance of the benefits and costs of enforcement. This is not only a point of theoretical interest, as shown in the analysis of actual enforcement policies.

May be it is possible to further integrate the different perspectives on compliance and enforcement. But possibly the best method to integrate is by investigating actual enforcement policies instead of trying to provide one theory.

16.4.2 The optimal enforcement style

My analysis shows that in the field of fire safety regulation in horeca establishments the use of a compliance style is questionable. Most arguments for applying this style are incorrect. Moreover, the analysis reveals two important general conclusions regarding the discussion on compliance styles. First, the scope of compliance styles is rather small. Only in rather exceptional cases will a properly applied deterrence style not work. Compliance styles are restricted to specific situations in which enforcement authorities have limited powers and limited information. In general, welfare will increase more by increasing these powers (provided that discretion can be controlled) than by using a compliance style.

Secondly, most arguments in the literature in favor of a compliance style are in fact arguments for why a deterrence style fails. However, the question should be whether a compliance style would really be better. For example, a compliance style might signal that norms are unimportant and hence destroy intrinsic motivation, or it may remove the incentive to acquire information

281 For instance, Van de Bunt et al. (2007*) provide critical comments to enforcement pyramid that are similar to mine concerning compliance strategies.

about the standards that apply. The arguments for a failure of a deterrence style are not sufficient to support a compliance style. It is important to investigate the failures of a compliance strategy. The optimal enforcement policy should balance the failures of a deterrence and a compliance strategy.

For economists there is an important lesson too. The conventional economic literature advises to trade off as much as possible the probability of detection and the magnitude of the sanction. In practice, however, enforcement authorities are often underfinanced making them unable to achieve the socially optimal level of deterrence. This has important implications for the enforcement problem. If there is a (serious) inefficient level of underdeterrence, the optimal enforcement policy might be different from that under the socially optimal level of (under)deterrence. In such situations so called compliance strategies can be more effective. Economic recommendations on optimal enforcement policies will improve if they take into consideration these actual constraints on policy.

16.4.3 *Remaining questions*

Further research should not only examine the failures of a compliance strategy, but should also focus on the relationship between administrative law enforcement and a compliance strategy. Sometimes, the use of administrative enforcement implies the use of a compliance strategy. But this might not be a necessary consequence. In some cases, administrative enforcement also uses direct fines instead of 'repair' sanctions. On the other hand, criminal law enforcers such as policemen are not only punishing, but often use warnings, cooperation and informal enforcement, for example when interacting with local youngsters in the neighborhood. Indeed, most literature on compliance strategies simply follows from the application of (criminal) fines. However, the usual metaphor implies a link between a policeman and a deterrence strategy. If the difference between administrative and criminal enforcement is not the enforcement style, what does characterize the difference? Why are there two types of public enforcement? Is there only a difference between the institutions that are responsible for enforcement (a public prosecutor versus administrative bodies)? Or are there more fundamental differences? Usual explanations focus on the difference between the magnitude of the sanction and the corresponding burden of proof. However, it is easy to provide examples of high administrative sanctions (for instance in competition law) or of low criminal sanctions (for instance in public order). There is no literature available that convincingly and in detail analyzes these questions (see chapter 10).

Furthermore, it became apparent that there remains much to be said on the relationship between norms and enforcement, despite the growing literature on norms and motivation. What is the relationship between public enforcement and (private) norms? Does public enforcement strengthen or destroy these norms?

Another important question is how to provide public (enforcement) officials with incentives to behave in the interests of society. These officials can make or break an enforcement policy. They should enforce in a professional, adequate way. How is this behavior guaranteed? Can governmental liability for insufficient enforcement help? Can fines on overdue decisions improve procedures? Should enforcement departments receive benefits from sanctions? What's more: to which extent is and should enforcement be affected by public fears for disasters, by lobby groups or by the enforcement officials themselves?

Finally, we need empirical analyses to improve our understanding of actual behavior patterns. There is still relatively little knowledge on the explanation for regulatory crime (Lott, 2000; Huisman and Beukelman, 2007*; Voermans, 2007*). Why do some firms comply? Why do others choose to non-comply? Moreover, there is much to be learned about the effects of enforcement on compliance. Economic textbooks and bibliographical works are full of basic theories on law enforcement. But to what extent can these be applied to actual enforcement problems? How strong is the influence of sanctions on (non)compliance? How does the number of inspections affect compliance? For such questions data collection and empirical analysis are essential.

16.4.4 *The importance of data collection*

The problem of data collection is to find or develop data on the benefits of enforcement: what is the level of compliance and how does it affect the expected damage? This problem is not easily solved for the field of administrative enforcement. Especially for situations of ex-ante enforcement, it is difficult to detect relationships between (ex-ante) enforcement actions and (ex-post) benefits from compliance. Even the relationship between (ex-ante) enforcement actions and (ex-ante) compliance is difficult because the level of (non)compliance cannot be measured using the victim reports. Moreover, when compliance concerns a level of precautions rather than an activity, it might be more difficult to measure compliance. This means that in analyzing the widespread use of administrative regulation and enforcement much work needs to be done as we have little actual knowledge on the relationship between enforcement, compliance and their benefits and costs.[282]

This thesis tries to collect data in different ways and applies different analyses. Of course, as sketched, these analyses do not answer all questions. Further research is required to provide unambiguous answers to the question

282 Similar observations have been made by Leeuw and Willemsen (2006*), Netherlands Court of Audit (2005*) and Voermans (2007*) for the Netherlands and by Macrory (2006b) for the United Kingdom.

of which level of compliance, and hence which level of enforcement resources, is desirable. However, I am convinced that it is worthwhile to begin with analyses such as the one in this thesis. It is important by now not only to improve the theoretical models of enforcement problems, but to demonstrate that these models actually provide insight into real enforcement problems. This is especially of interest since there is no abstract, optimal mix of enforcement instruments. Each problem requires its own solution. We know the basic conditions for the effectiveness of an enforcement method. But we know little concerning actual compliant behavior and the actual effects of enforcement actions.

This limited knowledge makes it all the more remarkable that the government has and should have so many and so clear ideas on what we should do. So many tools, instruments, programs and priority plans are produced, but these priorities are unlikely to have been chosen on the basis of an adequate balance of benefits and costs. Enforcement responses are usually inspired by disasters such as the one in Volendam or by the private ideas of bureaucrats, rather than as a result of a (tentative, best-available) balance of benefits and costs and of careful consideration on the effective way in which instruments can be used. But we do not have to reconcile to this situation. The analysis in this thesis shows that despite important limitations on the available data, policy-makers can actually investigate and plan their enforcement policies more thoroughly. Although these analyses do not provide decisive answers, they do provide insights into the line of reasoning and the direction of the conclusions.

Appendix 1 Interviews and interviewees

Section A1.1 of this appendix provides some background information on the interviews with enforcement officials in thirteen municipalities described in chapter 12. Section A1.2 provides a list of persons and organizations that were helpful in the cost-benefit analysis of chapter 13.

A1.1 Interviews in 13 municipalities

A1.1.1 The selection of municipalities

Sixteen municipalities were selected beforehand, from four Fire Department Regions, in order to create a representative selection of all Dutch municipalities with more than 15,000 and less than 250,000 inhabitants and with at least 20 horeca establishments.[283] The four largest municipalities were excluded because they have a different organizational structure (with more or less independent submunicipalities). The smallest municipalities with small horeca establishments were excluded because for them enforcing fire safety at horeca establishments is simply not important enough, or may be dealt with in a different way.

The selection of the Regions was made in such a way as to ensure a geographical spread that includes at least a region where Carnival is celebrated, one that is relatively rural and one that is relatively urban. In each Region, four municipalities were chosen: one larger city, one smaller village and two in between. The following thirteen municipalities were willingly to cooperate:

1. Amersfoort (Am)
2. Coevorden (Cv)
3. Driebergen – Rijsenburg (DR)
4. Haarlem (Ha)
5. 's Hertogenbosch (or: Den Bosch) (HB)
6. Hoogeveen (Ho)
7. Midden-Drenthe (MD)
8. Oss (Os)
9. Schijndel (Sc)

---

283 This excludes the four largest municipalities and the 176 very small ones. On January 1, 2003, the Netherlands contained 489 municipalities, of which 309 had between 25,000 and 250,000 inhabitants. Of these, 46 municipalities have fewer than 20 horeca establishments.

10. Sint-Oedenrode (Oe)
11. Soest (So)
12. Zandvoort (Za)
13. Zeist (Ze)

Table A1.1 provides summary information about these municipalities. The selection contains several (very) touristy municipalities (especially Za and Cv). Several city centers have a monumental character (Ha and especially HB). The municipalities in Drenthe especially consist of several smaller villages. Most apparent is that the selection does not contain real university or college towns.

Table A1.1 General information on the selected municipalities[a]

Mun.	Fire Department Region	Number of inhabitants	Number of horeca	Area (Km^2)	Urbanization[b]
MD	Drenthe	32,800	44	342	5
Cv	Drenthe	36,000	61	297	5
Ho	Drenthe	53,300	59	128	3
Za	Kennemerland	16,900	92	32	3
Ha	Kennemerland	147,100	294	29	1
Oe	NO-Noord-Brabant	17,000	26	64	4
Sc	NO-Noord-Brabant	23,400	32	42	4
Os	NO-Noord-Brabant	76,200	126	102	3
HB	NO-Noord-Brabant	132,500	263	85	2
DR	Utrechtsland	18,600	21	26	3
So	Utrechtsland	44,800	39	46	3
Ze	Utrechtsland	59,800	70	49	3
Am	Utrechtsland	131,200	174	63	2
Subtotal (13)		789,600	1301	1305	
NL 15-250 (309)[c]		12,375,400	17,821	25,090	
NL total (489)		16,192,500	25,762	33,780	

a. Source: *Horeca in numbers 2003* and CBS, Statline. Data concern January 1, 2003.
b. The extent of urbanization is measured by CBS based on the address density in the neighborhood on a scale of 1 (very urban) to 5 (rural).
c. Total for all municipalities in the Netherlands (NL) with inhabitants between 15,000 and 250,000. In parentheses the number of municipalities.

A1.1.2 The interviews

The interviews were conducted with a basically unstructured list of questions (but with several more structured questions).[284] The research covered the following subjects, with some questions to illustrate:[285]

1. *Basic figures* such as the number of inhabitants and horeca establishments, the political and financial situation, the regional (fire safety) networks, etc.
2. *Organisation and method of operation*. Which departments are involved with fire safety? Is there any help from outside? What is the number of personnel?
3. *Regulation and policy*. Is the model Bylaw of the VNG (Association of Dutch municipalities) applied? Are there policy and/or action plans? What are the policy priorities? Who developed the priorities and who is responsible for monitoring them? How are the rules made known? How long has the municipality worked in this way (Volendam disaster?)? What is the general impression of horeca proprietors? How do they characterize their style of enforcement?
4. *Licensing*. Which horeca establishments need a license? How many are licenced? Which requirements are added in the licence? What is the use of licensing? Is there an impression of the costs of licences for proprietors?
5. *Inspections and violations*. What is the policy and priority for the number of inspections? How often, in 2003, have inspections been carried out? How intense, at what time? How are establishments selected? Is there a presumption of violation? Which requirements were the focus of inspection? How large is the probability of detection? Are there differences between inspectors? How often is everything all right? How often do you detect a violation of one of the requirements identified? Are there differences between types of establishment? Why are the rules violated? How often is there discussion with the proprietor about a violation? Which standards are enforced?
6. *Sequential actions and sanctions*. What happens if a violation is detected? How is the proprietor informed? Can and do proprietors contest decisions? What is the role of informal sanctions? Which solutions are found? How often have formal sanctions been imposed? For each type of violation, how large is the probability and magnitude of sanctions?
7. *Other items*. To which extent should other effects (employment, municipal interests) be taken into account? Is there a problem with contradictory requirements? What is the role of insurance companies? Is it possible to introduce a certification system as for automobiles? Is is possible to say to which extent the (expected) damage decreases as a result of your actions? How many fires has the municipality observed? Do horeca establish-

284 See Hagan (2005), especially chapter six, on the methodology of interviews, including a discussion on the advantages and disadvantages of interviews. See also Bijleveld (2006*, chapter 7).

285 The interviews were conducted in Dutch. The translations are mine.

ments create a fire-safe profile? What is the importance of the administrative fine? Is it possible to enforce full compliance?
8 *Eight propositions* on which the interviews were asked to indicate whether they agree on a five-point scale.

The basic figures for topic 1 can be found in the written sources. For topics 2 and 3, there are usually also some reports and other written materials available. Of course, the main focus of the interviews was on topics 3, 5 and 6.

The questions asked depend on the person interviewed. Sometimes an interview dealt with all topics because the interviewee(s) was involved in policymaking, granting licenses, inspection and sanctioning. In other municipalities, I interviewed different persons with different tasks so that the questions of course varied.

Furthermore, all interviews were recorded.[286] Respondents were told that I would report on who had been interviewed, but the results will not be related to a specific municipality. This was no problem for anyone. Most of them had no objection to full disclosure. The municipalities that did not cooperate gave time-constraints as the reason.

Before beginning the interviews, I accompanied a fire safety inspector of the Fire Brigade in Leiden one afternoon visiting a number of horeca establishments. I also had contact with Mr. R. (Rene) Hagen of the Netherlands Institute Physical Safety Nibra, and Mrs. Y. (Yolanda) van Setten of the VNG (Association of Dutch Municipalities).

A1.1.3 List of persons interviewed in 13 municipalities

The following persons were interviewed in the thirteen municipalities. I am very grateful to them for their cooperation.

Amersfoort

30 March 2004	ing. J. (Jan) Hazeleger Hoofd afdeling Veiligheid, Brandweer Amersfoort [head safety department, Fire Brigade Amersfoort]

Coevorden

9 March 2004	dhr. J. (Jaap) de Koning plaatsvervangend commandant Brandweer Coevorden [deputy chief Fire Brigade Coevorden] dhr. A. (Anthony) Bergman medewerker preventie Brandweer Coevorden [assistant prevention Fire Brigade Coevorden]

286 Although one recording is of bad quality.

Driebergen-Rijsenburg

24 May 2004	dhr. T. (Ton) Leppers medewerker preventie [assistant prevention]

Haarlem

11 March 2004	dhr. T. (Tom) Bersee Handhavingsregisseur (concernstaf) [enforcement director]
22 March 2004	dhr. W. (Wim) de Zwart Hoofd Bureau Preventie (en: wnd. Hoofd preventie en preparatie), Brandweer Haarlem [head prevention office, Fire Brigade Haarlem] Dhr. C. (Carlo) Giezen Inspecteur preventie, Brandweer Haarlem [inspector prevention, Fire Brigade Haarlem] Mevr. B. (Bettina) Zevenbergen Juridisch medewerker afdeling Preventie, Brandweer Haarlem [legal assistant prevention office, Fire Brigade Haarlem]

's-Hertogenbosch

25 March 2004	dhr. J. (Jo) Peters handhavingsregisseur [enforcement director] dhr. M. (Michel) Thijssen hoofd afdeling Preventie Brandweer (fungerend commandant) [head prevention department fire brigade, acting chief] dhr. N. (Nico) de Feiter hoofd Bouwen (sector Stedelijke Ontwikkeling) [head building department]
7 July 2004	mevr. C.T. (Carolien) Reitsma officier preventie Brandweer 's Hertogenbosch [officer prevention, Fire Brigade 's Hertogenbosch]

Hoogeveen

24 March 2004	dhr. R. (Rob) Tiemes handhavingsregisseur [enforcement director] dhr. K. Lubbinge dhr. Peters Inspecteurs Preventie, Brandweer Hoogeveen [inspectors prevention, Fire Brigage Hoogeveen]

Midden-Drenthe

5 April 2004	dhr. J. (Jan) Zinger medewerker Brandweer / Brandpreventie [assistant Fire Brigade / Fire Prevention]

Oss

16 March 2004	dhr. P.J.J.M. Verspeek Brandweercommandant [fire chief] dhr. F.A.M (Freddy) Peters dhr. N. (Niek) van Boven medewerkers preventie / proactie [assistants prevention / proaction]

Schijndel

30 June 2004	dhr. B. (Bert) van Doorn dhr. L.A.M. (Marcel) Vervoort preventiemedewerkers Brandweer Schijndel [assistants prevention Fire Brigade Schijndel]

Sint-Oedenrode

23 June 2004	dhr. H. (Harrie) van Dijk medewerker Bouwen en Wonen [assistant Building and Living]

Soest

1 July 2004	Dhr. H. (Harrie) Tolboom medewerker preventie Brandweer Soest [assistant prevention Fire Brigade Soest]

Zandvoort

8 March 2004	Dhr. S. (Sander) Boon Officier Brandweer Zandvoort, afdeling Preventie [officer Fire Brigade Zandvoort, prevention department]

Zeist

10 June 2004	dhr. J. (Jouke) van Dijk Hoofd preventie Brandweer Zeist [head prevention Fire Brigade Zeist] dhr. J. (Johan) van der Bruggen coördinator T&H Brandweer Zeist [supervisor inspection and enforcement Fire Brigade Zeist]

A1.2 List of persons with additional information

The following persons and organizations provided information that was used in the analysis of chapter 13 (a cost-benefit analysis of the enforcement policies after Volendam). I thank them for this information.

dhr. J.L.M. Boek	Ministry of Justice, Expert Centre Law Enforcement
dhr. W. van Oppen	NCP, National Centre for Prevention
dhr. A. Stuivenberg	IOOV, Public Order and Safety Inspectorate
dhr. M. van der Velden	AEF, Andersson Elffers Felix
dhr. H. Veldkamp and	
dhr. J. Weges	Netherlands Institute Physical Safety Nibra
dhr. P. de Wolf	Koninklijk Horeca Nederland (Royal Horeca NL)

Samenvatting [Summary in Dutch]

Handhaving van brandveiligheid in de horeca.
Een economische analyse.

INLEIDING

Tijdens de nieuwjaarsnacht van 1 januari 2001 breekt er brand uit in café "'t Hemeltje" in de gemeente Volendam. Op dat moment zijn er zo'n 300 mensen aanwezig, voornamelijk jongeren. Het is een korte maar hevige brand. Uiteindelijk overlijden er 14 personen en zijn er vele gewonden, waarvan ongeveer 200 ernstig. De commissie Alders concludeert dat de naleving en de handhaving van de brandveiligheid ernstig tekortschoot. De uitbater Veerman had gezorgd voor een gevaarlijke situatie doordat nooduitgangen en vluchtwegen waren geblokkeerd, er veel te veel bezoekers waren toegelaten en de kerstversiering niet voldeed aan de brandveiligheidseisen. Maar ook de gemeente droeg schuld. Zij had onvoldoende toezicht uitgeoefend op de naleving van de brandveiligheidsvoorschriften. In 1998 al stelde de Commissie Michiels vast dat de overheid ernstig tekort schoot in het handhaven van de door haar gestelde regels. In reactie hierop hebben gemeenten de afgelopen jaren veel geïnvesteerd in het professionaliseren van hun handhavingsbeleid. Met name voor de handhaving van de brandveiligheid is er extra personeel aangesteld om vergunningen te verlenen en controles uit te voeren.

Vraagstelling

Dit proefschrift probeert de effectiviteit en efficiëntie te bepalen van deze handhavingsacties. De centrale vraag is: Wat is een effectief en efficiënt handhavingsbeleid voor de brandveiligheid in de horeca? De handhaving van de brandveiligheid kan in beginsel op drie manieren worden uitgevoerd: privaatrechtelijk, bestuursrechtelijk of strafrechtelijk. In deel I analyseer ik op grond van de economische literatuur onder welke voorwaarden de verschillende handhavingsmethodes effectief zijn in het vergroten van de naleving en efficiënt in het terugdringen van de totale kosten van naleving en handhaving. Meer in het bijzonder presenteer ik een overzicht van de economische literatuur over het gebruik van informele, coöperatieve handhavingsacties. Onder welke voorwaarden zijn advies, overtuiging en waarschuwing effectiever dan bestraffen?

Het afwegen van de kosten en baten, het bepalen van de relevante omstandigheden en restricties, het vormgeven van de juiste mix van handhavingsmethodes e.d. is uiteindelijk een empirische kwestie waarvoor concrete data nodig zijn. In deel II van dit proefschrift analyseer ik de handhaving van de brandveiligheid in de horeca in Nederland. Horeca is in dit proefschrift gedefinieerd als restaurants, cafés, bars, discotheken, partycentra e.d., en omvat geen hotels en ook geen snackbars of kleine afhaaltentjes. De horecabaas moet voldoen aan de gebruiksvoorschriften voor de brandveiligheid, zoals het vrijhouden van (nood)uitgangen en vluchtwegen, het correct ophangen van versiering, het goed laten functioneren van de vluchtwegaanduiding, blusmiddelen, etc., en het beperken van het aantal bezoekers tot het maximum toelaatbaar aantal personen.

Achtergrond

Dit proefschrift analyseert het handhavingsbeleid vanuit een economisch perspectief in navolging van Becker (1968). Die heeft zich gebogen over een verstandige inzet van de middelen aan criminaliteitsbestrijding, waarbij de kosten van handhaving worden afgewogen tegen de bate van de afschrikking die daarmee bereikt wordt. Een risiconeutraal persoon wordt afgeschrikt van het plegen van een delict door een verwachte boete die minstens zo groot is als zijn persoonlijk gewin. Als opsporing geld kost en het opleggen van een sanctie niet of nauwelijks (zoals bij een boete), dan worden de handhavingskosten geminimaliseerd door een zo hoog mogelijke sanctie op te leggen (beperkt door bijvoorbeeld het vermogen van de crimineel) en een (zo laag mogelijke) pakkans die voldoende afschrikking genereert.

Deze analyse is in de literatuur ook toegepast op de handhaving van bestuursrechtelijke regelgeving. Denk bijvoorbeeld aan de regulering van markten, het milieu, arbeidsomstandigheden, consumentenveiligheid, etc. Desondanks is de analyse van bestuursrechtelijke handhaving beperkt van aard. De bestaande literatuur bestudeert voornamelijk de strafrechtelijke handhaving van de zwaarste overtredingen van de regels, of op zijn minst de bestraffing door middel van boetes. Maar de kleinere, minder serieuze overtredingen maken het leeuwendeel uit van de overtredingen. En (strafrechtelijke) sancties worden vaak voorafgegaan door pogingen om de overtreding ongedaan te maken en/of te beëindigen.

Het onderzoek naar bestuursrechtelijke handhaving verdient ook aandacht van de economische literatuur omdat deze wijze van handhaving op het eerste gezicht in tegenspraak is met het standaard economische model. Het standaard model benadrukt het belang van het bestraffen van bedrijven om hen af te schrikken van het plegen van een overtreding. Echter, in Nederland en diverse andere West-Europese landen zijn bestuursorganen als gemeenten niet bevoegd om boetes op te leggen, laat staan gevangenisstraf. De belangrijkste sancties zijn de last onder dwangsom, de last onder bestuursdwang en de intrekking van een vergunning. Dergelijke sancties bepalen dat een bedrijf in de gelegenheid wordt gesteld om de overtreding te beëindigen en de schade te herstellen. Als het bedrijf daar niet aan voldoet, wordt overgegaan tot executie van de dwangsom, bestuursdwang of intrekking van de vergunning. In de praktijk gaan er vaak diverse (informele) waarschuwingen aan deze sancties vooraf. Vanuit economisch perspectief lijken deze praktijken het afschrikkende effect van handhaving te schaden. Echter, in de niet-economische literatuur wordt vaak beargumenteerd dat een dergelijke handhavingsstijl, ook wel compliance strategy genoemd, effectiever is dan de deterrence strategy die door economen wordt voorgestaan. In een compliance strategie handelt de handhaver niet als een 'politieman', maar als een 'consultant'. Handhaving vindt niet plaats door bestraffing maar door het individu of het bedrijf te overtuigen van het belang van naleving en door te onderhandelen over de geëiste voorzorgsmaatregelen. Naleving wordt bereikt door coöperatief gedrag van zowel de handhaver als het bedrijf. Dit coöperatieve gedrag wordt bedreigd door het opleggen van (strenge) straffen. Formele sancties worden gereserveerd voor hardnekkige overtreders. Een compliance strategie veronderstelt niet dat er rationele keuzes worden gemaakt, maar erkent dat overtredingen het gevolg zijn van foutieve inschattingen, onwetendheid en/of incompetentie. De meeste mensen zijn ten principale bereid tot naleving, maar incompetent om dit te realiseren of van mening dat in bepaalde gevallen naleving onredelijk is. In dit proefschrift wordt gezocht naar een economische verklaring voor deze handhavingspraktijken. Waarom kiezen handhavers voor een coöperatieve handhavingsstijl en leidt dat inderdaad tot een hoger nalevingsniveau? Wat zijn de effecten van bestuursrechtelijke sancties op de naleving?

SAMENVATTING DEEL I - THEORIE

Waarom handhaving nodig is

Hoofdstuk 2 beschrijft het handhavingsprobleem. De maatschappelijke doelstelling van handhaving is de optimale preventie van ongevallen, oftewel het creeëren van een prikkel tot het nemen van voorzorgsmaatregelen. Die voorzorgsmaatregelen uiten zich in het naleven van de gebruikseisen.

Bij afwezigheid van handhaving zullen horecaondernemers uit zichzelf onvoldoende voorzorgsmaatregelen treffen. Horecabezoekers beschikken over te weinig informatie om bij hun locatiekeuze rekening te kunnen houden met de mate waarin de ondernemer de brandveiligheidsvoorschriften naleeft. Ook andere reputatie-effecten van (niet-)naleving zijn nauwelijks waarneembaar. Het is niet waarschijnlijk dat de intrinsieke motivatie en bereidheid tot naleving onverkort overeind blijft als er geen enkele publieke handhaving van de regels is. Afwezigheid van handhaving stimuleert een cultuur waarin niet-naleving onbelangrijk wordt gevonden. Verzekeringsmaatschappijen zijn primair geïnteresseerd in het beperken van het materiële verlies voor de ondernemer, en niet in het (immateriële) leed voor de bezoekers, en zijn niet in staat om alle voorzorgsmaatregelen onder controle te houden. Tot slot, horecaondernemers onderschatten de mogelijke schade omdat de informatie daarover complex van aard is en vanwege cognitieve beperkingen moeilijk te verwerken.

Aansprakelijkheid, private handhaving

Private handhaving door slachtoffers op grond van de aansprakelijkheidsregels is alleen effectief als de daders daadwerkelijk gedwongen worden de schade die zij veroorzaken in hun beslissing te internaliseren. Daarom gelden de volgende voorwaarden:

1. De schade bestaat (vooral) uit materiële schade. In geval van immateriële schade moeten rechters in staat en bereid zijn om deze schade op adequate wijze te compenseren.
2. De dader moet over voldoende vermogen beschikken om de schade te kunnen vergoeden, dan wel (volledig) zijn verzekerd, waarbij verzekeringsmaatschappijen de voorzorgsmaatregelen van de horecaondernemer kunnen controleren.
3. Rechters en daders moeten over voldoende informatie beschikken over de mogelijke schade, zodat rechters het efficiënte voorzorgsniveau kunnen handhaven en daders in staat zijn te anticiperen op de jurisprudentie.
4. Slachtoffers moeten voldoende toegang hebben tot de rechter, zodat de kans op schadevergoeding voldoende groot is. Slachtoffers moeten de dader gemakkelijk kunnen aanwijzen en de kosten van de rechtsgang moeten voldoende laag zijn.

Publieke handhaving: Een compliance of een deterrence strategie?

In hoofdstuk 3 tot en met 10 analyseer ik publiek handhavingsbeleid. De suggestie die uitgaat van de analyse van Becker (1968) is dat hoge verwachte straffen leiden tot een hoog nalevingsniveau. In de praktijk lijkt er sprake te zijn van wat (soms) bekend staat als de *Harrington Paradox* (naar Harrington, 1988): ondanks een lage pakkans, lage veroordelingskans en bescheiden sancties, lijkt het nalevingsniveau redelijk hoog. Ik analyseer welke argumenten de economische literatuur aandraagt voor de afwezigheid van sancties, of in elk geval van onmiddellijke oplegging van sancties. Wat is de waarde van een compliance strategie?

In de discussie over compliance en deterrence strategieën lopen vaak allerlei verschillende onderscheidingen door elkaar heen. Het kenmerkende verschil tussen beide strategieën is het verschil in de handhavingsstijl, waarmee bedoeld is de vraag hoe een handhaver reageert op de ontdekking van een overtreding (hoofdstuk 8). Een deterrence strategie is een stijl waarbij elke overtreding onmiddellijk wordt bestraft. Een compliance strategie is een coöperatieve stijl waarbij de handhaver bepaalde overtredingen (tijdelijk) kan gedogen om zo het bedrijf een prikkel te geven noodzakelijke informatie aan de handhaver bekend te maken. Welke handhavingsstijl optimaal is, hangt niet af van de rationaliteit, de moraliteit of de bereidheid tot naleving van de ondernemer. In dit proefschrift laat ik zien dat compliance strategieën ook voor rationele en/of winstmaximaliserende ondernemers efficiënt kunnen zijn. Aan de andere kant: de veronderstelling dat ondernemers morele overtuigingen hebben, is onvoldoende voor de stelling dat een deterrence strategie niet werkt. Het is van belang eerst het doel te onderscheiden dat met handhaving bereikt moet worden (hoofdstuk 10). Vervolgens is de vraag met welke handhavingsstijl dit doel het beste bereikt kan worden. Welke optimaal is, hangt af van de omstandigheden en kan niet in zijn algemeenheid omschreven worden.

Het doel van afschrikking

Bij afschrikking (de focus van de meeste economische literatuur) is het doel om naleving te bevorderen door overtreding onaantrekkelijk te maken. De dreiging met sancties geeft het signaal dat overtreding niet loont. Als afschrikking het doel is, dan is primair een deterrence strategie aangewezen. Het afschrikkende effect van handhaving is wezenlijk afhankelijk van de verwachting van potentiële overtreders dat een geconstateerde overtreding consequent wordt vervolgd en bestraft. Bij een compliance strategie is de ondernemer in de gelegenheid om naleving uit te stellen of een lager nalevingsniveau 'uit te onderhandelen'.

De wet is incompleet
Als de wet compleet is en in zijn toepassing perfect voorspelbaar, is een strikt handhavingsbeleid efficiënt. De regels kunnen dan letterlijk en tamelijk bureaucratisch worden gehandhaafd. In werkelijkheid is de regelgeving noodzakelijkerwijs incompleet (hoofdstuk 6). Open normen laten onduidelijkheid bestaan. Specifieke standaarden zijn nooit volledig. De regels bevatten zowel voorschriften die te streng zijn als voorschriften die niet ver genoeg gaan. Een deterrence strategie waarbij de wettelijke regels precies en strikt worden gehandhaafd, zal daarom niet leiden tot het optimale niveau van voorzorgsmaatregelen. Een vorm van compliance strategie kan dan effectief en efficiënt zijn. Twee kanttekeningen zijn daarbij van toepassing. Als de handhavers over voldoende discretionaire bevoegdheden en over voldoende informatie beschikken, dan kunnen zij per geval de juiste standaard bepalen en deze vervolgens strikt handhaven (hoofdstuk 6). Bovendien, (de plicht tot) een strikte handhaving van de standaard, dwingt de wetgever onnodig onduidelijke regelgeving te vermijden (hoofdstuk 9).

Het opleggen van sancties brengt kosten met zich mee
Zelfs als met zekerheid gesteld kan worden dat een overtreding inefficiënt is, kan een deterrence strategie contraproductief zijn omdat het opleggen van sancties geld en andere kosten met zich meebrengt (hoofdstuk 8). Daarbij gaat het uiteraard om de directe kosten van het opleggen en uitvoeren van een straf. Maar ook om de kosten en gevolgen van het veroordelen van onschuldigen en van het ontkennen of verbergen van overtredingen (par. 6.2)

of van het uithollen van de intrinsieke nalevingsbereidheid (par. 7.4). Een deterrence strategie is ook kostbaar als niet alle overtredingen gelijk zijn en het bestraffen van elke geconstateerde overtreding de totale afschrikking vermindert. Het verhogen van de sanctie voor lichte overtredingen is kostbaar als dit de afschrikking voor zwaardere overtredingen vermindert (marginale afschrikking, par. 3.2). En de sanctie voor zelf-gemelde overtredingen moet lager zijn dan die voor niet-gemelde (par. 3.4). Ook is in situaties van serieuze onderafschrikking het optimale handhavingsbeleid gebaseerd op de nalevingsgeschiedenis (par. 4.1 en 5.3).De handhaving is dan streng voor degenen met een slechte geschiedenis en/of degenen die een waarschuwing hebben gehad, maar relatief licht voor de anderen. Het verhogen van de sanctie voor de laatste groep vermindert de afschrikking voor de eerste groep.

Een deterrence strategie kan leiden tot hoge sanctiekosten omdat elke overtreding moet worden bestraft. Maar dit probleem ontstaat alleen als het handhavingsbudget te klein is om volledige naleving af te dwingen (zie bijvoorbeeld de paragrafen 7.5, 5.3, 4.1). Bij volledige naleving zijn er geen kosten van het opleggen van sancties. Als er geen maximum is voor de hoogte van de sanctie, kan er altijd volledige naleving worden afgedwongen met een minimale inzet aan handhavingsbudget langs de tradeoff van Becker (1968). Als de maximale sanctie wel gelimiteerd is, dan moet de pakkans voldoende groot gemaakt worden om afschrikking te realiseren. En dus is er voldoende handhavingsbudget nodig. Als de combinatie van de sanctie en het budget onvoldoende is om volledige naleving te realiseren, kan een compliance strategie de onderafschrikking verminderen. De handhavingsautoriteit profiteert als de regels beter worden nageleefd. Bedrijven profiteren als zij lagere sancties krijgen opgelegd of soms 'mogen' overtreden. Bij volledige naleving is een dergelijke tradeoff niet mogelijk.

Gebrek aan afschrikkingsbelang bij handhavers

Een deterrence strategie kan ook mislukken omdat handhavers onvoldoende belang hebben bij het realiseren van afschrikking en in het consequent bestraffen van elke overtreding (hoofdstuk 9). Dat kan leiden tot geloofwaardigheidsproblemen. Als de handhaver of de handhavingsautoriteit onvoldoende belang heeft bij het afschrikken van toekomstige overtredingen, zal het geen middelen willen opofferen voor de huidige handhavingszaken. Het kan ook zijn dat handhavers meer geïnteresseerd zijn in hun eigen carrière of in hun eigen (gemoeds)rust dan in het daadwerkelijk aanpakken van overtredingen.

Alternatieven voor een deterrence strategie

Als een deterrence strategie er onvoldoende in slaagt om overtredingen af te schrikken, zijn er diverse alternatieve handhavingsmogelijkheden, die in meer of mindere mate lijken op een compliance strategie.

1. Als slechts een gedeeltelijke naleving haalbaar is, is het optimaal om de handhavingsinzet te richten op een subgroep van de potentiële daders en de overtredingen in de resterende groep te gedogen. Daardoor zal ten-

minste deze eerste subgroep de regels volledig naleven. De beste aanpak is die waarbij de handhaving is gebaseerd op de nalevingsgeschiedenis, bijvoorbeeld door middel van waarschuwingen (paragraaf 5.3, 4.1 en 4.2). Bij een eerste overtreding volgt een waarschuwing waarna er sprake is van strikt toezicht op de naleving.

2. Als de wettelijke standaard niet gehandhaafd kan worden, is het doorgaans optimaal om de standaard naar beneden aan te passen en deze informele standaard te handhaven (paragraaf 6.2 en 5.2). Elk bedrijf leeft dan de regels gedeeltelijk na. Er is geen sprake van een strikte handhaving van de letter van de wet, maar wel is het model van Becker (1968) toepasbaar op het handhaven van de informele standaard.
3. Als er ten gevolge van incomplete regelgeving wederzijdse onzekerheid bestaat over de regels en de nalevingsmogelijkheden, dan kan een coöperatieve, flexibele handhavingsstijl efficient zijn. Van het bedrijf wordt geëist dat deze de grootste veiligheidsproblemen aanpakt in ruil voor lagere sancties op lichte vergrijpen. Deze flexibiliteit is wederzijds voordelig (paragraaf 6.1). Het bedrijf zal proberen de handhaver ervan te overtuigen dat naleving onredelijk is. De handhaver zal het bedrijf ervan proberen te overtuigen dat niet-naleving niet loont.

De andere handhavingsdoelen

Naast afschrikking kunnen andere handhavingsdoelen worden onderscheiden. Ook hiervoor geldt dat het mogelijk is om te bepalen onder welke omstandigheden een deterrence of een compliance strategie voor dit doel gewenst is.

Herstel

Herstel houdt in dat handhaving gericht is op het ongedaan maken van de overtreding, bijvoorbeeld door het repareren van een technisch mankement of het opruimen van de schade (zie hoofdstuk 4). Als het mogelijk is om onvoldoende herstel te observeren en te bestraffen, dan is een deterrence strategie efficiënt die sancties oplegt die afhankelijk zijn van de mate van herstel. Dit geeft het bedrijf de juiste prikkel om in herstel te investeren in plaats van af te wachten totdat de handhaver er iets van zegt. Bovendien geeft het de juiste prikkel om te investeren in het voorkómen van mankementen. Alleen als er sprake is van volledig onvoorzienbare mankementen én sancties niet afhankelijk gemaakt kunnen worden van de mate van het uitgevoerde herstel, is het beter om te beginnen met waarschuwingen. Maar het is moeilijk om daarbij een voorbeeld te bedenken.

Advies

Advies houdt in dat handhaving gericht is op het informeren van het bedrijf over alle relevante baten en kosten van naleving, de regels die van toepassing zijn en de beste manier om aan de regels te voldoen. Als de ondernemer in staat is om de relevante informatie tegen redelijke kosten te verzamelen, geeft het opleggen van de juiste sancties de ondernemer de efficiënte prikkel om informatie te verzamelen. Wanneer de ondernemer hiertoe niet in staat is, of alleen tegen hoge kosten, zoals bij technisch ingewikkelde regelgeving, is advies van belang. Als alle bedrijven min of meer dezelfde informatieproblemen hebben, dan zijn algemene voorlichtingscampagnes voldoende om bedrijven van informatie te voorzien. Als de regels te incompleet zijn, zodat de geëiste voorzorgsmaatregelen per bedrijf te veel verschillen, dan is het wellicht efficiënt als inspecties gebruikt worden om bedrijven van advies te voorzien. Dit vereist een soort compliance strategie waarin een inspectie – in elk geval in eerste instantie – wordt gekenmerkt door uitleg, overleg en samenwerking.

Opvoeding
Handhaving kan ook gericht zijn op het stimuleren van normen en waarden die de naleving vergroten. Publieke handhaving kan nodig zijn om het normbesef te handhaven, te versterken of zelfs te creëren (hoofdstuk 7). Om de kosten van informele, sociale sancties (reputatieschade) te realiseren, kan inspectie door overheidsorganen nodig zijn. De publieke bestraffing kan dan relatief laag blijven, maar over het algemeen moeten overtredingen bestraft worden, en wel meteen. Publieke handhaving kan nodig zijn om te voorkomen dat het geloof in de norm, en daarmee de bereidheid tot naleving, afneemt. Als goedwillende bedrijven zien dat deze overtredingen niet bestraft worden, zullen ze gaan denken dat naleving blijkbaar niet belangrijk gevonden wordt en zal hun geloof in de norm afbrokkelen. Daarvoor kan een deterrence strategie nodig zijn gericht op het bestraffen van overtreders, of een meer coöperatieve handhavingsstijl gericht op het uitspreken van waardering richting de nalevers.

Completering
Een ander doel van publieke handhaving is het aanvullen van private handhaving. Publieke handhaving kan gebruikt worden om de onzekerheid over de vraag wat 'voldoende' voorzorg is weg te nemen of om bedrijven en hun werknemers met hun cognitieve beperkingen bewust te maken van de gevolgen van private handhaving. Hierbij heeft publieke handhaving advies als doel, zoals boven beschreven. Ook is publieke handhaving nodig om private handhaving aan te vullen in het creëren van een efficiënt niveau van afschrikking. Private handhaving kan gehinderd worden door bijvoorbeeld beperkingen aan het vermogen of barrières voor slachtoffers. In deze gevallen is afschrikking het doel. In het algemeen is een strikt handhavingsbeleid nodig om de bedrijven die middels het privaatrecht worden onderafgeschrikt, te dwingen tot meer voorzorgsmaatregelen.

SAMENVATTING DEEL II – EMPIRISCHE ANALYSE

Publieke handhaving tussen 2001 en 2007

Handhavingstekorten en actieprogramma's
Gedurende de jaren negentig en aan het begin van deze eeuw zijn er serieuze handhavingstekorten vastgesteld op vele gebieden van de bestuursrechtelijke handhaving, als gevolg van wetenschappelijke analyses, overheidsrapporten en ernstige rampen. Vanuit de landelijke overheid werden gemeenten gestimuleerd om hun handhaving te professionaliseren, bijvoorbeeld door het toepassen van 'programmatisch handhaven'. Dit moet garanderen dat handhaving volgens plan verloopt, in plaats van ad hoc en incidentgericht. Verschillende commissies, zoals de commissie Alders, hebben geleid tot actieprogramma's om het handhavingsbeleid en de regelgeving te verbeteren. Deze initiatieven zijn ook van grote invloed geweest op de handhaving van de brandveiligheid door gemeenten (hoofdstuk 11).

Het gemeentelijke handhavingsbeleid brandveiligheid
In 2001 was er een achterstand in het verlenen van gebruiksvergunningen voor meer dan 70% van de gebouwen (IOOV, 2002). Deze achterstand ging gepaard met een personeelstekort voor zowel het verlenen als het handhaven van de vergunningen. In bijna geen enkele gemeente vond systematische en adequate inspectie van de vergunningen plaats. In de meeste gemeenten ontbrak een goede onderbouwing van het beleid. Na 2001 hebben gemeenten fors geïnvesteerd in het handhaven van de brandveiligheid, met name ook in horecagelegenheden. Gemeenten zijn begonnen om de achterstanden in gebruiksvergunningen weg te werken en bedrijven periodiek te inspecteren.

Ik heb interviews afgenomen met handhavingsambtenaren in dertien gemeenten om inzicht te krijgen in de daadwerkelijke handhavingsacties. Ik gebruik deze interviews om te onderzoeken of het beleid van de laatste jaren effectief is in het afdwingen van naleving; en met name of het gebruik van een compliance strategie effectief en efficiënt is volgens de literatuur uit deel I. Zoals beschreven in hoofdstuk 12 worden cafés en restaurants doorgaans jaarlijks gecontroleerd (afgezien van hercontroles). Horecagelegenheden in het centrum of in een uitgaansgebied krijgen vaak ook nog een controle tijdens een speciale gebeurtenis of tijdens de feestdagen of carnaval. Gemeenten beginnen doorgaans met informele waarschuwingen en hercontroles. Formele sancties zoals een dwangsom, bestuursdwang of een (tijdelijke) intrekking van de vergunning worden zelden opgelegd.

Een Harrington Paradox?
Deze waarnemingen betekenen niet dat er sprake is van een Harrington Paradox. Het nalevingsniveau is niet onverklaarbaar hoog in relatie tot de verwachte sanctie. Handhavers vinden vele kleine(re) overtredingen bij een eerste controle. In minder dan 10% van de gevallen wordt er bij een controle

geen enkele tekortkoming geconstateerd. In meer dan 50% van de gevallen kondigen de handhavers hercontrole aan. Dit is een logisch gevolg van het feit dat horecaondernemers niet worden bestraft, maar alleen worden gewaarschuwd dat in de toekomst mogelijk sancties volgen. Het handhavingsbeleid is in overeenstemming met het targeting schema dat Harrington (1988) aanbeveelt. In eerste instantie worden overtreders niet bestraft, maar verplaatst naar een target groep waarin ze strikt worden gemonitord door hercontroles net zolang tot ze (weer) naleven. In reactie kiezen horecaondernemers ervoor om naleving uit te stellen totdat ze gecontroleerd worden. Uiteindelijk slagen gemeenten erin om naleving af te dwingen. Voor horecaondernemers is het van belang om geen grove handelingen na te laten of te plegen, waarvan iedereen weet dat deze een brandgevaarlijke situatie creëren. In dergelijke gevallen is er de geloofwaardige dreiging van voldoende grote sancties (mogelijk strafrechtelijk) zodat horecaondernemers hiervan worden afgeschrikt. Hierdoor komt de daadwerkelijke oplegging van sancties nauwelijks voor.

Een compliance strategie?
De handhavers van de brandveiligheid in de horeca zelf geven aan dat zij gebruikmaken van advies, overtuiging en waarschuwingen (een compliance strategie) en betogen zij dat dit ook de beste aanpak is. Ze beweren (1) dat inspecties nodig zijn om aan de horecaondernemer uit te leggen wat er mis is en wat het gevaar is van niet-naleving (vanwege onwetendheid); (2) dat inspecties nodig zijn om overtredingen te beëindigen en daarmee de brandveiligheid te herstellen (en niet zozeer om overtredingen te bestraffen); (3) dat inspecties nodig zijn om de horecaondernemer ervan te overtuigen dat naleving gewenst is (omdat de meeste horecaondernemers bereid zijn tot naleving); (4) dat uitleg en samenwerking effectiever zijn dan sancties (vanwege de juridische procedures daarvan); (5) dat ze zich flexibel opstellen zolang de ondernemer dat ook doet; en (6) dat inspecties maar een momentopname zijn en dat brandveiligheid vooral de eigen verantwoordelijkheid is van de ondernemer.

Nader onderzoek naar deze beweringen aan de hand van de literatuur uit deel I (hoofdstuk 12) laat zien dat deze argumenten over het algemeen niet juist zijn. Sterker nog, de meeste houden in dat in de huidige situatie meer nadruk zou moeten liggen op een deterrence strategie. De genoemde beweringen maken onvoldoende onderscheid tussen het doel van inspecties en de handhavingsstijl. Dat horecaondernemers bijvoorbeeld bereidwillig zijn of dat ze overtreden uit onwetendheid of per ongeluk, is op zichzelf onvoldoende reden om een compliance strategie te kiezen. Binnen de (juridische) mogelijkheden is de gemeente in staat om strikter te handhaven door sneller over te gaan tot het opleggen van formele sancties in plaats van te (blijven) waarschuwen. Tijd kan bespaard worden door minder nadruk te leggen op advies, uitleg en overtuiging tijdens een (her)controle.

Een kosten-batenanalyse

In hoofdstuk 13 onderzoek ik de baten en kosten van de intensivering van het handhavingsbeleid in horecagelegenheden na / in reactie op de Volendam-ramp. De jaarlijkse kosten van de additionele handhavingsacties en de additionele naleving worden geschat op €13,0 tot €23,7 miljoen. Het bedrag dat maximaal bespaard kan worden, is de gemiddelde totale schade door brand in de horeca voorafgaand aan de beleidsintensivering. Deze schade kan geschat worden op €30,4 miljoen per jaar. Om tot een positief saldo van baten en kosten te komen is dus een reductie in de verwachte schade vereist van minstens 43%. Deze 43% is een ondergrens, uitgaande van de laagste schatting van de kosten van naleving en handhaving. De minimaal vereiste effectiviteit kan oplopen tot 78% als de werkelijke kosten hoger zijn. Het is onwaarschijnlijk dat de verwachte schade van brand met meer dan 40% is gedaald.

Conclusies over de actieprogramma's

De analyse van het gemeentelijke handhavingsbeleid laat zien dat een goede onderbouwing van het huidige beleid ontbreekt. Er wordt uitgebreid gefilosofeerd over taken, organisatie, coördinatie en inzet van fte's, maar in het midden blijft wat nu precies het beoogde doel is. Zelfs van een begin van een afweging van middelen en doelen, en van kosten en baten, is dan ook geen sprake. Ondanks het (inmiddels) wijdverspreide gebruik van programmatisch handhaven lijkt het brandveiligheidsbeleid primair een reactie op de Volendam-ramp. Introductie van het Gebruiksbesluit direct na de Volendam-ramp zou bijvoorbeeld €100 miljoen hebben bespaard. Nu worden gebruiksvergunningen afgeschaft juist op het moment dat we een grote inspanning hebben geleverd om alle gebouwen van een vergunning te voorzien. Het is opvallend dat juist het actieprogramma van de commissie Alders leidt tot zoveel beleidsdocumenten en ander bureauwerk. Aandacht voor daadwerkelijke handhavingsacties gaat verloren aan beleidsontwikkeling, gemeentelijke reorganisaties en het afgeven van vergunningen.

Evaluatie van de schadeclaims van de Volendam-slachtoffers

In hoofdstuk 14 onderzoek ik de afwikkeling van de schadeclaims van de slachtoffers van de Volendam-ramp. Deze casestudy is waardevol omdat de afwikkeling van de Volendam-ramp relatief veel aandacht heeft getrokken en daarom waarschijnlijk ook van grote invloed is op de perceptie van horecaondernemers en andere betrokkenen.

De slachtoffers van de Volendam-ramp hebben naar schatting financiële ondersteuning gekregen ter grootte van €60 miljoen, grotendeels betaald door de Rijksoverheid. De horecabaas Veerman heeft hooguit ruim €3 miljoen betaald. De totale schade voor de slachtoffers, inclusief de immateriële schade, kan geschat worden op bijna €120 miljoen. Er is geen goede reden waarom slachtoffers niet geprobeerd hebben een grotere schadevergoeding van Veerman te krijgen. Blijkbaar verwachten zij geen hogere schadevergoedingen meer omdat de overheid daar al ruimschoots in voorzien heeft, daarbij gestimuleerd door de Volendamse gemeenschap om de zaak af te sluiten. Het gevolg is echter dat het signaal richting horecaondernemers is dat zij zich niet zoveel zorgen hoeven te maken over het betalen van de schade van brand in hun gelegenheid. De financiële rekening van de Volendam-ramp wordt grotendeels betaald door de belastingbetaler. Bovendien, omdat de immateriële verliezen grotendeels niet gecompenseerd worden, hebben zowel de horecaondernemers als de overheid geen goede prikkel om deze verliezen te voorkomen.

Handhaving in een representatieve gemeente

De analyse van bestaande handhavingsactiviteiten laat zien dat er belangrijke tekortkomingen zijn. De vraag is of andere beleidsopties effectiever en efficienter zijn. In hoofdstuk 15 verzamel en combineer ik de beschikbare gegevens om de jaarlijkse kosten van handhaving in een representatieve, hypothetische gemeente te onderzoeken. In deze gemeente worden de laagste kosten van verwachte schade en nalevingskosten behaald bij een nalevingsniveau van 113 (op een schaal van 0 tot 200) met totale kosten van €176.800 (nalevingskosten €57.500 en verwachte schade €119.400). Ik simuleer de effecten van diverse handhavingsmethodes onder de veronderstelling dat de horecaondernemer een risiconeutrale winstmaximaliseerder is. Op basis van deze veronderstelling kiezen alle ondernemers ervoor om de regels niet na te leven als er geen handhaving is. Er zijn dan geen kosten van handhaving of naleving, maar de verwachte schade is tamelijk hoog: €663.300. Alle drie de basale handhavingsmethodes zijn in staat om een beter resultaat te realiseren, als het beleid tenminste goed wordt uitgevoerd.

De analyse van aansprakelijkheidsclaims door slachtoffers laat zien dat via private handhaving een redelijk maatschappelijk welvaartsniveau bereikt kan worden: de totale kosten variëren van €176.800 tot €191.600. De handhavingskosten zijn beperkt omdat handhaving alleen plaatsvindt als er brand is geweest. Slachtoffers hebben voldoende toegang tot de rechter. Of private handhaving werkelijk tot efficiëntie leidt hangt af van (1) een adequate vergoeding die op de volledige schade is gebaseerd, (2) voldoende vermogen bij horecaondernemers, en (3) efficiënte toepassing van het criterium van 'voldoende' voorzorgsmaatregelen door de rechters.

Het is ook mogelijk om de verwachte schade te reduceren door middel van bestuursrechtelijke handhaving. Echter, het nadeel van deze aanpak is dat horecaondernemers naleving zullen uitstellen totdat er de werkelijke dreiging van executie van sancties is. Bestuursrechtelijke handhaving vereist daarom aanzienlijke handhavingskosten en resulteert in totale kosten van €260.000 tot €331.500.

Strafrechtelijke handhaving van de brandveiligheid is een goed alternatief. Met relatief lage handhavingskosten kunnen goede resultaten worden geboekt (totale kosten €176.800 tot €223.700). Hogere boetes maken het mogelijk het aantal inspecties te verminderen. In elk geval, boetes moeten hoog genoeg zijn om naleving af te dwingen (hier: minstens €250). Anders is het verstandig om helemaal af te zien van handhaving of om naleving te bereiken door niet elke overtreding onmiddellijk te bestraffen.

De analyse laat zien dat er een voorkeur is voor schadegerelateerde handhaving (ex-post), zowel met behulp van privaatrecht als van strafrecht. Een strafrechtelijke handhaving is daarbij het beste alternatief als het vermogen van de horecaondernemer onvoldoende is en vervangende gevangenisstraf nodig is. De andere problemen met schadegerelateerde handhaving zijn niet zo groot of kunnen opgelost worden door aanvullend een beetje te investeren in bestuursrechtelijke handhaving ex-ante. De rol van bestuursrechtelijke handhaving is vooral informerend en aanvullend. De voornaamste prikkel tot naleving kan worden gerealiseerd door middel van ex-post handhaving.

HANDHAVING VAN DE BRANDVEILIGHEID IN DE HORECA

Het optimale handhavingsbeleid

De hoofdvraag van dit proefschrift is: Wat is een effectief en efficiënt handhavingsbeleid voor de brandveiligheid in de horeca? Mijn antwoord op deze vraag laat zich in vijf punten samenvatten.

Ten eerste: enige vorm van handhaving is nodig om horecaondernemers te stimuleren tot het nemen van voorzorgsmaatregelen. Zonder handhaving zullen zij kiezen voor onvoldoende naleving (hoofdstuk 2, 12, 15). De prikkel die uitgaat van bezoekers, verzekeringsmaatschappijen of ondernemers zelf, is onvoldoende.

Ten tweede: om horecaondernemers te stimuleren tot voorzorgsmaatregelen zijn publieke middelen nodig, maar wel in beperkte omvang. Publieke middelen zijn nodig om de horecaondernemers te informeren over het belang van brandveiligheid en uit te leggen welke regels gelden (hoofdstuk 12, 15). Dit geld moet niet aan inspecties worden besteed, maar vooral aan voorlichtingscampagnes om horecaondernemers bewust te maken van bra ndveiligheid(svoorschriften). Bovendien, gelet op een doelmatige besteding van de publieke middelen is een beperkte investering in brandveiligheid in de horeca verstandig (hoofdstuk 13).

Ten derde: het handhavingsbeleid moet bestaan uit een slimme combinatie van alle handhavingsmethodes (hoofdstuk 14, 15). Ex-ante is er actie nodig om horecaondernemers bewust te maken van brandveiligheid. Bestuursrechtelijke handhaving ex-ante is nodig voor die ondernemers die de regels ernstig of vaak blijken te overtreden. De horecagelegenheden van deze ondernemers kunnen gesloten worden door de gemeente op het moment dat zij signalen krijgt van bijvoorbeeld andere handhavers (milieu, openbare orde, etc.). Daarnaast kan de gemeente gezamenlijke handhavingsacties organiseren waarbij brandveiligheid onderdeel is van een (vlugge) inspectie.

Ten vierde: afschrikking kan met name gerealiseerd worden door ex-post handhaving met een combinatie van privaatrechtelijke en strafrechtelijke handhaving (hoofdstuk 15). Private handhaving werkt in principe goed, omdat slachtoffers de dader kennen en voldoende schadevergoeding verwachten om de kosten van het indienen van een claim te rechtvaardigen, en omdat de meeste horecaondernemers een substantieel deel van de schade kunnen betalen. Aanvullende gevangenisstraf middels het strafrecht is nodig als de ondernemer niet in staat is te betalen of als het aandeel van de immateriële verliezen te groot is.

Ten slotte is het goed om expliciet op te merken dat de handhavingsstijl voor publieke handhaving (zowel ex-ante als ex-post) primair een deterrence strategie is (hoofdstuk 12). Dit volgt uit een analyse van de genoemde doelen van handhaving. Een compliance strategie geeft de horecaondernemer de gelegenheid om met de naleving te wachten totdat er inspectie plaatsvindt of zelfs tot er daadwerkelijk sancties worden aangekondigd. Dit blijkt met name uit het uitstellen van het herstel van mankementen aan de technische installaties. Voor advies zijn algemene voorlichtingscampagnes voldoende. Een strikt handhavingsbeleid geeft de horecaondernemer een prikkel om zichzelf te informeren over brandveiligheidsregels. Bij opvoeding is het probleem dat de huidige compliance strategie een cultuur stimuleert waarbij naleving niet belangrijk wordt gevonden en tot de verantwoordelijkheid van de overheid wordt gemaakt. Wel dient een deterrence strategie zodanig uitgevoerd te worden dat de motivatie van de bereidwillige horecaondernemers om de regels na te leven niet wordt ondermijnd. Ook voor de aanvullende rol van (bestuursrechtelijke) handhaving is vooral een deterrence strategie nodig als het gaat om het ingrijpen bij horecaondernemers die middels privaatrechtelijke en strafrechtelijke handhaving onvoldoende worden afgeschrikt. Een kanttekening bij het gebruik van een deterrence strategie is van belang. Gegeven de beperkte middelen en sancties moet de gemeente zijn inspecties en sancties richten op de meest ernstige overtredingen en op herhaalde overtredingen. Dat kan door het handhaven van een informele standaard en (met name bij ex-ante handhaving) door te beginnen met een waarschuwing.

Opmerkingen bij het handhaven van de brandveiligheid in de horeca

In het huidige beleid speelt bestuursrechtelijke handhaving ex-ante een centrale rol. Die uit zich in de frequentie en relatief lange duur van inspecties. Er zijn echter twee problemen met deze inspecties. Ten eerste, deze inspecties doen een relatief groot beroep op de schaarse publieke middelen (hoofdstuk 13 en 15). Het is efficiënter om te vertrouwen op ex-post handhaving en/of te accepteren dat er brandschade ontstaat, door het terugdringen van de intensiteit van de inspecties. Ten tweede, deze inspecties en bijbehorende sancties creëren weinig afschrikking vanwege hun informele, herstellende karakter (hoofdstuk 12).

Het huidige beleid kan verder verbeterd worden door de juridische mogelijkheden uit te breiden (hoofdstuk 12 en 15), bijvoorbeeld door toepassing van hogere boetes of van bestuursrechtelijke boetes. Dergelijke mogelijkheden vergroten wel het gevaar dat de handhavers hun discretionaire bevoegdheden gebruiken voor hun eigen belangen. Bovendien, een kleine boete van bijvoorbeeld €100 werkt alleen maar contraproductief, omdat deze te laag is om substantieel bij te dragen aan de naleving. Een optie die wel veelbelovend is om de naleving te vergroten, is het vergroten van de vergoeding voor immaterieel leed. Dit verbetert de werking van het privaatrecht.

Uiteraard zijn de beleidsaanbevelingen beperkt door de reikwijdte van het onderzoek. De beschrijving van het effectieve en efficiënte beleid is van toepassing op de horeca en kan niet zomaar worden toegepast op andere sectoren. Voor andere sectoren is mogelijk een ander soort beleid efficiënt. De vraag is echter of verschillen in handhavingsbeleid haalbaar en geloofwaardig zijn. Verder is er slechts beperkte aandacht besteed aan bijvoorbeeld de analyse van een verplichte verzekering tegen brandschade voor horecaondernemers, het gebruik van certificering en private keuringen, de optimale wijze van vergoeding van slachtoffers, de complicaties van principaal-agent problemen in ingewikkeldere horecabedrijven, beperkte aansprakelijkheid, etc. Maar deze zaken zijn in beginsel van ondergeschikt belang. Als het bijvoorbeeld gaat om een verplichte verzekering, dan is deze alleen effectief en efficiënt als de horecaondernemer tevens verplicht wordt werkelijk de geleden schade te vergoeden, met name waar het immaterieel leed betreft.

AFSLUITING

De (basale) economische theorie over handhaving staat in ieder tekstboek. Maar in hoeverre kan deze worden toegepast op actuele handhavingsproblemen? Empirische analyses zijn nodig om ons begrip van naleving en handhaving in de praktijk te verbeteren. Er is nog relatief weinig kennis om de regelovertreding door bedrijven te verklaren. Welke achtergrondkenmerken verklaren het nalevingsgedrag van bedrijven? Hoe groot is de invloed van sancties op de naleving? Hoe beïnvloedt het aantal inspecties de naleving? Voor zulke vragen is dataverzameling en empirische analyse essentieel. We weten inmiddels tamelijk veel over de *output* van handhavingsorganisaties (hoeveel inspecties, sancties etc.), maar weinig over de *outcome*: In hoeverre leidt die output tot betere naleving en tot een lagere verwachte schade? In dit proefschrift heb ik geprobeerd met diverse methodes bij te dragen aan het vergroten van deze kennis. Deze analyses geven geen definitief antwoord op alle vragen. Verder onderzoek is nodig om eenduidige antwoorden te krijgen op de vraag welk nalevingsniveau en daarmee welk niveau van handhaving(skosten) gewenst is. Ik ben ervan overtuigd dat het waardevol is deze analyses verder uit te bouwen. De analyse in dit proefschrift laat zien dat, ondanks belangrijke beperkingen aan de beschikbare data, beleidsmakers in staat zijn om hun handhavingsbeleid beter te onderbouwen. Alhoewel die analyses geen doorslaggevend bewijs opleveren, laten ze wel zien in welke richting de conclusies gaan.

References

References with an asterix (*) are in Dutch.

AEF (Andersson Elffers Felix), 2002*. *Staat van handhaving - Analyse verlening en handhaving vergunningen in 13 casusgemeenten [The enforcement state - Analysis of licenses in 13 municipalities]*. By order of the WODC. Utrecht: Andersson Elffers Felix.

AEF (Andersson Elffers Felix), 2005*. *Inventarisatie bevoegdheden handhavingsorganisaties [Inventory of enforcement powers of enforcement authorities]* (report gj82/004). By order of the Ministry of Justice. Utrecht: Andersson Elffers Felix.

Akerlof, G.A., 1980. A theory of social custom, of which unemployment may be one consequence. *Quarterly Journal of Economics* 94 (4), pp. 749-775.

Alders Committee (in full: Committee of Inquiry Cafe Fire New Year's Eve 2001), 2001*. *Cafebrand Nieuwjaarsnacht. Eindrapport [Cafe Fire New Year's Eve]* (het rapport bestaat uit een eindrapport, bestaande uit de delen A t/m E plus twee deelonderzoeken (A en B) en verder een aparte publieksversie en een Engelse samenvatting). The Hague.

Alexander, C.R., 2004. *Corporate crime, markets and enforcement: a review*. In: Sjögren, H., and G. Skogh (Eds.), pp. 20-41.

Arguedas, C., 2005. Bargaining in environmental regulation revisited. *Journal of Environmental Economics and Management* 50, pp. 422-433.

Arguedas, C., and H. Hamoudi, 2004. Controlling pollution with relaxed regulations. *Journal of Regulatory Economics* 26 (1), pp. 85-104.

Arlen, J., 2000. Tort damages. In: Bouckaert, B., and G. de Geest (Eds.), *Encyclopedia of law and economics* (volume III), pp. 682-734.

AVD (Adviesbureau Van Dijke), 2004*. *Versterking inrichting van het toezicht brandveilig gebruik bouwwerken [Reinforcing the organization of supervising a fire safe use of buildings]*. By order of the Ministry of Internal Affairs. Best: Adviesbureau Van Dijke.

AVV (Adviesdienst Verkeer en Vervoer), 2006*. *Kosten verkeersongevallen in Nederland. Ontwikkelingen 1997-2003 [Costs of traffic accidents in the Netherlands. Trends 1997-2003]*. Rotterdam.

Axelrod, R., 1984. *The evolution of cooperation*. New York: Basic Books, Inc., Publishers.

Ayres, I., and J. Braithwaite, 1992. *Responsive regulation: transcending the deregulation debate*. New York: Oxford University Press.

Baker, M.J., and T.J. Miceli, 2005. Credible criminal enforcement. *European Journal of Law and Economics* 20, pp. 5-15.

Bardach, E., and R.A. Kagan, 1982. *Going by the book: The problem of regulatory unreasonableness*. Philadelphia: Temple University Press.

Barendrecht, M., 2002. Aansprakelijkheid en welzijn [Liability and well-being]. *Nederlands Juristenblad* (12), pp. 605-617.

Bar-Gill, O., and A. Harel, 2001. Crime rates and expected sanctions: The economics of deterrence revisited. *Journal of Legal Studies* 30, pp. 485-501.

Becker, G.S., 1968. Crime and punishment: An economic approach. *Journal of Political Economy* 76 (2), pp. 169-217.

Bekkers, V.J.J.M., et al., 2003*. *Integrale handhaving: een onderzoek naar vormen van integrale handhaving van het beleid ten aanzien van openbare inrichtingen [Intregal enforcement with respect to public establishments]*. Rotterdam: Erasmus Universiteit Rotterdam, Faculteit der Sociale Wetenschappen, Opleiding Bestuurskunde.

Benson, B.L., D.D. Rasmussen, and I. Kim, 1998. Deterrence and public policy: Trade-offs in the allocation of police resources. *International Review of Law and Economics* 18, pp. 77-100.

Benson, B.L., D.D. Rasmussen, and D.L. Sollars, 1995. Police bureaucracies, their incentives, and the war on drugs. *Public Choice* 83, pp. 21-45.

Bergkamp, L., 2001. *Liability and environment. Private and public law aspects of civil liability for environmental harm in an international context.* The Hague: Kluwer Law International.

Bianco, W.T., P.C. Ordeshook, and G. Tsebelis, 1990. Crime and punishment: Are one-shot, two-person games enough? *American Political Science Review* 84 (2), pp. 569-586.

Bijleveld, C.C.J.H., 2006. *Methoden en technieken van onderzoek in de criminologie [Methods of research in criminology]* (2nd edition). Den Haag: Boom Juridische uitgevers.

Bishop, W., 1990. A theory of administrative law. *Journal of Legal Studies* 19, pp. 489-530.

Blomberg, A.B., and F.C.M.A. Michiels, 1997*. *Handhaven met effect. Een empirisch-juridische studie naar de mogelijkheden voor een effectieve handhaving van het milieurecht [Effective enforcement. An empirical-legal study of the possibilities of effective enforcement of environmental law].* 's Gravenhage: VUGA Uitgeverij.

Boadway, R., N. Marceau, and M. Marchand, 1996. Time-consistent criminal sanctions. *Public Finance / Finances Publiques* 51 (2), pp. 149-165.

Boardman, A.E., et al., 2001. *Cost-benefit analysis. Concepts and practice* (2nd edition). Upper Saddle River, New Yersey: Prentice Hall.

Boek, J.L.M., 2006*. Programmatisch handhaven [Programmed enforcement], in: Michiels and Muller (2006*), pp. 207-218.

Boek, J.L.M., and F.C.M.A. Michiels, 2006*. Handhaven op Niveau [High Quality Enforcement], in: Michiels and Muller (2006*), pp. 199-206.

Bouckaert, B., and G. de Geest (Eds.), 2000. *Encyclopedia of law and economics.* Cheltenham, UK: Edward Elgar.

Boyd, J., and D.E. Ingberman, 1999. Do punitive damages promote deterrence? *International Review of Law and Economics* 19, pp. 47-68.

Braithwaite, J., 1985. *To punish or to persuade.* Albany: State University of New York Press.

Brehm, J., and J.T. Hamilton, 1996. Noncompliance in environmental reporting: Are violators ignorant, or evasive, of the law? *American Journal of Political Science* 40 (2), pp. 444-477.

Burrows, P., 1999. Combining regulation and legal liability for the control of external costs. *International Review of Law and Economics* 19, pp. 227-244.

Calabresi, G., 1970. *The costs of accidents. A legal and economic analysis.* New Haven, CT: Yale University Press.

Calcott, P., and S. Hutton, 2006. The choice of a liability regime when there is a regulatory gatekeeper. *Journal of Environmental Economics and Management* 51 (2), pp. 153-164.

Calfee, J.E., and P.H. Rubin, 1992. Some implications of damage payments for non-pecuniary losses. *Journal of Legal Studies* 21, pp. 371-411.

Cason, T.N., and L. Gangadharan, 2006. An experimental study of compliance and leverage in auditing and regulatory enforcement. *Economic Inquiry* 44 (2), pp. 352-366.

Chu, C.Y.C., S.-C. Hu, and T.-Y. Huang, 2000. Punishing repeat offenders more severely. *International Review of Law and Economics* 20, pp. 127-140.

Clark, J., L. Friesen, and A. Muller, 2004. The good, the bad, and the regulator: An experimental test of two conditional audit schemes. *Economic Inquiry* 42 (1), pp. 69-87.

Cohen, M.A., 1999. Monitoring and enforcement of environmental policy, pp. 44-106. In: Tietenberg, T., and H. Folmer (Eds.), *International Yearbook of Environmental and Resource Economics 1999/2000*: Edward Elgar publishers.

Coase, R.H., 1960. The problem of social cost. *Journal of Law and Economics* 3, pp. 1-44.

Cohen, M.A., 2000. Measuring the costs and benefits of crime and justice, pp. 263-315, *Criminal Justice 2000* Vol. 4: Washington DC: National Institute of Justice.

Cohen, M.A., 2005. *The costs of crime and justice.* London / New York: Routledge.

Conley, J.P., and P. Wang, 2006. Crime and ethics. *Journal of Urban Economics* 60, pp. 107-123.

Cook, P.J., and D.A. Graham, 1977. The demand for insurance and protection: The case of irreplaceable commodities. *Quarterly Journal of Economics* 91 (1), pp. 143-156.

Cooter, R.D., 1984. Prices and sanctions. *Columbia Law Review* 84, pp. 1523-1560.
Cooter, R.D., and D.L. Rubinfeld, 1989. Economic Analysis of Legal Disputes and Their Resolution. *Journal of Economic Literature* 27, pp. 1067-1097.
Curran, C., 2000. The endowment effect. In: Bouckaert, B., and G. de Geest (Eds.), *Encyclopedia of law and economics* (volume I) , pp. 819-835.
Daughety, A.F., 2000. Settlement,. In: Bouckaert, B., and G. de Geest (Eds.), *Encyclopedia of Law and Economics* (volume V), pp. 95-158.
Decker, C.S., 2003. Corporate environmentalism and environmental statutory permitting. *Journal of Law and Economics* 46, pp. 103-129.
De Blaeij, A.T., 2003. *The value of a statistical life in road safety* (thesis). Amsterdam: Free University.
De Jong, P.O., and M. Herweijer, 2004*. *Alle regels tellen [Counting all rules]*. WODC Onderzoek en Beleid no. 225. The Hague: Boom Juridische uitgevers.
DHV Huisvesting en Vastgoed, 2002*. *Knelpuntenonderzoek veiligheidsregelgeving. Inventarisatie en analyse van knelpunten in de regelgeving die de publieksveiligheid raken [Inventory and analysis of bottlenecks in safety regulation]* (report V0028). By order of the Ministry of Justice. Amersfoort: DHV Huisvesting en Vastgoed.
Den Hertog, J., 2000. General theories of regulation. In: Bouckaert, B., and G. de Geest (Eds.), *Encyclopedia of law and economics* (volume III), pp. 223-270.
Eckert, H., 2004. Inspections, warnings, and compliance: the case of petroleum storage regulation. *Journal of Environmental Economics and Management* 47, pp. 232-259.
Ehrlich, I., 1996. Crime, punishment, and the market for offenses. *Journal of Economic Perspectives* 10 (1), pp. 43-67.
Eide, E., 1994. *Economics of crime: deterrence and the rational offender*. Amsterdam: Elsevier Science.
Eide, E., 2000. Economics of criminal behavior. In: Bouckaert, B., and G. de Geest (Eds.), *Encyclopedia of law and economics* (volume VIII), pp. 345-389.
Eijgenraam, C.J.J., et al., 2000*. *Evaluatie van infrastructuurprojecten; leidraad voor kosten-baten-analyse [Evaluation of projects concerning the infrastructure; guideline for cost-benefit analysis]*. The Hague: Sdu Uitgevers.
Elffers, H., 2005*. *De rationele regelovertreder* (inaugural lecture Antwerp University). The Hague: Boom Juridische uitgevers.
Elffers, H., P. Verboon and W. Huisman, 2006. *Managing and maintaining compliance*. Den Haag: Boom Legal publishers.
Ellickson, R.C., 1991. *Order without law. How neighbors settle disputes*. Cambridge, Massachusetts: Harvard University Press.
Emons, W., 2003. A note on the optimal punishment for repeat offenders. *International Review of Law and Economics* 23, pp. 253-259.
European Commission, 2000. *Priorities in EU road safety. Progress report and ranking of actions*. Brussels.
Ewerhart, C., and P.W. Schmitz, 1998. Ex post liability for harm vs. ex ante safety regulation: Substitutes or complements? Comment. *American Economic Review* 88 (4), pp. 1027-1028.
Expert Centre Law Enforcement (Ministry of Justice), 2006*. *Gids programmatisch handhaven voor decentrale overheden [Guidebook enforcing programmedly for local authorities]*. The Hague: Expert Centre Law Enforcement (Handhaven met Effect).
Faure, M.G., 2000. Environmental regulation. In: Bouckaert, B., and G. de Geest (Eds.), *Encyclopedia of law and economics* (volume II) , pp. 443-520.
Faure, M., 2007. Economic analysis of tort and regulatory law, pp. 399-415. In: Van Boom, W.H., M. Lukas, C. Kissling (eds.), *Tort and regulatory law*. Vienna: Springer-Verlag.
Faure, M.G., and M. Visser, 2004. *Law and economics of environmental crime*. In: Sjögren, H., and G. Skogh (Eds.), pp. 57-75.
Feld, L.P., and S. Voigt, 2003. Economic growth and judicial independence: cross-country evidence using a new set of indicators. *European Journal of Political Economy* 19, pp. 497-527.
Fenn, P., and C.G. Veljanovski, 1988. A positive economic theory of regulatory enforcement. *Economic Journal* 98 (393), pp. 1055-1070.

Firestone, J., 2002. Agency governance and enforcement: The influence of mission on environmental decisionmaking. *Journal of Policy Analysis and Management* 21 (3), pp. 409-426.

Franckx, L., 2002. The use of ambient inspections in environmental monitoring and enforcement when the inspection agency cannot commit itself to announced inspection probabilities. *Journal of Environmental Economics and Management* 43, pp. 71-92.

Frey, B.S., 1997. *Not just for the money: an economic theory of personal motivation*. Cheltenham, UK: Edward Elgar.

Friedman, D.D., 1982. What is 'fair compensation' for death or injury. *International Review of Law and Economics* 2, pp. 81-93.

Friedman, D.D., 2000. *Law's order. What economics has to do with the law and why it matters*. Princeton, NJ: Princeton University Press.

Friesen, L., 2003. Targeting enforcement to improve compliance with environmental regulations. *Journal of Environmental Economics and Management* 46, pp. 72-85.

Friesen, L., 2006. The social welfare implications of industry self-auditing. *Journal of Environmental Economics and Management* 51, pp. 280-294.

Garoupa, N., 2004. *The economics of business crime*. In: Sjögren, H., and G. Skogh (Eds.), pp. 5-19.

Garoupa, N., and F. Gomez-Pomar, 2004. Punish once or twice: A theory of the use of criminal sanctions in addition to regulatory penalties. *American Law and Economics Review* 6 (2), pp. 410-433.

Garvie, D., and A. Keeler, 1994. Incomplete enforcement with endogenous regulatory choice. *Journal of Public Economics* 55, pp. 141-162.

Glaeser, E.L., B. Sacerdote, and J.A. Scheinkman, 1996. Crime and social interactions. *Quarterly Journal of Economics* 111 (2), pp. 507-548.

Glaeser, E.L., S. Johnson, and A. Shleifer, 2001. Coase versus the Coasians. *Quarterly Journal of Economics* 116 (3), pp. 853-899.

Glaeser, E.L., and A. Shleifer, 2003. The rise of the regulatory state. *Journal of Economic Literature* 41 (2), pp. 401-425.

Gneezy, U., and A. Rustichini, 2000. A fine is a price. *Journal of Legal Studies* 29, pp. 1-17.

Gundelach, J., and F.C.M.A. Michiels, 2003*. *Ongegronde vrees: een onderzoek naar de rechtspraak inzake handhavingsbesluiten [Unfounded fear: a study of the jurisprudence concerning enforcement decisions* (by order of the Ministry of Internal Affairs and the Ministry of Justice). Deventer: Kluwer.

Hagan, F.E., 2005. *Research methods in criminal justice and criminology* (7th edition). Boston: Allyn and Bacon.

Harford, J.D., 1991. Measurement error and state-dependent pollution control enforcement. *Journal of Environmental Economics and Management* 21, pp. 67-81.

Harford, J.D., and W. Harrington, 1991. A reconsideration of enforcement leverage when penalties are restricted. *Journal of Public Economics* 45, pp. 391-395.

Harrington, W., 1988. Enforcement leverage when penalties are restricted. *Journal of Public Economics* 37 (1), pp. 29-53.

Harrington, W., and A.G. Heyes, 2001. The theory of penalties: 'Leverage' and 'dealing', pp. 185-197. In: Heyes, A. (Ed.), *The law and economics of the environment*. Cheltenham, UK, Northampton, MA, USA: Edward Elgar.

Hawkins, K., 1983. Bargain and bluff. Compliance strategy and deterrence in the enforcement of regulation. *Law & Policy Quarterly* 5 (1), pp. 35-73.

Hawkins, K., 1984. *Environment and enforcement. Regulation and the social definition of pollution*. Oxford: Clarendon Press.

Hawkins, K., and J.M. Thomas (Eds.), 1984. *Enforcing Regulation*. Boston: Kluwer-Nijhoff Publishing.

Heineke, J.M. (Ed.), 1978. *Economic models of criminal behavioral*. Amsterdam: North-Holland Publishing Company.

Helland, E., 1998. The enforcement of pollution control laws: Inspections, violations and self-reporting. *Review of Economics and Statistics* 80 (1), pp. 141-153.

Helsloot, I., 2006*. Handhaving van brandveiligheidsvoorschriften [Enforcement of fire safety regulation], in: Michiels and Muller (2006*), pp. 491-524.

Helsloot, I., E.R. Muller, J.D. Berghuijs, 2007*. *Brandweer. Studies over organisatie, functioneren en omgeving*. Deventer: Kluwer.
Heyes, A.G., 1994. Environmental enforcement when 'inspectability' is endogenous: A model with overshooting properties. *Environmental and Resource Economics* 4, pp. 479-494.
Heyes, A.G., 2000. Implementing environmental regulation: Enforcement and compliance. *Journal of Regulatory Economics* 17 (2), pp. 107-129.
Heyes, A.G., 2002. A theory of filtered enforcement. *Journal of Environmental Economics and Management* 43, pp. 34-46.
Heyes, A.G., and N. Rickman, 1999. Regulatory dealing - revisiting the Harrington paradox. *Journal of Public Economics* 72 (3), pp. 361-378.
Hiriart, Y., D. Martimort, and J. Pouyet, 2004. On the optimal use of ex ante regulation and ex post liability. *Economics Letters* 84, pp. 231-235.
Hoitink, J.E., and F.C.M.A. Michiels, 1993*. *Handhaven met beleid. Een empirisch-juridische studie over de handhaving van bestemmingsplannen*. Zwolle: W.E.J. Tjeenk Willink.
Huisman, W., 2001*. *Tussen winst en moraal. Achtergronden van regelnaleving en regelovertreding door ondernemingen [Between profit and morals. Backgrounds of compliance and violation by corporations]* (thesis). Amsterdam: Free University.
Huisman, W., and A.M. Beukelman, 2006*. Regulering en regelnaleving door bedrijven [Regulation and regulatory compliance by corporations], pp. 95-117. In: Huisman, W., L.M. Moerings, and G. Suurmond (Eds.), *Veiligheid en Recht. Nieuwe doelwitten en strategieën [Security and law. New targets and strategies]* (Meijers-reeks nr. 113). Den Haag: Boom Juridische uitgevers.
Huisman, W., and A.M. Beukelman, 2007*. *Invloeden op regelnaleving door bedrijven. Inzichten uit wetenschappelijk onderzoek [Effects on regulatory compliance. Scientific evidence]*. The Hague: Boom Juridische uitgevers.
Hutter, B.M., 1997. *Compliance: Regulation and environment*. Oxford: Clarendon Press.
Innes, R., 1999a. Optimal liability with stochastic harms, judgement-proof injurers, and asymmetric information. *International Review of Law and Economics* 19, pp. 181-203.
Innes, R., 1999b. Remediation and self-reporting in optimal law enforcement. *Journal of Public Economics* 72, pp. 379-393.
Innes, R., 2004. Enforcement costs, optimal sanctions, and the choice between ex-post liability and ex-ante regulation. *International Review of Law and Economics* 24, pp. 29-48.
IOOV (Inspectie Openbare Orde en Veiligheid, Public Order and Safety Inspectorate), 2002*. *Gemeentelijke actieprogramma's brandveiligheid [Municipal action programs concerning fire safety]*. The Hague: Ministry of Internal Affairs.
Jolls, C., 2004. On law enforcement with boundedly rational actors, *The Harvard John M. Ohlin Discussion Paper Series* no. 494. Cambridge, MA.
Jost, P.-J., 1997a. Regulatory enforcement in the presence of a court system. *International Review of Law and Economics* 17, pp. 491-508.
Jost, P.-J., 1997b. Monitoring, appeal, and investigation: The enforcement and legal process. *Journal of Regulatory Economics* 12, pp. 127-146.
Kagan, R.A., 1984. *On Regulatory Inspectorates and Police*. In: Hawkins, K., and J.M. Thomas (Eds.), pp. 37-64.
Kagan, R.A., and J.T. Scholz, 1984. *The "Criminology of the Corporation" and Regulatory Enforcement Strategies*. In: Hawkins, K., and J.M. Thomas (Eds.), pp. 67-95.
Kaplow, L., and S. Shavell, 1994a. Optimal law enforcement with self-reporting of behavior. *Journal of Political Economy* 102 (3), pp. 583-606.
Kaplow, L., and S. Shavell, 1994b. Accuracy in the determination of liability. *Journal of Law and Economics* 37 (1), pp. 1-16.
Kaplow, L., and S. Shavell, 2001. *Fairness versus welfare*. Cambridge MA/London: Harvard University Press.
Karpoff, J.M., and J.R. Lott Jr., 1993. The reputational penalty firms bear from committing criminal fraud. *Journal of Law and Economics* 36, pp. 757-802.
Karpoff, J.M., J.R. Lott Jr., and E.W. Wehrly, 2005. The reputational penalties for environmental violations: empirical evidence. *Journal of Law and Economics* 48, pp. 653-675.

Kennan, J., and R. Wilson, 1993. Bargaining with private information. *Journal of Economic Literature* 31 (1), pp. 45-104.

Khambu, J., 1989. Regulatory standards, noncompliance and enforcement. *Journal of Regulatory Economics* 1, pp. 103-114.

Khambu, J., 1990. Direct controls and incentive systems of regulation. *Journal of Environmental Economics and Management* 18, pp. S72-85.

Kobayashi, B.H., and J.S. Parker, 2000. Civil procedure: General. In: Bouckaert, B., and G. de Geest (Eds.), *Encyclopedia of Law and Economics* (volume V), pp. 1-26.

Kolstad, C.D., T.S. Ulen, and G.V. Johnson, 1990. Ex post liability for harm vs. ex ante safety regulation: Substitutes or complements? *American Economic Review* 80 (4), pp. 888-901.

Kraakman, R.H., 2000. Vicarious and corporate civil liability. In: Bouckaert, B., and G. de Geest (Eds.), *Encyclopedia of law and economics* (volume II), pp 669-681.

Kramer, L., and A.O. Sykes, 1987. Municipal liability under §1983: a legal and economic analysis. *Supreme Court Review* (1987), pp. 249-301.

Kydland, F.E., and E.C. Prescott, 1977. Rules rather than discretion: The inconsistency of optimal plans. *Journal of Political Economy* 85 (3), pp. 473-491.

Lai, C.-C., C.-Y. Yang, and J.-J. Chang, 2003. Environmental regulations and social norms. *International Tax and Public Finance* 10, pp. 63-75.

Lando, H., 2006. Does wrongful conviction lower deterrence? *Journal of Legal Studies* 35, pp. 327-337.

Lando, H., and S. Shavell, 2004. The advantage of focusing law enforcement effort. *International Review of Law and Economics* 24, pp. 209-218.

Landsberger, M., and I. Meilijson, 1982. Incentive generating state dependent penalty system. *Journal of Public Economics* 19, pp. 333-352.

Law, M.T., 2006. How do regulators regulate? Enforcement of the Pure Food and Drugs Act, 1907-38. *Journal of Law, Economics and Behaviour* 22 (2), pp. 459-489.

Leeuw, F.L., and F. Willemsen, 2006*. Toezicht en inspectie: trends, kosten en baten [Supervision and inspection: trends, costs and benefits]. *Nederlands Juristenblad* 2006/37, pp. 2108-2116.

Lin, C.-C., and C.C. Yang, 2006. Fine enough or don't fine at all. *Journal of Economic Behavior & Organization* 59, pp. 195-213.

Lipsky, M., 1980. *Street-level bureaucracy. Dilemmas of the individual in public services*. New York: The Russell Sage Foundation.

Livernois, J., and C.J. Mckenna, 1999. Truth or consequences: Enforcing pollution standards with self-reporting. *Journal of Public Economics* 41, pp. 415-440.

Lomwel, A.G.C. van, and J.H.M. Nelissen, 2005*. *Handhaven onder de druk der omstandigheden. Het gebruik van schriftelijke waarschuwingen door gemeenten [Enforcement by force of circumstances. The use of written warnings by municipalities]*. WODC / CentER Applied Research.

Lott, J.R., Jr., 2000. Corporate Criminal Liability. In: Bouckaert, B., and G. de Geest (Eds.), *Encyclopedia of Law and Economics* (volume V), pp. 492-501.

Van Maanen, , G.E., 2007*. Overheidsaansprakelijkheid voor gebrekkig toezicht [Government liability for insufficient inspection]. *Rechtsgeleerd Magazijn Themis* (nr. 4), pp. 127-140.

Macrory, R.B., 2006a. *Regulatory justice: sanctioning in a post-Hampton world* (consultation document), www.berr.gov.uk

Macrory, R.B., 2006b. *Regulatory justice: making sanctions effective* (final report), www.berr.gov.uk

Malik, A.S., 1990. Avoidance, screening and optimum enforcement. *Rand Journal of Economics* 21 (3), pp. 341-353.

Malik, A.S., 1993. Self-reporting and the design of policies for regulating stochastic pollution. *Journal of Environmental Economics and Management* 24, pp. 241-257.

Marette, S., and J.M. Crespi, 2005. The financing of regulatory agencies. *Journal of Regulatory Economics* 27 (1), pp. 95-113.

Mason, R., 2004. Dividends, safety and liquidation when liabilities are long-term and stochastic. *European Economic Review* 48, pp. 1179-1210.

May, P.J., and R.S. Wood, 2003. At the regulatory front lines: Inspectors' enforcement styles and regulatory compliance. *Journal of Public Administration Research and Theory* 13 (2), pp. 117-139.

Meijer, F.M., and H.J. Visscher, 2004*. Bouwtoezichtsystemen in acht Europese landen vergeleken [Systems of building supervision in eight European countries]. *Bouwrecht* (12), pp. 1005-1015.

Miceli, T.J., 1997. *Economics of the law. Torts, contracts, property, litigation.* Oxford: Oxford University Press.

Miceli, T.J., and K. Segerson, 2003. A note on optimal care by wealth-constrained injurers. *International Review of Law and Economics* 23, pp. 273-284.

Michiels Committee (in full: Committee Administrative and Private Law Enforcement), 1998*. *Handhaven op niveau [High quality enforcement].* Deventer: W.E.J. Tjeenk Willink.

Michiels, F.C.M.A., 2006*. *Houdbaar handhavingsrecht [Tenable enforcement law]* (inaugural lecture). Deventer: Kluwer.

Michiels, F.C.M.A., and E.R. Muller, 2006*. *Handhaving. Bestuurlijk handhaven in Nederland [Enforcement. Administrative enforcement in the Netherlands].* Deventer: Kluwer.

Miller, G.P., 1997. *Class actions.* In: Newman, P. (Ed.), *The New Palgrave Dictionary of Economics and the Law* (volume I), pp. 257-262.

Ministry of Internal Affairs, 1997*. *Handleiding Prevap [Guideline Prevap (Prevention Action Plan)].* The Hague: Ministry of Internal Affairs, directorate Fire Service and Crisis Management.

Ministry of Justice, 2002*. *Risicomatrix; prioriteren op basis van risico-inschatting [Risk Matrix; priorities based on risk assessment.* The Hague: Ministry of Justice, Expert Centre Law Enforcement, second concept.

Ministry of Justice, 2004. *The 'table-of-eleven'. An aid to formulating policy with a high compliance factor.* The Hague: Ministry of Justice, Expert Centre Law Enforcement.

Mookherjee, D., and I.P.L. Png, 1992. Monitoring vis-a-vis investigation in enforcement of law. *American Economic Review* 82 (3), pp. 556-565.

Mookherjee, D., and I.P.L. Png, 1994. Marginal deterrence in enforcement of law. *Journal of Political Economy* 102 (5), pp. 1039-1066.

Mulder, L.B., et al., 2005. The effect of feedback on support for a sanctioning system in a social dilemma. The difference between installing and maintaining the sanction. *Journal of Economic Psychology* 26, pp. 443-458.

Nadeau, L.W., 1997. EPA effectiveness at reducing the duration of plant-level noncompliance. *Journal of Environmental Economics and Management* 34, pp. 54-78.

Netherlands Court of Audit, 1996. *Vergunningen [Licenses].* The Hague: Sdu Uitgevers (presented to the Dutch Second Chamber, *Kamerstukken II* 1995/96, 24 656, no.2).

Netherlands Court of Audit, 2002*. *Handhaving door rijksinspecties [Enforcement by national inspectorates].* The Hague: Sdu Uitgevers (presented to the Dutch Second Chamber, *Kamerstukken II* 2001/02, 28 271, no.2).

Netherlands Court of Audit, 2005*. *Handhaven en gedogen [Enforcement and tolerance].* The Hague: Sdu Uitgevers (presented to the Dutch Second Chamber, *Kamerstukken II* 2004/05, 30 050, no. 1-2.

Newman, P. (Ed.), 1998. The new Palgrave dictionary of economics and the law. London: Macmillan Reference Limited.

Nowell, C., and J. Shogren, 1994. Challenging the enforcement of environmental regulation. *Journal of Regulatory Economics* 6 (3), pp. 265-282.

Nyborg, K., and K. Telle, 2004. The role of warnings in regulation: keeping control with less punishment. *Journal of Public Economics* 88 (12), pp. 2801-2816.

Nyborg, K., and K. Telle, 2006. Firms' compliance to environmental regulation: Is there really a paradox? *Environmental and Resource Economics* 35, pp. 1-18.

Ogus, A., 2001. Regulation: The public interest and the private interest, pp. 141-145. In: Deffains, B., and T. Kirat (eds.), *Law and economics in civil law countries.* Amsterdam: Elsevier Science.

Ogus, A., 2007. The relationship between regulation and tort law: goals and strategies, pp. 377-389. In: Van Boom, W.H., M. Lukas, C. Kissling (eds.), *Tort and regulatory law.* Vienna: Springer-Verlag.

Ogus, A., and C. Abbot, 2002. Sanctions for pollution: do we have the right regime? *Journal of Environmental Law* 14 (3), pp. 283-298.
Ott, C., and H.-B. Schäfer, 1997. Negligence as untaken precaution, limited information, and efficient standard formation in the civil liability system. *International Review of Law and Economics* 17, pp. 15-29.
Persson, M., and C.-H. Siven, 2006. Incentive and incarceration effects in a general equilibrium model of crime. *Journal of Economic Behavior & Organization* 59, pp. 214-229.
Pfaff, A.S.P., and C.W. Sanchirico, 2000. Environmental self-auditing: Setting the proper incentives for discovery and correction of environmental harm. *Journal of Law, Economics and Behaviour* 16 (1), pp. 189-208.
Pistor, K., and C. Xu, 2003. Incomplete law. *N.Y.U. Journal of International Law and Politics* 35, pp. 931-1013.
Polinsky, A.M., 2006. The optimal use of fines and imprisonment when wealth is unobservable. *Journal of Public Economics* 90, pp. 823-835.
Polinsky, A.M., and D.L. Rubinfeld, 1991. A model of optimal fines for repeat offenders. *Journal of Public Economics* 46, pp. 291-306.
Polinsky, A.M., and S. Shavell, 1979. The optimal trade-off between the probability and magnitude of fines. *American Economic Review* 69, pp. 880-891.
Polinsky, A.M., and S. Shavell, 1984. The optimal use of fines and imprisonment. *Journal of Public Economics* 24, pp. 89-99.
Polinsky, A.M., and S. Shavell, 1992. Enforcement costs and the optimal magnitude and probability of fines. *Journal of Law and Economics* 35, pp. 133-148.
Polinsky, A.M., and S. Shavell, 1993. Should employees be subject to fines and imprisonment given the existence of corporate liability? *International Review of Law and Economics* 13, pp. 239-257.
Polinsky, A.M., and S. Shavell, 1998. On offense history and the theory of deterrence. *International Review of Law and Economics* 18, pp. 305-324.
Polinsky, A.M., and S. Shavell, 2000. The economic theory of public enforcement of law. *Journal of Economic Literature* 38, pp. 45-76.
Polinsky, A.M., and S. Shavell, 2000b. Public enforcement of law. In: Bouckaert, B., and G. de Geest (Eds.), *Encyclopedia of law and economics* (volume VIII), pp. 307-344.
Ponsaers, P., and B. Hoogenboom, 2004. Het moeilijke spel van wortel en stok. Organisatiecriminaliteit en handhavingstragieen van bijzondere inspectie- en opsporingsdiensten [The difficult game of carrot and stick. Organized crime and enforcement strategies of special investigation departments]. *Tijdschrift voor Criminologie* 46 (2), pp. 165-181.
Posner, R.A., and E.B. Rasmusen, 1999. Creating and enforcing social norms, with special reference to sanctions. *International Review of Law and Economics* 19, pp. 369-382.
Raymond, M., 1999. Enforcement leverage when penalties are restricted: a reconsideration under asymmetric information. *Journal of Public Economics* 73, pp. 289-295.
Raymond, M., 2004. Regulatory compliance with costly and uncertain litigation. *Journal of Regulatory Economics* 26 (2), pp. 165-176.
Reinganum, J.F., 1988. Plea bargaining and prosecutorial discretion. *American Economic Review* 784 (4), pp. 713-728.
Ricketts, M., and A. Peacock, 1996. Bargaining and the regulatory system. *International Review of Law and Economics* 6, pp. 3-16.
Ringleb, A.H., and S.N. Wiggins, 1990. Liability and large-scale, long-term hazards. *Journal of Political Economy* 98 (3), pp. 574-595.
RIVM, 2003*. *Nuchter omgaan met risico's [Sensible management of risks]* (report 251701047/2003). Bilthoven: Milieu- en Natuurplanbureau - RIVM.
Rose-Ackerman, S., 1991. Regulation and the law of torts. *American Economic Review Papers and Proceedings* 81 (2), pp. 54-58.
Rubinstein, A., 1979. An optimal conviction policy for offenses that may have been committed by accident, pp. 406-413. In: Brams, S.J., A. Schotter, and G. Schwodiauer (Eds.), *Applied game theory*. Wuerzburg: Physica-Verlag.

Savage, I., 1993. An empirical investigation into the effects of psychological perceptions on the willingness-to-pay to reduce risk. *Journal of Risk and Uncertainty* 6, pp. 75-90.

Schäfer, H.-B., and C. Ott, 2004. *The economic analysis of civil law*. Cheltenham, UK: Edward Elgar.

Schmitz, P.W., 2000. On the joint use of liability and safety regulation. *International Review of Law and Economics* 20, pp. 371-382.

Scholz, J.T., 1984. Cooperation, deterrence, and the ecology of regulatory enforcement. *Law and Society Review* 18 (2), pp. 179-224.

Scholz, J.T., 1991. Cooperative regulatory enforcement and the politics of administrative effectiveness. *American Political Science Review* 85 (1), pp. 115-136.

Segerson, K., and T. Tietenberg, 1992. The structure of penalties in environmental enforcement: An economic analysis. *Journal of Environmental Economics and Management* 23, pp. 179-200.

SER (Sociaal Economische Raad), 1999*. *Investeren in verkeersveiligheid [Investments in traffic safety]* (SER-advies 99/13), Den Haag.

Shavell, S., 1979. On moral hazard and insurance. *Quarterly Journal of Economics*, pp. 541-562.

Shavell, S., 1980. Strict liability versus negligence. *Journal of Legal Studies* 9, pp. 1-25.

Shavell, S., 1982. On liability and insurance. *Bell Journal of Economics*, pp. 120-132.

Shavell, S., 1984a. Liability for harm versus regulation of safety. *Journal of Legal Studies* 13, pp. 357-374.

Shavell, S., 1984b. A model of the optimal use of liability and safety regulation. *Rand Journal of Economics* 15 (2), pp. 271-280.

Shavell, S., 1987. *Economic analysis of accident law*. Canbridge, Massachusetts: Harvard University Press.

Shavell, S., 1991. Specific versus general enforcement of law. *Journal of Political Economy* 99 (5), pp. 1088-1108.

Shavell, S., 1992a. Liability and the incentive to obtain information about risk. *Journal of Legal Studies* 21, pp. 259-270.

Shavell, S., 1992b. A note on marginal deterrence. *International Review of Law and Economics* 12, pp. 345-355.

Shavell, S., 1993. The optimal structure of law enforcement. *Journal of Law and Economics* 36, pp. 255-287.

Shavell, S., 2004. *Foundations of economic analysis of law*. Cambridge, Massachusetts: Harvard University Press.

Shavell, S., 2005. Optimal discretion in the application of rules, *The Harvard John M. Olin Discussion Paper Series* no. 03/2005. Cambridge, MA.

SIRA Consulting, 2004*. *Administratieve lasten gemeenten. Eindrapportage voor het onderzoek nulmeting administratieve lasten van gemeenten [Administrative burden municipalities. Final report of the starting measurement]*. Nieuwegein.

Silver, C., 2000. Class actions - Representative proceedings. In: Bouckaert, B., and G. De Geest (Eds.), *Encyclopedia of Law and Economics* (volume V), pp. 194-240.

Sjögren, H., and G. Skogh (Eds.), 2004. *New Perspectives on Economic Crime*. Cheltenham: Edward Elgar.

Sloan, F.A., et al., 1998. Alternative approaches to valuing intangible health losses: the evidence for multiple sclerosis. *Journal of Health Economics* 17, pp. 475-497.

Spence, M., 1977. Consumer misperceptions, product failure and producer liability. *Review of Economic Studies* 44 (3), pp. 561-572.

Stigler, G.J., 1970. The optimum enforcement of laws. *Journal of Political Economy* 78 (3), pp. 526-236.

Sugden, R., 1989. Spontaneous order. *Journal of Economic Perspectives* 3 (4), pp. 85-97.

Sunstein, C.R., 1997. Bad deaths. *Journal of Risk and Uncertainty* 14, pp. 259-282.

Sunstein, C.R., 2001a. The laws of fear, *Chicago John M. Olin Law & Economics Working Paper* no. 128, Chicago.

Sunstein, C.R., 2001b. Probability neglect: Emotions, worst cases, and law, *Chicago John M. Olin Law & Economics Working Paper Series* no. 138, Chicago.

Suurmond, G., 2004*. Brandveiligheid in de horeca. Een economische analyse van de handhaving in een representatieve gemeente [Fire safety in horeca establishments. An economic analysis of enforcement is a representative municipality]. *Department of Economics Research Memorandum* no. 2004.03. Leiden.

Suurmond, G., 2005*. Handhaving brandveiligheid in de horeca: een economische analyse [Enforcement of fire safety in horeca establishments: an economic analysis], pp. 157-200. In: Barkhuysen, T., W. den Ouden and J.E.M. Polak (eds.), *Recht realiseren. Bijdragen rond het thema adequate naleving van rechtsregels [Realizing justice. Contributions on the theme of adequate compliance with the law]* (Meijers-reeks nr. 89). Deventer: Kluwer.

Suurmond, G., 2006*. Rapport Schipholbrand: waar het echt om gaat ... [Report Schiphol Airport fire: the real point of interest...] *Nederlands Juristenblad* 39 (3 november 2006), pp. 2233-2236.

Suurmond, G., 2007. Compliance to fire safety regulation. The effects of the enforcement strategy. *Department of Economics Research Memorandum* no. 2007.03 (paper presented at the EALE 24th Annual Conference in Copenhagen). Leiden.

Suurmond, G., and B.C.J. van Velthoven, 2003*. Brandveiligheid in de horeca. Toepassing van kosten-batenanalyses op justitieel terrein [Fire safety in horeca establishments. Application of cost-benefit analysis to legal issues]. *Justitiële Verkenningen* 29 (9), pp. 8-29.

Suurmond, G., and B.C.J. van Velthoven, 2004*. *Handhaven: eerst kiezen dan doen. Economische mogelijkheden en beperkingen [Enforcement: choosing before acting. Economic possibilities and constraints* (report section IV). The Hague: Ministry of Justice, Expert Centre Law Enforcement.

Suurmond, G., and B.C.J. van Velthoven, 2005*. Vergoeding van affectieschade. Te weinig met het oog op de daders en te veel met het oog op de slachtoffers [Hedonic damages. Too little for injurers, too much for victims]. *Nederlands Juristenblad* 37 (21 oktober 2005), pp. 1934-1936 (extensive version on the internet).

Suurmond, G., and B.C.J. van Velthoven, 2006*. Een beginselplicht tot handhaving: liever regels dan discretionaire vrijheid [A duty to enforce: rules rather than discretion], pp. 193-213. In: Huisman, W., L.M. Moerings, and G. Suurmond (Eds.), *Veiligheid en Recht. Nieuwe doelwitten en strategieën [Security and law. New targets and strategies]* (Meijers-reeks nr. 113). The Hague: Boom Juridische uitgevers.

Titmuss, R.M., 1970. *The gift relationship. From human blood to social policy*. London: George Allen & Unwin Ltd.

Tsebelis, G., 1989. The abuse of probability in political analysis: The Robinson Crusoe Fallacy. *American Political Science Review* 83 (1), pp. 77-91.

Tyler, T.R., 1990. *Why people obey the law*. New Haven: Yale University Press.

USFA (U.S. Fire Administration / National Fire Data Center), 2000. Nightclub Fires in 2000. *Topical Fire Research Series* 7 (3). Emmitsburg, Maryland.

Van de Bunt, H., J. van Erp and K. van Wingerde, 2007*. Hoe stevig is de piramide van Braithwaite? [How solid is the pyramid of Braithwaite?] *Tijdschrift voor Criminologie* 49 (4), pp. 386-399.

Van de Peppel, R., 2006*. Het handhavingstekort [The enforcement deficit], in: Michiels and Muller (2006*), pp. 163-174.

Van Rooij, B., 2006. *Regulating land and pollution in China. Lawmaking, compliance, and enforcement; theory and cases* (thesis). Leiden: Leiden University Press.

Van Stokkum, B., 2004*. *Handhaven: eerst kiezen dan doen. Sociaal-wetenschappelijke mogelijkheden en beperkingen [Enforcement: choosing before acting. Social scientific possibilities and constraints* (report section I). The Hague: Ministry of Justice, Expert Centre Law Enforcement.

Van der Tak, T., 1988*. *Vergunning verleend. Een bestuurskundige studie naar vergunningen op grond van de Wet inzake de luchtverontreiniging en de Wet algemene bepalingen milieuhygiëne [License granted. A study of public administrative perspective to environmental licenses]* (thesis). Rotterdam: Erasmus University Rotterdam.

Van Velthoven, B.C.J., 1994*. De economie van misdaad en straf. Een overzicht en evaluatie van de literatuur [The economics of crime and punishment. An overview and evaluation of the literature], *Department of Economics Research Memorandum, Leiden University* no. 94.08. Leiden.

Van Velthoven, B.C.J., 2007*. Kosten-batenanalyse van criminaliteitsbeleid. Over de methodiek in het algemeen en Nederlandse toepassingen in het bijzonder [Cost-benefit analysis of criminal justice policy. On the method in general and on Dutch applications in particular]. *Department of Economics Research Memorandum, Leiden University* no. 2007.05. Leiden.

Van Velthoven, B.C.J., and G. Suurmond, 2003*. Doelmatigheid in de rechtshuishouding. Een beschouwing naar aanleiding van het WRR-rapport over de nationale rechtsstaat [Efficiency in the legal system. A consideration of the WRR report on the national constitutional state]. *Rechtsgeleerd Magazijn Themis* 164 (3), pp. 132-144.

Veljanovski, C.G., 1984. *The economics of regulatory enforcement*. In: Hawkins, K., and J.M. Thomas (Eds.), pp. 171-188.

Viscusi, W.K., 1993. The value of risks to life and health. *Journal of Economic Literature* 31 (4), pp. 1912-1946.

Viscusi, W.K., 1998. Valuing life and risks to life. In: Newman, P. (Ed.), *The New Palgrave Dictionary of Economics and the Law* (volume 3), pp. 660-669.

Viscusi, W.K., and J.E. Aldy, 2003. The value of a statistical life: A critical review of market estimates throughout the world. *Journal of Risk and Uncertainty* 27 (1), pp. 5-76.

Viscusi, W.K., and W.N. Evans, 1990. Utility functions that depend on health status: Estimates and economic implications. *American Economic Review* 80 (3), pp. 353-374.

Viscusi, W.K., and R.J. Zeckhauser, 1979. Optimal standards with incomplete enforcement. *Public Policy* 27 (4), pp. 437-456.

Visscher, L.T., 2005*. *Een rechtseconomische analyse van het Nederlandse onrechtmatigedaadsrecht [A law and economics analysis of the Dutch law on unlawful behaviour]* (thesis). Rotterdam: Erasmus University Rotterdam.

VNG (Vereniging van Nederlands Gemeenten), 2002*. *Model-bouwverordening [Model Building Bylaw]*. The Hague: VNG Uitgeverij.

Voermans, W., 2007*. De aspirinewerking van sanctioneren, in: L. Loeber (ed.), *Bruikbare wetgeving*, preadviezen van Ph. Eijlander en P. Popelier aan de Vereniging voor wetgeving en wetgevingsbeleid. Nijmegen: Wolff Legal Publishers, pp. 57-64.

Weerkamp, J.W., 2004*. Waarheen met de gebruiksvergunning? [What to do with the use license?] *Bouwrecht* (9), pp. 740-746.

Wesemann, P, A.T. de Blaeij and P. Rietveld, 2005*. De waardering van bespaarde verkeersdoden: covernota bij het proefschrift De Blaeij (2003) [The valuation of saved lives in traffic: a note to the thesis of De Blaeij (2003)]. Leidschendam: SWOV (report R-2005-4).

Weterings, W.C.T., 1999*. *Vergoeding van letselschade en transaktiekosten. Een kwalitatieve en kwantitatieve analyse [Compensation of injury and transaction costs. A qualitative and quantitative analysis]*. Deventer: W.E.J. Tjeenk Willink.

Wiering, M.A., 1999*. *Controleurs in context [Inspectors in context]* , Lelystad: Koninklijke Vermande

Wittberg, L., 2006. Can communication activities improve compliance? In: Elffers et al., 2006, pp. 25-43.

WRR (Wetenschappelijke Raad voor het Regeringsbeleid), 2003*. *Waarden, normen en de last van het gedrag [Values, norms and the burden of behavior]*. Amsterdam: Amsterdam University Press.

Xu, C., and K. Pistor, 2002. Law enforcement under incomplete law: Theory and evidence from financial market regulation, *London School of Economics and Political Science, Suntory and Toyota Centres for Economics and Related Disciplines* no. Working Paper No. TE/02/442.

Zeckhauser, R.J., 1996. The economics of catastrophes. *Journal of Risk and Uncertainty* 12, pp. 113-140.

Zoomers, G.M., 2001*. *Handboek gebruiksvergunningen [Handbook use licenses]*. Arnhem: Nederlands Instituut voor Brandweer en Rampenbestrijding

Index

Curriculum vitae

Guido Suurmond was born in Gouda on September 6, 1979. He completed pre-university education at the Gereformeerde Scholengemeenschap Rotterdam in 1997. After that, he studied economics at the Erasmus University Rotterdam. He worked as a research assistant during the *Erasmus Advies Project 2000/2001* for a study on the effects of age on savings (joint with Martin Ruiter). This project won the *prof. dr. Lambers Prize* and an honorable mention at the *Grote Financiën Prijs 2001*. In 2001-2002 he worked as a research assistant of prof. dr. O.H. Swank. He graduated in 2002 with an undergraduate thesis on the influence of reputational concerns on decision-making ("vanity as a driving force"). Since then, he has been working as a Ph.D.-fellow at the Department of Economics of the Faculty of Law at Leiden University. In addition to writing this thesis, he has realized several publications and taught courses in the Economic Analysis of Law. As a member of the research programme Security and Law of the E.M. Meijers Institute for Legal Research he also participated in the Research School Safety and Security in Society (OMV).

In de boekenreeks van het E.M. Meijers Instituut voor Rechtswetenschappelijk Onderzoek van de Faculteit der Rechtsgeleerdheid, Universiteit Leiden, zijn in 2007 en 2008 verschenen:

MI-117 H.J.Th.M. van Roosmalen, *Overheidsaansprakelijkheid in Engeland en Nederland* (diss. Leiden), Den Haag: Sdu 2007, ISBN 978 90 12 11846 0

MI-118 R.W.J. Crommelin, *Het aanvullen van de rechtsgronden* (diss. Leiden), Alphen aan den Rijn: Kluwer 2007, ISBN 978 90 13 04635 9

MI-119 L.A.R. Siemerink, *De overeenkomst van Internet Service Providers met consumenten* (diss. Leiden), Deventer: Kluwer 2007, ISBN 90 13 04357 7

MI-120 I.S.J. Houben, K.J.O. Jansen, P. Memelink, J.H. Nieuwenhuis & L. Reurich (red.), *Europees contractenrecht. Techniek en moraal*, Deventer: Kluwer 2007, ISBN 90 13 04036 5

MI-121 S. Hillebrink, *Political Decolonization and Self-Determination. The Case of the Netherlands Antilles and Aruba* (diss. Leiden), 2007, ISBN 978 90 9021470 2

MI-122 B.W. Schermer, *Software agents, surveillance, and the right to privacy: a legislative framework for agent-enabled surveillance* (diss. Leiden), Leiden: Leiden University Press 2007, ISBN 978 90 8728 021 5

MI-123 C.G. Breedveld-de Voogd, *Vorm, vrijheid en gebondenheid bij de koop van een woning* (diss. Leiden), Den Haag: Boom Juridische uitgevers 2007, ISBN 978 90 5454 412 8

MI-124 T. Barkhuysen, W. den Ouden & E. Steyger (red.), *Europees recht effectueren. Algemeen bestuursrecht als instrument voor de effectieve uitvoering van EG-recht*, Alpen aan den Rijn: Kluwer 2007, ISBN 978 90 13046 28 1

MI-125 A.C. Beunen, *Protection for databases, The European Database Directive and its effects in the Netherlands, France and the United Kingdom* (diss. Leiden), Nijmegen: Wolf Legal Publishers 2007, ISBN 978 90 5850 267 4

MI-126 Z. Deen-Racsmány, *Active Personality and Non-extradition of Nationals in International Criminal Law at the Dawn of the Twenty-first Century: Adapting Key Functions of Nationality to the Requirements of International Criminal Justice* (diss. Leiden), Leiden: Leiden University Press.

MI-127 C.P.L. van Woensel, *MERK, GOD EN VERBOD, Oneigenlijk gebruik en monopolisering van tekens met een grote symbolische waarde* (diss. Leiden), Amstelveen: Uitgeverij deLex BV 2007, ISBN 978 90 8692 010 5

MI-128 J.A.J. Peter, *Levering van roerende zaken* (diss. Leiden), Deventer: Kluwer 2007, ISBN 978 90 13043 81 5

MI-129 A.C. Hendriks & H.-M.Th.D. ten Napel (red.), *Volksgezondheid in een veellagige rechtsorde. Eenheid en verscheidenheid van norm en praktijk*, Alphen a/d Rijn: Kluwer 2007, ISBN 978 90 13 04720 2

MI-130 H.R. Kranenborg, *Toegang tot documenten en bescherming van persoonsgegevens in de Europese Unie. Over de openbaarheid van persoonsgegevens* (diss. Leiden), Deventer: Kluwer 2007, ISBN 978 90 13 04886 5

MI-131 M. Kaeding, *Better regulator in the European Union – Lost in translation or Full Steam ahead?* (diss. Leiden), Leiden University Press, ISBN 978 90 8728 026 0

MI-132 S. van Wingerden, M. Moerings & J. van Wilsem, *De praktijk van schadevergoeding voor slachtoffers van misdrijven*, Den Haag: Boom Juridische uitgevers 2007, ISBN 978 90 5454 948 2

MI-133 Jubileumbundel Meijers Instituut, *Ik zou het zo weer doen. Terugblikken op proefschriften uit het 10-jarig bestaan van het Meijers Instituut*, Leiden: Leiden University Press

MI-134 C.L.J. Caminada, A.M. Haberham, J.H. Hoogteijling en H. Vording (red.), *Belasting met beleid*, Den Haag: Sdu 2007, ISBN 978 90 12 12470 6

MI-135 A. Vermeer-Künzli, *The Protection of Indificudals by means of Diplomatic Protection. Diplomatic Protection as a Human Rights Instrument (*diss. Leiden) 2007, ISBN 978 90 9022 487 9

MI-136 I.S.J. Houben, K.J.O. Jansen, P. Memelink, J.H. Nieuwenhuis & L. Reurich (red.), *Samenloop*, Deventer: Kluwer 2007, ISBN 978 90 13 05033 2

MI-137 T.C. Leemans, *De toetsing door de bestuursrechter in milieugeschillen. Over rechterlijke toetsingsintensiteit, bestuurlijke beslissingsruimte en deskundigenadvisering* (diss. Leiden), Den Haag: Boom Jurdische uitgevers 2008, ISBN 978 90 5454 986 4

MI-138 P. Kuypers, *Forumkeuze in het Nederlands internationaal privaatrecht* (diss. Leiden), Deventer: Kluwer 2008, ISBN 978 90 13 04797 4
MI-139 A. Meuwese, *Impact Assessment in EU Lawmaking* (diss. Leiden), Zutphen: Wöhrmann Printing Service 2008
MI-140 P.C. Adriaanse e.a., *Implementatie van EU-handhavingsvoorschriften*, Den Haag: Boom Juridische uitgevers 2008, ISBN 978 90 5454 862 1
MI-141 S.D. Dikker Hupkes, *What Constitutes Occupation? Israel as the occupying power in the Gaza Strip after the Disengagement*, Leiden: Jongbloed 2008
MI-142 R.A. Visser, E. van Gemerden, P.A. More & R.C.J. de Roon, *Sturing en samenwerking in handhavingsprojecten*, Leiden: Leiden University Press, ISBN 978 90 8728 0383
MI-143 B.M. Dijksterhuis, *Rechters normeren de alimentatiehoogte. Een empirisch onderzoek naar rechterlijke samenwerking in de Werkgroep Alimentatienormen (1975-2007)*, Leiden: Leiden University Press, ISBN 978 90 8728 045 1
MI-144 F.P. Ölçer, *Eerlijk proces en bijzondere opsporing*, Nijmegen: Wolf Legal Publishers 2007, ISBN 978 90 5850 376 3
MI-145 J.H. Crijns, P.P.J. van der Meij & J.M. ten Voorde, *De waarde van waarheid. Opstellen over waarheid en waarheidsvinding in het strafrecht*, Den Haag: Boom Juridische uitgevers 2008, ISBN 978 90 8974 020 5
MI-146 G.K. Schoep, *Straftoemetingsrecht en strafvorming* (diss. Leiden), Deventer: Kluwer 2008, ISBN 978 90 1306 018 8
MI-147 A.R. Spanjer, Structural and regulatory reform of the European natural gas market – Does the current approach secure the public service obligations? (diss. Leiden), Wageningen: Ponsen & Looijen BV, ISBN 978 90 6464 300 2
MI-148 E.-J. Zippro, Privaatrechtelijke handhaving van mededingingsrecht (diss. Leiden), Deventer: Kluwer 2008
MI-149 G. Suurmond, Enforcing fire safety in the catering industry. An economic analysis, Leiden: Leiden University Press 2008, ISBN 978 90 8728 061 1
MI-150 J.P. Loof (red.), Juridische ruimte voor gewetensbezwaren?, Leiden: NJCM-Boekerij 2008, ISBN 978 90 6750 048 7

Zie voor de volledige lijst van publicaties: www.law.leidenuniv.nl/onderzoek

Printed by Amazon Italia Logistica S.r.l.
Torrazza Piemonte (TO), Italy

92438570R00227